Initiation of Polymerization

Initiation of Polymerization

Frederick E. Bailey, Jr.,
EDITOR
Union Carbide Corporation

ASSOCIATE EDITORS
**E. J. Vandenberg, A. Blumstein,
M. J. Bowden, Jett C. Arthur,
Joginder Lal, R. M. Ottenbrite**

Based on a symposium sponsored
by the Macromolecular Secretariat,
at the 183rd ACS National Meeting,
Las Vegas, Nevada,
March 28–April 2, 1982

ACS SYMPOSIUM SERIES 212

AMERICAN CHEMICAL SOCIETY
WASHINGTON, D.C. 1983

Library of Congress Cataloging in Publication Data

American Chemical Society. National Meeting
(183rd: 1982: Las Vegas, Nev.)
Initiation of polymerization.

(ACS symposium series, ISSN 0097-6156-83/0; 212)

Includes bibliographies and index.

1. Polymers and polymerization—Congresses.
I. Bailey, Frederick E. (Frederick Eugene), 1927- . II. American Chemical Society. Macromolecular Secretariat. III. Title. IV. Series.

QD380.A4 1983 668'.9 83-2613
ISBN 0-8412-0765-8

PRINTED IN THE UNITED STATES OF AMERICA

ACS Symposium Series

M. Joan Comstock, *Series Editor*

FOREWORD

The ACS SYMPOSIUM SERIES was founded in 1974 to provide a medium for publishing symposia quickly in book form. The format of the Series parallels that of the continuing ADVANCES IN CHEMISTRY SERIES except that in order to save time the papers are not typeset but are reproduced as they are submitted by the authors in camera-ready form. Papers are reviewed under the supervision of the Editors with the assistance of the Series Advisory Board and are selected to maintain the integrity of the symposia; however, verbatim reproductions of previously published papers are not accepted. Both reviews and reports of research are acceptable since symposia may embrace both types of presentation.

CONTENTS

BIOLOGICAL INFLUENCE OF INITIATION

POLYMER CHEMISTRY

PREFACE

Although the initiation step in polymerization is of fundamental importance in polymer synthesis, no in-depth assessment of its status has been made in recent years—years in which great strides were made in understanding and controlling polymer architecture by exploiting knowledge of initiation mechanisms. Such an assessment, therefore, is long overdue. The topic of the initiation step in polymer synthesis fit well into the general format of Macromolecular Secretariat symposia because of its highly interdisciplinary character.

The Macromolecular Secretariat is a consortium of five ACS divisions: the Division of Cellulose, Paper, and Textile Chemistry; the Division of Colloid and Surface Chemistry; the Division of Organic Coatings and Plastics Chemistry; the Division of Polymer Chemistry; and the Rubber Division. Each division contributed its particular focus and expertise to the symposium presented here by selecting chairmen who are eminent authorities in their fields, and who also served as associate editors of this volume: Jett C. Arthur, Jr.; Alexandre Blumstein; Murrae J. Bowden; Edwin J. Vandenberg; Joginder Lal; and Ray M. Ottenbrite. The associate editors selected papers for inclusion in the symposium to reflect the most recent advances in control of polymerization processes. Knowledge of the mechanisms and kinetics of initiation extends the range of use and applications of polymers. For their assistance, I am most indebted and grateful.

The international character of this symposium, evident from the significant contributions by speakers from outside the United States, was made possible by a Special Educational Opportunity Grant from the Petroleum Research Fund, which provided the registration and travel assistance that enabled many foreign guests to participate. Special thanks are due to the Petroleum Research Fund for this grant, and also to Union Carbide Corporation for sponsoring an informal opening of the symposium prior to the plenary session.

Frederick E. Bailey, Jr.
Union Carbide Corporation
South Charleston, WV 25303

October 29, 1982

PLENARY LECTURES

New Syntheses of Functional and Sequential Polymers by Exploiting Knowledge of the Mechanism of Initiation

JOSEPH P. KENNEDY

The University of Akron, Institute of Polymer Science, Akron, OH 44325

The first part of this presentation concerns a brief overview of the mechanism of initiation of carbocationic polymerizations; in particular it addresses head group control by protonation, cationation, and the controlled initiation concept which led to a large new family of block, graft, and bigraft copolymers. Subsequently the first carbocationic macromer synthesis yielding polyisobutenylstyrene is described. The copolymerization of the latter macromer with acrylates gave rise to new graft copolymers. Then head group control with inorganic moieties, i.e., Cl-, NO_2- and $Ø_3$Si-, is outlined. The first part concludes with a discussion of the similarity between the mechanisms of initiation and chain transfer, the appreciation of which led to the inifer concept, which in turn yielded new telechelics, networks, sequential copolymers, etc. The second part of this presentation focuses on practical consequences of understanding details of the mechanism of initiation. The synthesis of a new family of telechelic linear and tri-arm star polyisobutylenes will be described. Among the new prepolymers are telechelic olefins, epoxides, aldehydes, alcohols, and amines. The preparation of new ionomers and polyisobutylene-based polyurethanes will be outlined and some fundamental properties of these new materials will be discussed.

0097-6156/83/0212-0003$06.00/0

The surest way toward desirable new polymer structures is by a systematic exploitation of the detailed mechanistic understanding of polymerization processes. The aim of this presentation is to examine in some depth the mechanism of carbocationic polymerizations and subsequently to apply this insight toward the preparation of new polymeric materials possessing advantageous combinations of processing and/or physical-mechanical characteristics.

Initiation in Carbocationic Polymerization

Early Developments till the Discovery of Controlled Initiation. Under suitable conditions any electrophilic species may induce cationic polymerizations (1). As a practical matter, the most convenient cationogens are Bronsted acids alone or in conjunction with Friedel-Crafts acids (1). Systematic research on the initiation of carbocationic polymerization became possible by the discovery of coinitiation by British investigators (2-4). These workers found that the strong Lewis acid BF_3 alone is unable to initiate isobutylene polymerization but in the presence of suitable cationogens, i.e., H_2O, immediate and vigorous polymerization ensues. Their formalism:

$$BF_3 + H_2O \rightleftharpoons H^{\oplus}BF_3OH^{\ominus}$$

$$H^{\oplus}BF_3OH^{\ominus} + CH_2{=}C(CH_3)_2 \rightarrow CH_3\text{-}\overset{\oplus}{C}(CH_3)_2BF_3OH^{\ominus}$$

provided valuable guidance for subsequent research on the mechanism of protic initiation and in the context of this presentation is regarded as the first proposition of head group control, i.e., the incorporation of a proton as the "head group" of a macromolecule. Not much later Pepper (5) suggested that certain alkyl halides RX in conjunction with Friedel-Crafts acids MX_n may also act as initiators:

$$RX + MX_n \rightleftharpoons R^{\oplus}MX_{n+1}^{\ominus}$$

$$R^{\oplus} + C{=}C \rightarrow R\text{-}C\text{-}C^{\oplus}$$

Although the formalism of cationation by alkyl halides was in principle correct, the significance of this mechanism for head group control could not be exploited because the conventional Friedel-Crafts acids (BF_3, $AlCl_3$, $SnCl_4$, $TiCl_4$) used by the early workers were extremely moisture sensitive (i.e., induced polymerization by protonation in the presence of putative traces of moisture impurities) and because large quantities of CH_3- head groups also arose by unavoidable chain transfer to monomer.

In 1965 it was discovered (6,7) that under conventional "open" laboratory conditions (i.e., in the presence of moisture) olefin polymerizations can be readily initiated by protic acids or by "active" organic halides (t-butyl, benzyl, allyl) in conjunction with certain alkylaluminum compounds (Me_3Al, Et_3Al, Et_2AlCl, etc.). Since these di- and trialkylaluminum compounds are effective moisture scavengers and their reaction products with small amounts of H_2O are polymerization inactive, initiation can occur only upon the purposeful addition of suitable cationogens. For reasons beyond the scope of this presentation, polymerizations induced by RX/R_2AlX combinations are chain transferless (transferless polymerizations are discussed in ref. 1) so that the head groups of polymers formed by these initiating systems can be readily controlled. For example, (M = monomer):

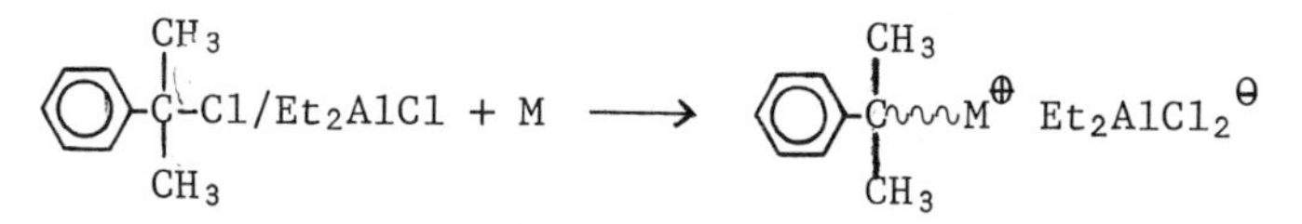

The head groups of polymers formed by such "controlled initiation" systems are determined by the nature of the initiator molecule. A large number of controlled initiation systems giving rise to polymers bearing desirable functional head groups e.g., allyl, benzyl, cyclopentadienyl, silyl, have been described (1).

Possibilities offered by controlled initiation have been exploited for the preparation of new sequential copolymers and macromers and are discussed below.

Sequential (Block and Graft) Copolymers by Controlled Initiation. Controlled initiation by active high molecular weight halides (i.e., macromolecules containing tertiary alkyl or benzylic halides) in conjunction with di- and trialkylaluminum coinitiators led to the synthesis of a large variety of block, graft and bigraft copolymers. Schematically (X and Y = halide functions):

$$\text{AAAAAAAA-X} \xrightarrow{+\ nB} \text{AAAAAAAA-BBBBB}$$

diblocks

$$\text{X-AAAAAAAA-X} \xrightarrow{+\ nB} \text{BBBBB-AAAAAAAA-BBBBBB}$$

triblocks

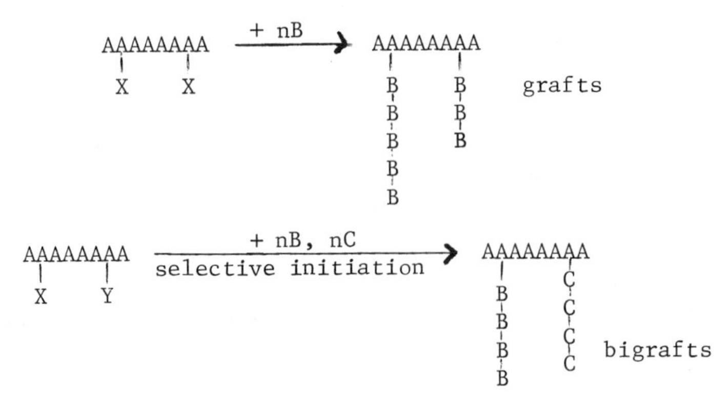

A large number of sequential copolymers comprising many combinations of glassy and elastomeric segments have been prepared (1). The block and graft copolymers prepared by controlled initiation are remarkably well defined and many are free of homopolymers. Sequential copolymers obtained by controlled cationic initiation have been recently comprehensively surveyed (1,8).

Macromers by Controlled Initiation. New and unique graft copolymers can be prepared by copolymerizing macromers (macromolecular monomers) with conventional monomers. The synthesis of poly(butyl acrylate-*g*-isobutylene), i.e., the first graft synthesis that involves carbocationic controlled initiation, has recently been accomplished by the following route (9):

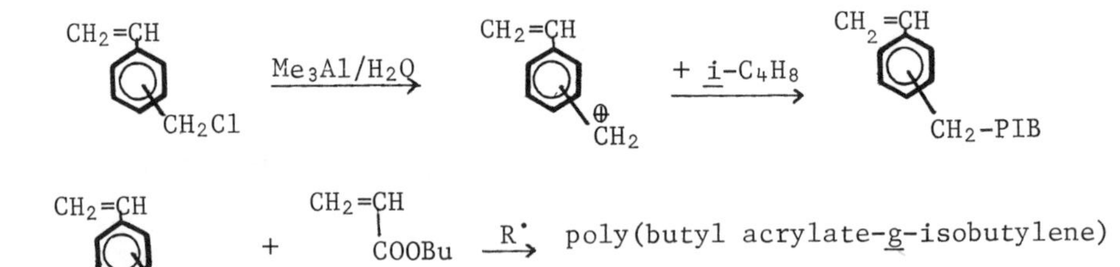

The copolymerization of the polyisobutenylstyrene macromer with methyl methacrylate and styrene gave further interesting new materials (10,11).

Inorganic Head Groups by Controlled Initiation. In addition to organic functional groups, inorganic head groups, e.g., Cl-, Br-, O_2N-, can also be introduced into polymers by controlled initiation. For example, the incorporation of Cl- head group into polyisobutylene has been accomplished by the Cl_2/BCl_3 system (12):

$$Cl_2 + BCl_3 \rightleftharpoons Cl^{\oplus}\ BCl_4^{\ominus} + \xrightarrow{i\text{-}C_4H_8} Cl\text{-}CH_2\text{-}\overset{\oplus}{C}(CH_3)_2 \rightrightarrows Cl\text{-}PIB$$

Similarity between Initiation and Chain Transfer:

1. The Inifer Concept. A detailed analysis of the respective mechanisms of initiation and various chain transfer processes showed that the fundamentals of initiation by alkyl halides and chain transfer to alkyl halides are indeed very similar. It is postulated that an alkyl halide that is able to initiate carbocationic polymerizations, may also be able to function as a chain transfer agent. Schematically (1):

Initiation:

$$RX + MX_n \rightleftharpoons R^{\oplus}MX_{n+1}^{\ominus} \xrightarrow{+\ C=C} R\text{-}C\text{-}C^{\oplus}\ MX_{n+1}^{\ominus}$$

Chain Transfer

$$\sim\!\!\sim C\text{-}C^{\oplus}MX_{n+1}^{\ominus} + RX \rightleftharpoons \sim\!\!\sim C\text{-}CX + R^{\oplus}MX_{n+1}^{\ominus} \xrightarrow{+C=C} \sim\!\!\sim C\text{-}CX + R\text{-}C\text{-}C^{\oplus}MX_{n+1}^{\ominus}$$

Alkyl halides that perform simultaneously as initiators and chain transfer agents are in general termed inifers. Thus in the absence of chain transfer to monomer i.e., when $R_{tr,i} > R_{tr,M}$ (rate of chain transfer to inifer is larger than that to monomer), the termini of polymers can be readily controlled by inifers. Monofunctional inifers termed minifers can be used to control one terminus; bifunctional inifers XRX, termed binifers, are useful to control two termini; trifunctional inifers XR(X)X, termed trinifers, can control the three termini of three-arm star molecules, etc. The following scheme helps to visualize the gist of the mechanism leading to end group control i.e., to telechelic polymers, be these linear or star-shaped, by inifers (1):

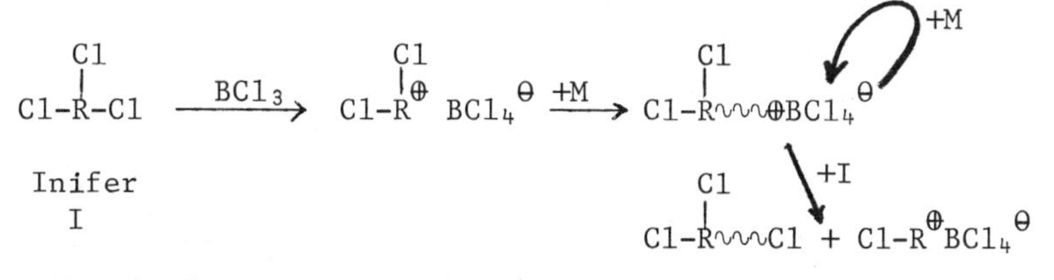

2. New Products Prepared by the Inifer Technique. A new family of linear and three-arm star polyisobutylenes have been prepared by the inifer technique. The inifer systems used were cumyl chloride (minifer) p-dicumyl chloride (binifer) and sym-tricumyl chloride (trinifer) always in combination with BCl_3 coinitiator (1):

$$C_6H_5\text{-}C(CH_3)_2\text{-}Cl \xrightarrow[BCl_3]{iC_4H_8} C_6H_5\text{-}C(CH_3)_2\sim\!\!\sim PIB \sim\!\!\sim CH_2\text{-}C(CH_3)_2\text{-}Cl$$

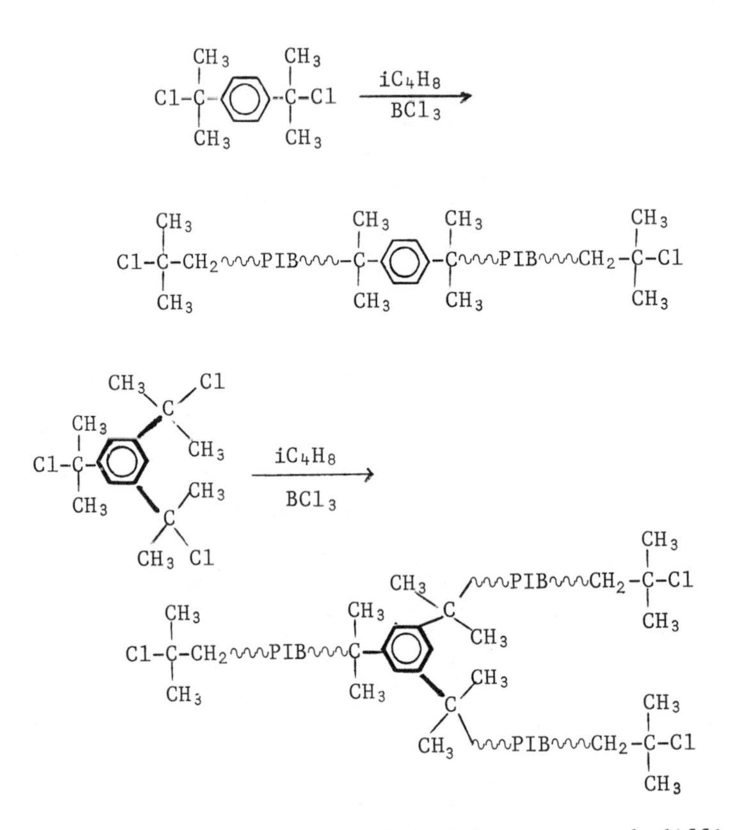

Side-reactions can be avoided without too much difficulty so that virtually theoretical end group functionalities can be obtained, i.e., $\overline{F}_n = 1.0$, 2.0 or 3.0. Molecular weight control can be readily accomplished by adjusting reagent concentrations and products ranging from liquids to rubbery solids can be prepared.

The tertiary chlorine termini are useful starting points for a variety of derivatizations, e.g., diblock, triblock and tri-star block syntheses with styrene, α-methylstyrene (by alkylaluminum chemistry) and/or cyclic ethers (by silver salt chemistry). For example, the synthesis of the triblock PαMeSt-*b*-PIB-*b*-PαMeSt, a thermoplastic elastomer whose physical-mechanical properties are determined by the central saturated rubbery PIB block sandwiched between two glassy PαMeSt blocks, is outlined by the following set of equations:

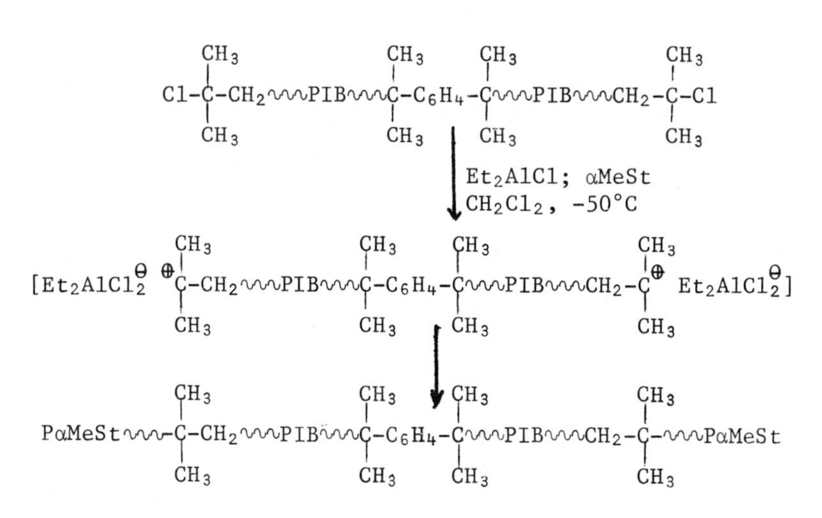

Details on the synthesis and characterization of three-arm star thermoplastic elastomer Ø(PIB-b-PαMeSt)$_3$ are given in a presentation delivered in another Section of this A.C.S. Meeting (13):

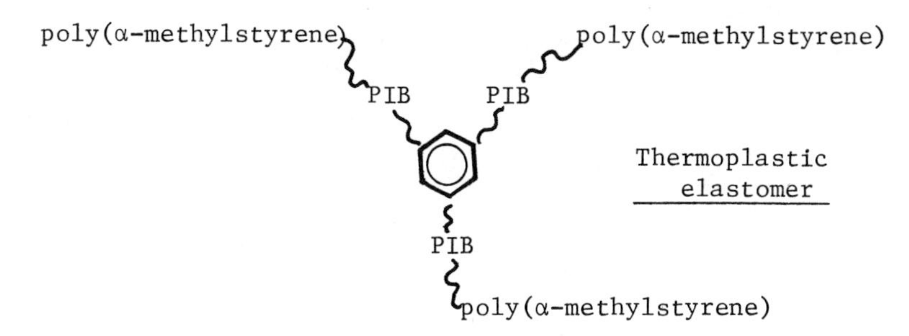

Regioselective quantitative dehydrochlorination (t-BuOK, THF) of the chlorine-terminated products led to olefin telechelic derivatives which in turn yielded many potentially useful materials (14). Thus hydroboration (a quantitative reaction) gave rise to new polyisobutylene-based diols and triols which in combination with isocyanates gave unique polyurethanes:

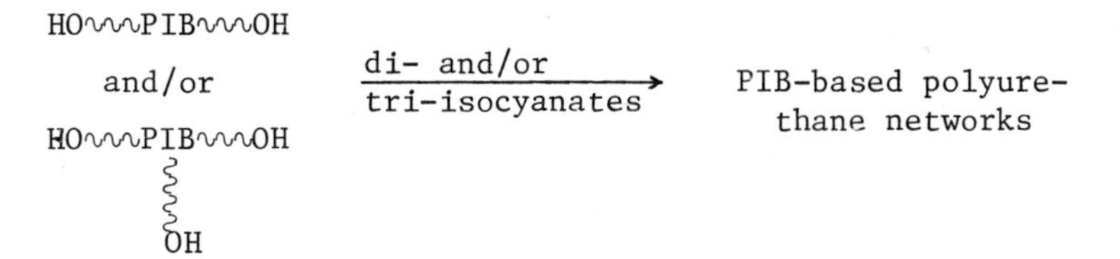

Polyurethane sheets containing saturated rubbery (Tg ∿ -70°C) polyisobutylene segments exhibit excellent barrier properties and environmental resistance.

An interesting group of new ionomers have been prepared from liquid olefin telechelic polyisobutylenes. Syntheses involved the quantitative sulfonation (by acetyl sulfate) of linear and/or tri-arm star polyisobutylenes carrying exactly two or three $-CH_2-C(CH_3)=CH_2$ termini, respectively, followed by neutralization with various bases:

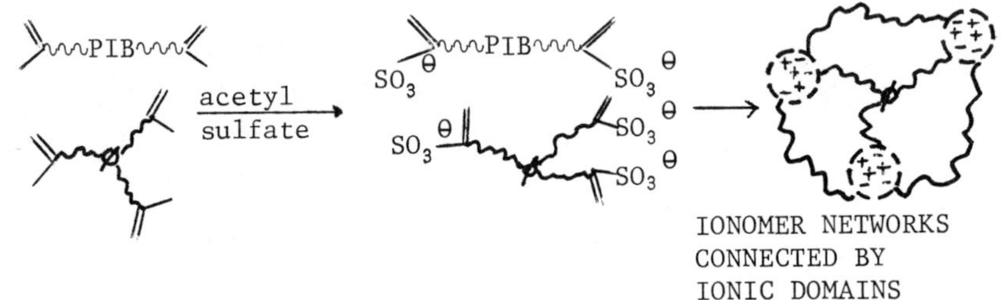

These ionomers are optically clear tough, rubbery materials that can be readily formed and reformed for example by casting. Since every terminus carries a functional group, dangling ends are largely absent and every network element participates in load-bearing. Some details of the synthesis and characterization of these new easily processable polyisobutylene-based ionomers are given in a presentation delivered in another Section of this ACS Meeting (15).

Quantitative epoxidation of olefin-telechelic polyisobutylenes yielded interesting new epoxy-telechelic products. Isomerization of the linear epoxy-telechelic polyisobutylene quantitatively yielded aldehyde-telechelic chains:

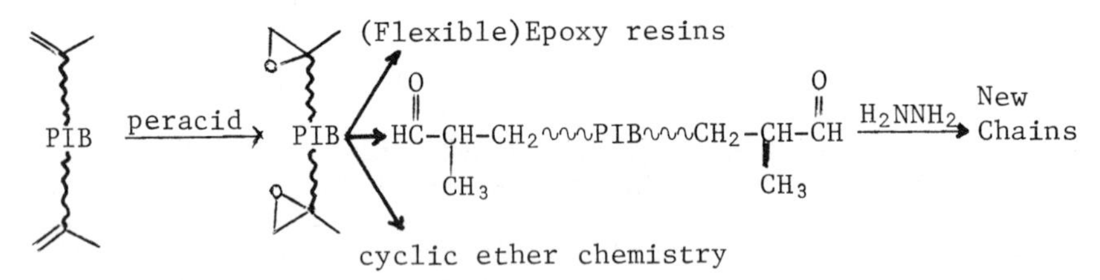

A large variety of quantitative conversions and derivatizations have been accomplished and/or are under intensive investigation involving polyisobutylenes carrying exactly one, two or three (three-arm star) OH- termini. Some of the reactions that have been accomplished together with the products obtained are summarized:

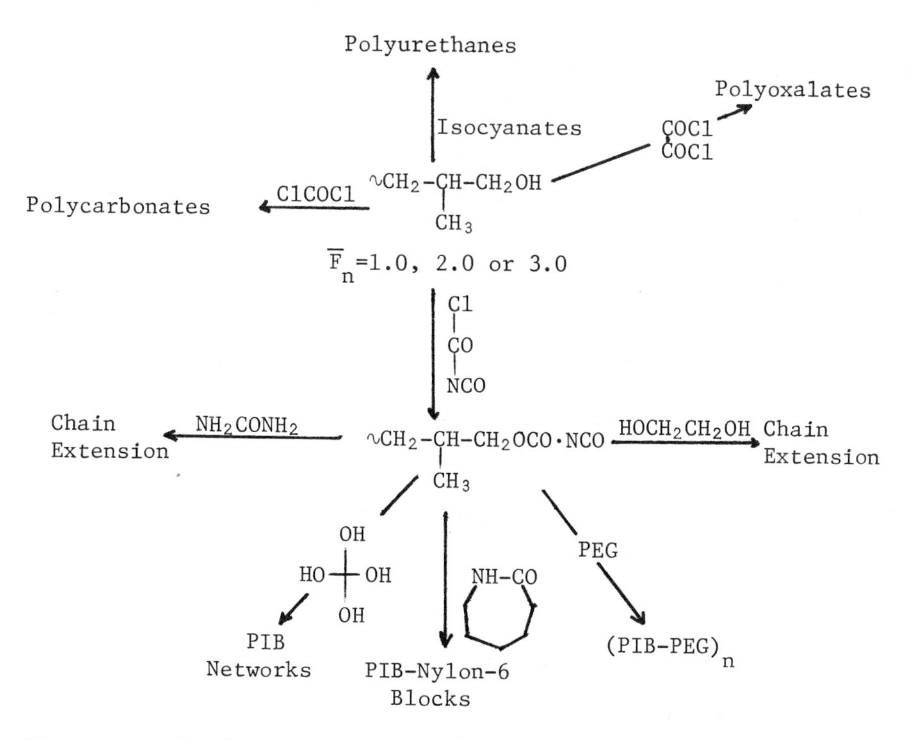

Literature Cited

1. J. P. Kennedy and E. Maréchal, "Carbocationic Polymerizations", J. Wiley-Interscience, New York, 1982.
2. A. G. Evans, D. Holden, P. H. Plesch, M. Polányi, H. A. Skinner and M. A. Weinberg, Nature, 157, 102 (1946).
3. A. G. Evans and M. Polányi, J. Chem. Soc., 252 (1947).
4. P. H. Plesch, M. Polányi, and H. A. Skinner, J. Chem. Soc., 257 (1947).
5. D. C. Pepper, Trans. Faraday Soc. 45, 404 (1949).
6. J. P. Kennedy, Belgian Patent 663,319 (April 30, 1965).
7. J. P. Kennedy and F. P. Baldwin, Belgian Patent 663,320 (April 30, 1965).
8. Cationic Graft Copolymerization, J. P. Kennedy, ed., J. Appl. Polym. Sci., Appl. Polym. Symp., 30 (1978).
9. J. P. Kennedy and K. C. Frisch, Jr., Preprints, Macro Florence IUPAC, Vol. 2, 162 (1980).
10. J. P. Kennedy and K. C. Frisch, unpublished, Akron, 1980.
11. J. P. Kennedy and C. Y. Lo, unpublished, Akron, 1982.
12. J. P. Kennedy and F. J.-Y. Chen, Polymer Prepr., 20, 310 (1979).

13. J. P. Kennedy, L. R. Ross and S. C. Guhaniyogi, Presentation 38, Div. of Org. Coating and Plastics Chemistry, 183rd Natl. ACS Meeting, Las Vegas, 1982.
14. J. P. Kennedy, V. S.-C. Chang, R. A. Smith and B. Iván, Polym. Bull., 1, 575 (1979).
15. J. P. Kennedy and R. F. Storey, Presentation 39, Div. Org. Coating and Plastics Chemistry, 183rd Natl. ACS Meeting, Las Vegas, 1982.

RECEIVED August 24, 1982

2

Catalysis by Water-Soluble Imidazole-Containing Polymers

C. G. OVERBERGER and RICHARD TOMKO

The University of Michigan, Department of Chemistry and the Macromolecular Research Center, Ann Arbor, MI 48109

A summary of our recent publications describing water-soluble imidazole containing polymers is presented. Copoly[1-alkyl-4- or 5-vinylimidazole/4(5)-vinylimidazole] (I), copoly[vinylamine/4(5)-vinylimidazole] (II), and dodecane-block-poly[ethylenimine-graft-4(5)-methylimidazole] (III) have been used to investigate the hydrophobic interaction in esterolytic reactions. A brief survey of related work is presented along with our current work.

Several reviews have been published describing our work on imidazole catalysis.(1-4) Other investigators in this field have published comprehensive review articles comparing synthetic catalysts to enzymes.(5-9) The esterolytic activity of the three catalyst systems reviewed in this paper have been compared to poly[4(5)-vinylimidazole] (PVIm) in 28.5% ethanol-water solutions since PVIm is not soluble in water. However, within each system, minor changes in catalyst apolarity illustrate the effect of hydrophobic interactions on the rate of esterolysis. The substrates consisted of a series of neutral _p_-nitrophenyl esters (Sn) and negatively charged 4-acyloxy-3-nitrobenzoic acids (Sn^-), where n denotes the number of carbons in the acyl chain.

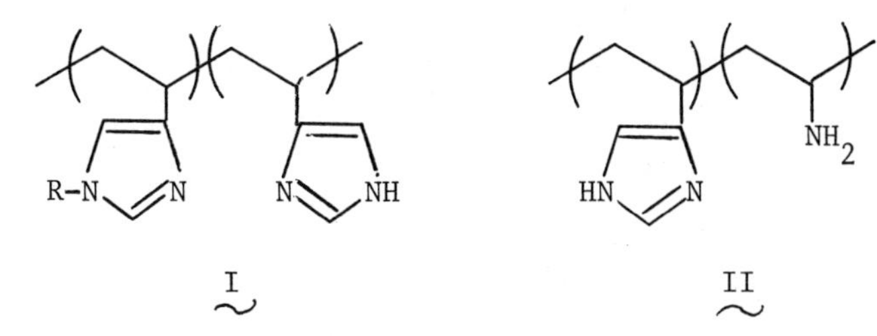

0097-6156/83/0212-0013$06.00/0

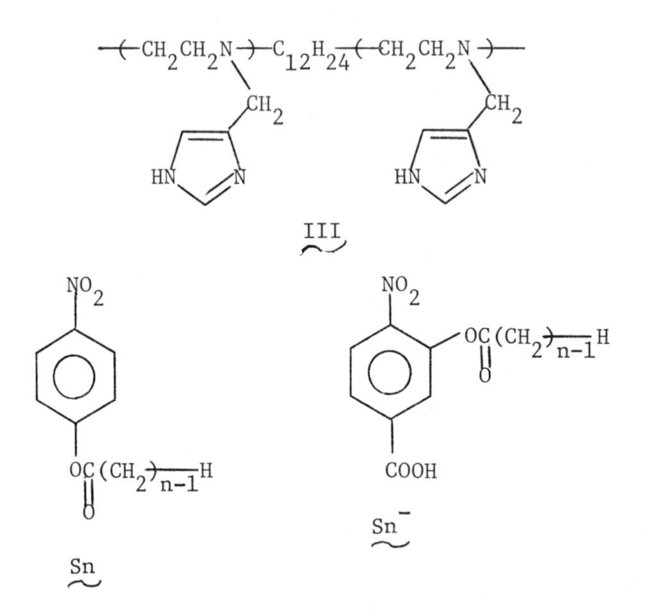

Copolymers of 1-alkyl-4- or 5-vinylimidazole and 4(5)-vinylimidazole (I) were synthesized with variations in the alkyl chain length and the 4(5)-vinylimidazole content.(10) The incorporation of alkylated imidazole residues into the backbone of poly[4(5)-vinylimidazole] not only provided efficient regeneration of the catalysts, but also imparted water solubility to the copolymers. (11) These copolymers were compared with mixtures of their component homopolymers having the same vinylimidazole content.

The pKa values of the imidazole residues in the copolymers decreased with decreasing vinylimidazole content. The increased acidity was attributed to a destabilization of the protonated imidazoles by intramolecular interaction with the alkylated imidazole residues. There was also a slight decrease in pKa as the length of the alkyl imidazole side chain was increased. The imidazole pKa values were higher in water (4.99-5.92) than in 28.5% ethanol-water (4.94-5.80)(12) due to enhanced stability of the charged residues in the more polar medium. The pKa values for the copolymers were lower than the pKa values for PVIm in 28.5% ethanol-water (5.78-6.20). The pH-rate profiles for the copolymers in the hydrolysis of S_2 in 28.5% ethanol-water were similar to PVIm.(13) The copolymers were more efficient catalysts than a mixture of the component homopolymers due to the increased basicity of the copolymers. Lengthening the alkyl side chain from one to three carbons had little effect on the rate of hydrolysis.

Bell-shaped pH-rate profiles similar to PVIm were observed in the hydrolysis of the Sn^- series.(14) There was a maximum

Effect of Variations in Alkyl Chain Length on Copolymers of Vinylimidazoles

Catalyst	pH	Solvent [a]	$k_{cat}(M^{-1}min^{-1})$ [b] S_2^-	$k_{cat}(M^{-1}min^{-1})$ [b] S_{12}^-	pKa [c]	$k_{cat}(M^{-1}min^{-1})$ [b] S_2 (pH)
Imidazole [d]	8.0	Water	70	40	6.95 [e]	16 (7)
Imidazole [f]	8.0	20% EtOH-water	58	26	7.00 [g]	11.5 (7) [h]
PVIm [f]	8.0	20% EtOH-water	338	10,316		
PVIm [i]	7.1	28.5% EtOH-water	110	1,409	6.00 [g]	7 (7) [g] 18.2 (8)
$P(VIm_{75}-VAm)$ [j]	8.0	Water	1,101	10,706	5.40	84 (8) 47 (7)
$P(VIm_{67}-VAm)$ [j]	8.0	Water	984	11,409	5.11	77 (8) 51 (7.2)
P(EIIm) [k]	7.0	Water	88	1,734	5.35	- -
$P(EIIm_{85}-C_{12})$ [l]	7.0	Water	107	3,566	5.13	1.89 (7)
$P(EIIm_{25}-C_{12})$ [l]	7.0	Water	52	1,325	5.02	- -
$P(VIm-1,4Me)_{50}$ [m]	7.1	28.5% EtOH-water	58	1,264	5.45	10 (7)
50% Homo Mix	7.1	28.5% EtOH-water	58	643	- -	- -
$P(VIm-1,5Me)_{50}$ [m]	7.1	28.5% EtOH-water	71	1,183	5.37	10 (7)
50% Homo Mix	7.1	28.5% EtOH-water	79	686	- -	- -
P(1,4Me) [n]	6.85	28.5% EtOH-water	- - -	- - -	5.2	0
P(1,5Me) [n]	6.85	28.5% EtOH-water	25.4	- - -	4.7	4.1 (7)

Footnotes on next page

Footnotes to table

[a] Tris buffer, $\mu = 0.02$, water solvent contained 3.3% acetonitrile, 26°C.

[b] [Catalyst] = 5×10^{-4} $\underline{M}$, [substrate] = 5×10^{-5} $\underline{M}$.

[c] Determined at 25 or 26°C, $\mu = 0.02$.

[d] Data taken from Reference 17.

[e] Data taken from Reference 34.

[f] Data taken from Reference 14.

[g] In 25% EtOH-water; data taken from References 11 and 13.

[h] In 28.5% EtOH-water; data taken from Reference 35.

[i] Data taken from Reference 10.

[j] Structure II contains 67 or 75% vinylimidazole residues; data taken from Reference 15.

[k] PEI-graft-4(5)-methylimidazole; data taken from Reference 18.

[l] Structure III with PEI blocks of DP 25 or 85; data taken from Reference 18.

[m] Structure I 50:50 Copolymer of 4(5)-vinylimidazole and 1-methyl-4- or 5-vinylimidazole compared to a 50% mixture of homopolymers. Reference 10.

[n] Homopolymers of 1-methyl-4- or 5-vinylimidazole; data taken from Reference 11.

around pH 7 where there were significant amounts of both protonated imidazoles for electrostatic attraction and catalytically active neutral imidazoles. PVIm was a better catalyst than the copolymers and the mixtures of component homopolymers in 28.5% ethanol-water through the Sn^- series. There was little difference in reactivity between the copolymers and the mixture of their component homopolymers in the hydrolysis of S_2^-. However, the copolymers were better catalysts from S_4^- to S_{18}^-. In general, the reactivity of the copolymers increased with 4(5)-vinylimidazole content, the substrate chain length, and the 1-alkyl imidazole side chain length.

Finally, the second order rate constant (k_{cat}) exhibited a greater increase with Sn^- and the catalyst side chain length in water than in 28.5% ethanol-water, where hydrophobic interactions are weaker. The copolymers hydrolyzed S_{12}^- about five times faster in water than in 28.5% ethanol-water.

Copolymerization of 4(5)-vinylimidazole with vinylamine increased the water solubility of the catalyst and, therefore its esterolytic activity.(15) Copoly[4(5)-vinylimidazole-vinylamine] (II) had considerably lower pKa values for the imidazole residues in water (5.11-5.40) than the imidazole residues for PVIm in 28.5% ethanol-water (5.78-6.20).(14) Apart from the difference in solvent, the polyelectrolyte effect from the strongly basic amine residues (pKa = 7.35-7.74 in water) was responsible for the lower pKa values.

The pH-rate profiles for the hydrolysis of S_2 with these copolymers in water are similar to PVIm in 28.5% ethanol-water. However, the copolymers are very efficient catalysts even at the lower pH values compared to PVIm which is less efficient than imidazole as a catalyst.(10,13) This greater catalytic efficiency was due to the availability of more neutral imidazole residues (lower pKa) and the more compact coil in the aqueous media. Both of these factors increase the imidazole-imidazole cooperative interactions.(5) There was no evidence for either cooperative interactions involving vinylamine units or vinylamine catalyzed aminolysis of the substrate.(16)

The pH-rate profiles for the copolymers with S_2^- did not result in the bell-shaped curves found in previous studies.(10,14) The continuous rate increase with pH (similar to the PVIm and copoly[1-alkyl-4(5)-vinylimidazole-4(5)-vinylimidazole] hydrolysis of S_2) was rationalized by an electrostatic attraction of the substrate to the primary ammonium residues in the backbone while the imidazole residues remained catalytically active. The copolymers hydrolyzed S_2^- about three times faster in water than PVIm in 20% ethanol-water at pH 8.(14) The rate constants increased continuously in going to longer chain substrates (S_2^- to S_{12}^-). Therefore, hydrophobic interactions begin to play a dominant role even with S_4^- as opposed to PVIm where the rate constant decreased from S_2^- to S_4^-.(11,14,17) The best copolymer catalyst for S_7^- was the more tightly coiled copolymer with the lesser amine content. The hydrophobic contribution of the substrate eventually over-rides all other factors in the hydrolysis of S_{12}^- where the rates of hydrolysis for the copolymers are nearly equivalent. In the hydrolysis of S_{12}^- at pH 8, the copolymers were nearly 300 times as efficient as imidazole in water which makes them just slightly better catalysts than PVIm in 20% ethanol-water.(14)

The more apolar substrates exhibited decelerative behavior as seen in other systems where the hydrophobicity of the catalyst increased due to acylation of the imidazole residues.(14) Solvolysis in excess substrate revealed that the steady state concentration of acylated imidazoles is much less for the copolymers than for PVIm.

The copolymers in 20% ethanol-water were more effective catalysts than PVIm with S_2^- where electrostatic interactions are more important than hydrophobic interactions. However, with S_{12}^- where apolar interactions are predominant, the copolymers were less efficient catalysts than PVIm.(14)

Dodecane-block-poly[ethylenimine-graft-4(5)-methylimidazole] copolymers (III) were efficient catalysts for the hydrolysis of p-nitrophenyl esters in water.(18) These copolymers differ from previous poly[ethylenimine] (PEI) catalysts(8,21-24) because they are derived from linear PEI, incorporate a well-defined apolar binding site (dodecane block) as a portion of the polymer backbone, and they contain a high weight percent of pendant imidazole groups. Since all of copolymers contained the same dodecane block, a greater apolar weight was associated with the lower-molecular-weight polymers (fewer PEI units).

The pKa values of the pendant imidazoles (5.02-5.13 in water) and the ethylenimine groups (7.00-7.08 in water) decreased with decreasing molecular weight of the copolymers. This decrease in pKa with increasing apolarity is further evidenced by the largest imidazole pKa value (5.35 in water) belonging to the homopolymer graft, poly[ethylenimine-graft-4(5)-methylimidazole], which contains no apolar block. Another factor lowering the pKa values may be the close proximity of the protonated ethylenimine units.(15)

The pH-rate profile for the hydrolysis of S_2 showed that imidazole in water was a 3-fold more effective catalyst than the PEI grafts at every pH. At pH less than 7.25, the copolymers were less efficient than the catalyst containing no apolar block. At pH greater than 7.25, the reactivities of the copolymers increased which indicated a more efficient use of the apolar binding site. This was due to the greater concentration of pendant imidazoles associated with the apolar domain. Since the lower-molecular-weight copolymers (greater apolar weight) exhibited the greater rate enhancements, cooperative interactions were ruled out in favor of hydrophobic effects.

There was a continuous increase in the second order rate constant (k_{cat}) from 20- to 100-fold at pH 9.15 through the Sn series. The nearly equivalent catalytic efficiencies of the copolymer and homopolymer grafts suggest that the major hydrophobic interaction involved the N-acylimidazole intermediate. However, throughout this series, the copolymers were still better catalysts than the homopolymer.

Analogous to other polymeric imidazole catalysts,(10,14) the copolymer and homopolymer grafts displayed bell-shaped pH-rate profiles for the hydrolysis of S_2^-. The more efficient copolymer catalysts contained ethylenimine blocks of higher molecular weight which possessed a stronger electrostatic attraction to the substrate. This attraction makes these copolymers in water almost as efficient catalysts as PVIm in 28.5% ethanol-water with S_2^- at pH 7.11.(10) For the Sn^- series, again the largest rate enhancements were displayed by the higher-molecular weight copolymers with the lesser apolar weight. The highest-molecular weight copolymer at pH 7 with S_{12}^- is 2.5 times more reactive than PVIm in 28.5% ethanol-water at pH 7.11,(14) however the homopolymer graft is only slightly more reactive than PVIm. Therefore, the highest molecular-weight copolymer catalyst made the most effective use of both electrostatic and hydrophobic interactions.

Saturation kinetics with the highest molecular-weight copolymer in water indicated that both substrate binding and the rate of intracomplex breakdown increased through the Sn^- series. (5,19) The Michaelis-Menten kinetics for PVIm(11,14) in ethanol-water suggested that increased substrate binding has generally correlated with decreases in the rate of intracomplex breakdown. Therefore, the results from the block copolymers indicate that the PEI backbone provided extra stabilization for N-acylimidazole hydrolysis within the apolar environment.(5)

Our current work focuses on grafting imidazole pendants onto water soluble polymers and the synthesis of optically active poly[4(5)-vinylimidazole] catalysts. In our continuing effort to investigate the hydrophobic effect, ω-(1-imidazolyl)alkanoic acids have been grafted onto poly(vinylamine) through an amide bond. These grafting reactions allow us to vary the apolarity of the catalyst while, ensuring water solubility and locating the imidazole groups away from the polymer backbone. 1-(S)-(1-Methylbutyl)-4-vinylimidazole and 1-(S)-(2-methylbutyl)-5-vinylimidazole have been copolymerized with various monomers and cross-linked with divinylbenzene. The hydrolysis of enantiomeric esters(17,20) is being studied in order to determine if these catalysts possess enantioselectivity.

Professor Irving Klotz and Professor Toyoki Kunitake are well-known contributors to the field of polymer catalysis. Klotz and his associates have continued their investigations of branched

PEI derivatives.(21-24) A series of substituted aminopyridines, covalently attached to laurylated PEI, hydrolyzed p-nitrophenylcaproate (S_6) 50-2000 times faster than the isolated aminopyridines. The aminopyridines are better nucleophiles than imidazole due to their higher pKa values.(25) Klotz intends to produce stereoselective PEI catalysts by coupling optically active amino acids to the aminopyridine nucleophile.(26,27) Klotz has also investigated salicylaldoxime as a nucleophile in quaternized PEI and analogous cationic micelles.(28)

Kunitake and his coworkers have investigated bifunctional polymer catalysts(29) in micelles(30) and polysoaps.(31) The bifunctional interaction between a hydroxamate nucleophile and a neighboring imidazole group at the catalytic site is similar to the charge relay system of serine proteases. This interaction leads to remarkably accelerated acylation and deacylation processes. In the hydrophobic environment of cationic micelles, where the reactivity of anionic nucleophiles are remarkably enhanced, the overall catalytic efficiency exceeded even that of α-chymotrypsin at pH 8.(30) Micellar monofunctional catalysts, nonmicellar bifunctional catalysts, and polysoap-bound bifunctional catalysts were less effective.

Kunitake has also studied the hydrolysis of p-nitrophenyl esters in the presence of aqueous bilayer vesicles. The rate of inter-vesicle S_2 hydrolysis by a cholesteryl derivative of 4(5)-imidazole carboxylic acid was much slower than the intra-vesicle counterpart.(32) Rate-limiting substrate transfer was responsible for the lower rates and was influenced by the crystalline-liquid crystalline transition.(33)

Acknowledgments: The authors are grateful for financial support from the National Science Foundation under Grants DMR78-13400 and DMR 81-06891, and the Sherwin-Williams Company.

Literature Cited

1. Overberger, C. G.; Salamone, J. C. Accts. Chem. Res., 1969, 2, 217.
2. Overberger, C. G.; Smith, T. W.; Dixon, K. W. J. Polym. Sci. Symp., 1975, 50, 1.
3. Overberger, C. G.; Guterl, A. C.; Kawakami, Y.; Mathias, Lon J.; Meenakshi, A.; Tomono, T. Pure Appl. Chem., 1978, 50, 309.
4. Pavlisko, J. A.; Overberger, C. G. "Biomedical and Dental Applications of Polymers," Polymer Science and Technology, Vol. 14, Gebelein, C. G. and Koblitz, F. F., Eds., Plenum Press, New York, 1981. p. 257.
5. Imanishi, Y. J. Polym. Sci., Macromol. Rev., 1979, 14, 1.
6. Kunitake, T.; Okahata, Y. Adv. Polym. Sci., 1976, 20, 159.
7. Kunitake, T.; Shinkai, S. Adv. Phys. Org. Chem., 1980, 17, 435.
8. Klotz, I. M. Adv. Chem. Phys., 1978, 39, 109.

9. Fendler, J. H.; Fendler, E. J. "Catalysis in Micellar and Macromolecular Systems," Academic Press, 1975, p. 387.
10. Overberger, C. G.; Kawakami, Y. J. Polym. Sci., Polym. Chem. Ed., 1978, 16, 1237, 1249.
11. Overberger, C. G.; Smith, T. W. Macromolecules, 1975, 8, 401, 407, 416.
12. Overberger, C. G.; St. Pierre, T.; Vorchheimer, N.; Lee, J.; Yaroslavsky, S. J. Am. Chem. Soc., 1965, 87, 296.
13. Overberger, C. G.; Morimoto, M. J. Am. Chem. Soc., 1971, 93, 3222.
14. Overberger, C. G.; Glowaky, R. C.; Vandewyer, P. H. J. Am. Chem. Soc., 1973, 95, 6008, 6014.
15. Overberger, C. G.; Mitra, S. Pure Appl. Chem., 1979, 51, 1391.
16. St. Pierre, T.; Vigee, G.; Hughes, A. R. "Reactions on Polymers," Moore, J. A., Ed., D. Reidel Publishing Co., Boston, 1973, p. 61.
17. Overberger, C. G.; Dixon, K. W. J. Polym. Sci., Polym. Chem. Ed., 1977, 15, 1863.
18. Pavlisko, J. A.; Overberger, C. G. J. Polym. Sci., Polym. Chem. Ed., 1981, 19, 1621, 1757.
19. Kunitake, T.; Shinkai, S. Makromol. Chem., 1972, 151, 127.
20. Overberger, C. G.; Cho, I. J. Polym. Sci., A-1, 1968, 6, 2741.
21. Klotz, I. M.; Spetnagel, W. J. Polym. Sci., Polym. Chem. Ed., 1977, 15, 621.
22. Klotz, I. M., Nango, M. J. Polym. Sci., Polym. Chem. Ed., 1978, 16, 1265.
23. Klotz, I. M.; Stryker, V. H. J. Am. Chem. Soc., 1978, 90, 2717.
24. Nango, M.; Gamson, E. P.; Klotz, I. M. J. Polym. Sci., Polym. Chem. Ed., 1979, 17, 1557.
25. Hierl, M. A.; Gamson, E. P.; Klotz, I. M. J. Am. Chem. Soc., 1979, 101, 6020.
26. Delaney, E. J.; Wood, L. E.; Klotz, I. M. J. Am. Chem. Soc., 1982, 104, 799.
27. Klotz, I. M. J. Polym. Sci., Polym. Lett., 1980, 18, 647.
28. Klotz, I. M.; Drake, E. N.; Sisido, M. Bioorg. Chem., 1981, 10, 63.
29. Kunitake, T.; Okahata, Y. J. Am. Chem. Soc., 1976, 98, 7793.
30. Kunitake, T.; Okahata, Y.; Sakamoto, T. J. Am. Chem. Soc., 1976, 98, 7799.
31. Kunitake, T.; Sakamoto, T. Polym. J., 1979, 11, 871.
32. Kunitake, T.; Sakamoto, T. J. Am. Chem. Soc., 1978, 100, 4615.
33. Kunitake, T.; Sakamoto, T. Chem. Lett., 1979, 1059.
34. "Handbook of Chemistry and Physics," The Chemical Rubber Company, 52nd Ed., 1971-1972.
35. Salamone, J. C. Ph.D. Thesis, Brooklyn Polytechnic Institute, 1967.

RECEIVED August 24, 1982

3

Initiation Reactions with Activated Monomer and/or Nucleophile in Ionic Polymerizations

TEIJI TSURUTA

Science University of Tokyo, Faculty of Engineering, Department of Engineering, Kagurazaka, Shinjuku-ku, Tokyo 162, Japan

This review article is concerned with chemical behavior of organo-lithium, -aluminum and -zinc compounds in initiation reactions of diolefins, polar vinyls and oxirane compounds. Discussions are given with respect to the following five topics: 1) lithium alkylamide as initiator for polymerizations of isoprene and 1,4-divinylbenzene; 2) initiation of N-carboxy-α-aminoacid anhydride(NCA) by a primary amino group; 3) activated aluminum alkyl and zinc alkyl; 4) initiation of stereospecific polymerization of methyloxirane; and 5) comparison of stereospecific polymerization of methyloxirane with Ziegler-Natta polymerization. A comprehensive interpretation is proposed for chemistry of reactivity and/or stereospecificity of organometallic compounds in ionic polymerizations.

This review article is concerned with chemical behavior of organo-lithium, -aluminum and -zinc compounds in initiation reactions of diolefins, polar vinyls and oxirane compounds. A comprehensive interpretation is proposed for metallic compounds in ionic polymerizations.

Lithium Alkyl Amide as Initiator for Polymerizations of Isoprene and 1,4-Divinylbenzene

Lithium dialkylamide having bulky alkyl groups, such as isopropyl groups, exhibits unique behavior in polymerization reactions of isoprene and divinylbenzene. It was previously reported by us that lithium dialkylamide underwent a stereospecific addition reaction with butadiene in the presence of an appropriate amount of dialkylamine in cyclohexane as solvent (1, 2). For instance, on reacting with butadiene, lithium diethylamide gave the sole adduct, 1-diethylamino-cis-butene-2, in a 98-99% purity. In the absence of free amine, on the other hand, no reaction took place under the same experimental conditions (50°C

0097-6156/83/0212-0023$06.00/0

in cyclohexane for 60 min.). By kinetic and spectroscopic studies, it was concluded that the reactive species in this addition reaction is the one to two complex of lithium dialkylamide and dialkylamine. In the one to two complex, lithium dialkylamide is activated enough to undergo the addition reaction with butadiene. A regioselective product, 1-diethylamino-3-methylbutene-2, was formed with isoprene as the diene reactant under reactions conditions similar to those for butadiene (3). In contrast, sodium alkylamide showed a lower degree of regioselectivity (4). As reported previously (2), lithium diisopropylamide exhibited only a slight reactivity in this addition reaction. Actually, none of addition products was observed to be formed in a 60-minute reaction which was carried out in cyclohexane at 50°C.

It has been found recently that lithium diisopropylamide is able to initiate isoprene polymerization at 80°C to an oligomer having the alkylamino group at the end of the polymer chain. The reaction conditions are more severe than those in previous study.

The degree of polymerization of the isoprene oligomers varied according to the ratio of amine to amide. This is understandable in terms of a transfer reaction which involves free amine. The writer and his coworkers focused their efforts on finding the best conditions for getting amino-containing isoprene oligomer having molecular weight of 1000 to 2000. In the course of study, it was noticed that "preformed oligomer" in the reaction system was one of the most suitable transfer reagents for molecular design of the amino-containing oligomers having a molecular weight of 1000. The reactions were carried out with lithium diisopropylamide as initiator in the absence of free amine because much lower molecular weight oligomers including members of the terpenoid family were formed in the presence of free amine.

The "preformed oligomer" has an allylamine structure at the chain-end. Examinations of chemical behavior were also carried out with low molecular weight tertiary amines having allylamine structure such as $(i\text{-}Pr)_2N\text{-}CH_2\text{-}CH{=}C(CH_3)_2$, $Et_2N\text{-}CH_2\text{-}CH{=}CH_2$ and others.

Results of examination of GPC patterns of oligomers prepared in the presence of these tertiary amines demonstrated that "the preformed oligomer" shows a unique behavior as chain transfer reagent; essentially no formation of the lower molecular weight oligomers is observed. Other allylamine homologues, in contrast, failed to suppress the formation of the lower molecular weight oligomers. Studies to elucidate mechanism for the unique behavior of "the preformed oligomer" are now in progress.

As the writer reported previously, lithium diethylamide, in the presence of diethylamine, underwent the one to one addition reaction with styrene (5) or divinylbenzene (DVB) (6). No reaction, however, took place between lithium diisopropylamide and styrene. A surprising result, in contrast, was obtained in the reaction of 1,4-DVB with lithium diisopropylamide, which initiated a prompt polymerization of 1,4-DVB at 20°C in tetrahydrofuran to form a soluble polymer (7). Remarkably enough, the propagation reaction proceeded to give a relatively high molecular weight (e.g. 100,000) polymer, even in the presence of a large excess of diisopropylamine (e.g. $i\text{-}Pr_2NH/i\text{-}Pr_2NLi$ molar ratio = 11/1).

The much greater reactivity of 1,4-DVB in the propagation step is in contrast to the lower reactivity of isoprene. Poly (1,4-DVB) obtained was soluble in THF, benzene and chloroform, and gave typical gel permeation chromatograms. When the polymerization was carried out with only a small excess of diisopropylamine, an insoluble crosslinked polymer was formed in a similar way to the reaction intitiated with butyllithium. In a nonpolar solvent such as benzene, lithium diisopropylamide exhibited little ability to add to 1,4-DVB even in the presence of excess amine. The nucleophilicity of the propagating carbanion should be an important factor determining the nature of the polymerization. The carbanion at the growing chain end of poly(1,4-DVB) is so stabilized owing to the extended conjugation with the para-vinyl group that the carbanion attacks predominantly the more reactive double bond of the DVB monomer. The pendant vinyl group of poly(DVB) was found to have a diminished reactivity toward nucleophilic reagents. This should be another reason for the absence of the cross-linking reaction. Under the same conditions, lithium isopropylamide, instead of lithium diisopropylamide, underwent the one to one addition reaction with 1,4-DVB to form 4-isopropylaminoethylstyrene. The soluble poly(DVB) reacts with various types of amine, so that it is possible to prepare a variety of amino-containing polymer.

Initiation of N-Carboxy-α-Aminoacid Anhydride(NCA) by a Primary Amino Group

Starting from a monomer or polymer having primary amino groups, Inoue and his coworkers have recently prepared a macromer (8) or graft copolymer (9) having polypeptide side chains. They found that the polymerization of N-carboxy-α-amino acid anhydride (NCA) was initiated by the primary amino groups of the original monomer or polymer to form the polypeptide side chain. Tertiary amino groups did not initiate the NCA polymerization.

The mechanism of NCA polymerization with amine as initiator has also been discussed earlier by many researchers (10). "The activated monomer mechanism" proposed by Bamford and Szwarc is one of the most widely accepted mechanisms. Proton abstraction reaction from -NH- group of NCA by tertiary amine is the essential step for the formation of activated monomer, which undergoes a nucleophilic attack at carbonyl carbon at 5th position of NCA to cleave the NCA-ring.

The activated monomer mechanism was based on the assumption that the major part of propagation reaction was carried out by $^{(-)}$NCA species rather than NHCOOH or $-NH_2$(with CO_2 elimination) group at the growing chain end.

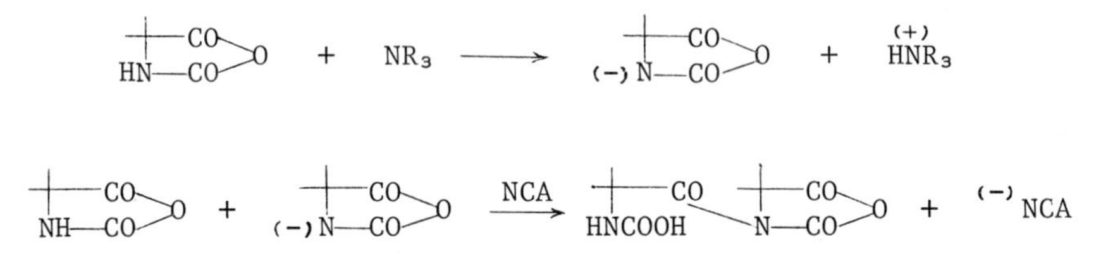

More details of the Inoue experiments are as follows: the poly (amino acid) macromer was synthesized in THF by reacting α-benzyl-L-glutamate (BLG)—NCA with N-methyl-N-(4-vinylphenethyl)-ethylenediamine.

In their experiments, the NCA polymerization must be initiated by the primary amino group in (I), because DP (number average) of the macromer (II) was almost equal to the mole ratio of the NCA to (I), and the gel permeation chromatogram showed a unimodal peak for (II).

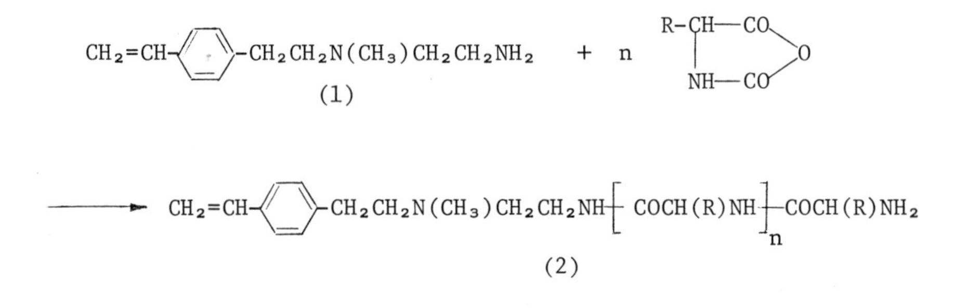

Inoue et al. also reported that copolymer of styrene with (I) could be converted to a novel polystyrene-poly(amino acid) graft copolymer by reacting the trunk copolymer with L-alanine-, BLG- or β-benzyl-L-aspartate-NCA. The formation of the graft copolymer was supported by results obtained in a turbidity titration of the graft copolymer in trifluoroacetic acid-chloroform mixture. The graft copolymer was perfectly soluble in the whole range of the solvent composition. The back-bone polymer, in contrast, was insoluble in the solvent containing more than 60% trifluoroacetic acid. They confirmed also that there was little evidence for the formation of homo(polyamino acid) in the reaction system.

Activated Aluminum Alkyl and Zinc Alkyl Initiators

Aluminum or zinc alkyls are known to form complexes with bypyridyl(Bipy), triphenylphosphine(PPh_3), sparteine(Spar), hexamethylphosphoric triamide(HMPT) and other Lewis bases (11). The nucleophilic reactivity of these metal alkyls was enhanced enough by the complexation to initiate the anionic polymerization of unsaturated esters and nitriles, e.g. methyl methacrylate (MMA) acrylonitrile(AN), etc. Among several $AlEt_3$-Lewis base complexes, the $AlEt_3$ complex with bidentate ligands or with a monodentate ligand of sufficiently high basicity showed high catalytic activity for the polymerization of AN and MMA in toluene at 30° C. $AlEt_3$ and rigid bidentate Lewis bases such as Bipy and Spar formed an unstable equimolar five-coordinated complex, and the complex was stabilized to a six-coordinated structure by the coordination of another complex or solvent to its sixth coordination site. When monomers were present along with the aluminum complex, some of the ligands at the sixth coordination site were replaced with monomer molecules, which was the preliminary stage for the initiation reaction of polymerization. Model reactions were studied with methyl α-isopropylacrylate to eludidate the initiation mechanism, because the α-isopropylacrylate was previously shown to be non-polymerizable. Michael-type addition products were isolated from a reaction mixture of the α-isopropylacrylate and $i\text{-}Bu_3Al$--Bipy complex. It also was confirmed that only one isobutyl group of $i\text{-}Bu_3Al$ had entered the model initiation reaction, even in the presence of excess (20 times) methyl α-isopropylacrylate.

Contrary to acrylic esters or acrylonitrile, a vinyl ketone could be polymerized with aluminum (or zinc) alkyl alone in the absence of any added Lewis bases. The mechanism of polymerization of phenyl vinyl ketone(PVK) was studied by using a model initiator, ethylzinc-1,3-diphenyl-1-pentene-1-olate (ZC), which possessed the same structure (see below) as the growing chain end in zinc alkyl-initiated polymerization of PVK (12).

$$C_6H_5(Et)CHCH = C(C_6H_5)\text{-}OZnEt$$

The polymerization system exhibited a "living character" with an initiator efficiency of 100%.

Kinetic studies showed that the initiation reaction proceeded according to the second order rate law with respect to [ZC] concentration, when excess PVK was present in the reaction system. Solubility data of a model binary system suggested the formation of a molecular complex between ZC and the vinyl ketone: the vinyl ketone by itself was slightly soluble in cyclohexane, while it was readily soluble in the presence of ZC.

UV spectra also supported complex formation in the binary system. In a model reaction between ZC and benzalacetophenone, it was concluded that the monomer was activated by ZC, while the residue of ZC was activated by the monomer coordination onto the zinc atom. The same mechanism could also be applied to the reaction with diethylzinc.

The chemical behavior of vinyl ketone is in sharp contrast to that of acrylonitrile. No polymerization of AN or methacrylonitrile (MAN) took place with aluminum alkyl in the absence of Lewis bases. In the presence of HMPT, on the other hand, AN or MAN was polymerized to produce colorless polymers as in the case of vinyl ketone. The last named monomer had to be activated by the coordination of aluminum alkyl, but HMPT, a strong base, took away the aluminum alkyl from the coordination site of vinyl ketone monomer.

Another type of activation of aluminum alkyl was found in the asymmetric-selective polymerization of epichlorohydrin (ECH) with an optically active cobalt-salen type complex [Co*(II)]. The structure of the salen-type cobalt complex was shown previously (13, 14). In a benzene solution of the binary system consisting of [Co*(II)] and $AlEt_3$, no evolution of ethane or ethylene was observed at room temperature. The NMR signals of the methyl protons for $AlEt_3$ shifted down field on mixing with [Co*(II)]. These observations together with a circular dichroism study indicated that $AlEt_3$ and [Co*(II)] formed a molecular complex in benzene, none of Al-Et bonds being cleaved by this complexation.

The shape of the GPC pattern of oligomers formed was not changed throughout the reaction process, which suggested repeated occurrence of a set of elementary reactions in the system. Some of these oligomers were isolated by preparative GPC technique. The isolated fractions were proven to be a series of oligomers having a oxirane ring at one terminal. On the basis of these observations, it was concluded that the ECH polymerization proceeded according to a cationic mechanism. A propagating species (see below) is formed at the initiation stage. Propagation takes place under the chiral circumstance of cobalt complex.

CH_2-Cl

O

H_2C

Cl

$[AlEt_3.Co]$

O

H_2C-O (+)

CH_2-Cl

$[ClAlEt_3.Co]^{(-)}$

The propagation will continue until chloride anion of the complex attacks the methylene-carbon of ECH monomer to cause a termination reaction, reproducing the catalyst species.

Inoue et al. found previously that a catalyst system, $\alpha,\beta,\gamma,\delta$ -tetraphenylporphyrin($TPPH_2$) with diethylaluminum chloride, exhibited an extremely high activity in oxirane polymerizations (15, 16). On reacting Et_2AlCl with $TPPH_2$ in dichloromethane or benzene, 2 moles of EtH were eliminated to form [TPP]-AlCl. Initiation reaction took place by the nucleophilic attack at β-carbon of methyloxirane with chloride anion to form [TPP]-OCH-(Me)CH_2-Cl. One of the significant features of the catalyst system is its "living nature" in terms of the capability of block copolymerization and of the very narrow molecular weight distribution observed for the polymer formed with this catalyst. Inoue and Aida reported recently that the living end was able to add carbon dioxide as well, forming a block copolymer (see below).

$$Cl-\left[CH_2CH(Me)O\right]_m-\left[\left(\underset{\displaystyle O}{\overset{\|}{C}}OCH_2CH(Me)O\right)_x-\left(CH_2CH(Me)O\right)_{1-x}\right]_n-H$$

prepolymer blocked chain

Initiation of Stereospecific Polymerization of Methyloxirane

It is widely recognized that Et_2Zn-H_2O system is one of the most active catalysts for the stereospecific polymerization of oxiranes. A variety of chemical species are formed in the following way: rapid formation of ethylzinc hydroxide, its aggregation, and elimination of ethane to form zinc oxide structure. The maximum catalyst activity was achieved when the mole ratio of zinc to water was one to one, where the predominant formation of a species, $Et(ZnO)_nH$, (III), was observed. If we use less amount of water, another species, $Et(ZnO)_nZnEt$, (IV), was also produced concurrently. Contrary to the anionic nature of the former species (III), the latter species (IV) exhibited a cationic nature. For instance, more than 95% of ring cleavage of methyloxirane takes place at $O-CH_2$ bond with species (III), while the cleavage at O-CH bond also takes place concurrently with species (IV).

A recent GPC study (17) of a poly(methyloxirane) sample prepared by Et_2Zn-H_2O (one to 0.1 system) clearly showed the cationic nature of this catalyst system. More than 50 weight % of the polymer obtained was found to be oligomers including pentamer and lower molecular wight compounds in contrast with the high molecular weight polymer obtained with Et_2Zn-H_2O, one to one, system.

^{13}C-NMR studies of these polymers showed that the spectrum of the oligomer fraction had very complicated features owing to the irregular structure admixed with head-to-head and tail-to-tail enchainments of the monomeric unit, which again is in sharp contrast with the clear-cut NMR patterns for the polymers prepared with the one to one catalyst system.

The active species of the diethylzinc-methanol system was previously proven to be zinc dimethoxide (18). Contrary to the zinc-water system, no trace of cationic nature was observed in the zinc-methanol system at any ratio of the two components.

When racemic methyloxirane is polymerized with zinc dimethoxide, D-and L-monomers are separately incorporated into growing chains to form an isotactic polymer consisting of poly(D-methyloxirane) and poly(L-methyloxirane). This stereoselective polymerization can be satisfactorily explained in terms of the enantiomorphic catalyst sites model (18). The d*-sites accept D-methyloxirane in preference to the L-monomer, resulting in the formation of -DDDD- isotactic sequences. The same situation is valid for the 1*-catalyst sites.

It was most desirable for us to elucidate the stereocontrol mechanism in terms of molecular level considerations rather than a phenomenological approach. No information, however, was available concerning the chiral structure of d*- and 1*-catalyst sites, because none of the active catalysts possesses a well-defined structure. The active zinc methoxide, for instance, was a disordered powdery substance and its catalyst efficiency was extremely low.

Seventeen years ago, we isolated (19) an organozinc complex in the form of single crystal, which had the composition:

$$[EtZnOCH_3]_6[Zn(OCH_3)_2]$$

It was soluble in benzene, and the benzene solution exhibited a catalytic activity for the oxirane polymerization at 80 °C, but no activity at room temperature.

According to the X-ray analysis by Kasai, the organozinc complex consisted of two enantiomorphic distorted cubes, d-cube and 1-cube (20). In the ^{13}C-NMR spectrum of the organozinc complex, four singlets were observed. Two of them were assigned to the inner and outer methoxy-carbons, respectively. We carried out NMR analysis of a reaction system in which the organozinc complex and racemic methyloxirane were allowed to react in benzene (21, 22, 23). A spectrum of the reaction system at 30° C was proven to be the simple overlapping of the individual spectrum of methyloxirane and the zinc complex. No reaction took place at 30° C. In a spectrum at 80° C, a number of new signals and shape changes in the original signals were observed owing to the occurrence of the polymerization reaction at 80°C. All of the observed signals in the reaction system could be explained reasonably by the initiation mechanism by one of the inner methoxy groups. Tacticity data of poly(methyloxirane) prepared by the organozinc complex was found to agree well with the predicted values from the enantiomorphic model. A possible mechanism is as follows: at 80° C, two of the longest bonds are loosened. If the bond-loosening takes place at d-cube, the steric course of an entering monomer will be influenced by the chiral structure around the central zinc atom. A molecular model study suggested that L-monomer will be preferentially accepted in this particular example. The initiation reaction takes place through the nucleophilic attack by the inner methoxy group. This may be the origin of the l*-catalyst site. The probability of the bond loosening in l-cube is exactly the same as that in d-cube, so that an equal number of d*-sites and l*-sites will be established in the reaction system. This explains the experimental results that poly(D-methyloxirane) and poly(L-methyloxirane) are formed in equimolar amounts in the reaction system.

Comparison of Stereospecific Polymerization of Methyloxirane with Ziegler-Natta Polymerization

The last topic that is to be discussed is the comparison of our organozinc complex with Ziegler-Natta catalysts.

It is widely accepted that the formation of isotactic polypropylene is brought about by the chiral structure, d* or l*, of the titanium species in the catalyst system. The concept of the steric control in propylene polymerization is the same as that in the methyloxirane polymerizations which have been discussed in the foregoing sections. An active center having d*- or l*-chirality is formed along the crystal surface of δ*-titanium trichloride.

One of the most important points of argument about the nature of the "super active Ziegler catalyst" is the role of ethyl benzoate. The writer will discuss this in reference to a catalyst system which was reported (24) recently by Kashiwa of Mitsui Petrochemical Co.

As the first step for the catalyst preparation, the ball-milled magnesium chloride was allowed to stand with titanium tetrachloride at 80° C for two hours. The $MgCl_2(TiCl_4)$, i.e., $TiCl_4$ adsorbed on $MgCl_2$ was then treated with $AlEt_3$ and ethyl benzoate (EB) at 60° C. The catalyst (24)) thus obtained gave the following analytical results:

	Ti	Cl	Mg	Al	EB
weight percent	0.6	71.0	28.0	0.3	1.2
mole ratio	1.2	200	115	1.1	0.8

It is to be noted that Ti, Al and ethyl benzoate are present in nearly equal molar ratio. Results of propylene polymerization with this type of catalyst showed that ethyl benzoate enhanced the rate of formation of isotactic polypropylene, while it decreased the rate for atactic polypropylene. Ethyl benzoate seems to make the catalytic sites sterically more specific.

According to Kashiwa, $TiCl_4$ chemisorbs loosely at the non-specific site so that some of $TiCl_4$ molecules may be extracted by ethyl benzoate, which causes the deactivation of the non-specific site. Furthermore, non-specific sites having the stronger acidity will be poisoned more easily by ethyl benzoate.

Results of study on the molecular wight distribution of the "isotactic fraction" of polypropylene showed that isotactic polymer having higher molecular weight was produced in the presence of ethyl benzoate. Termination reaction is believed to be a transfer of "growing chain" from Ti-center to Al-center and the coordination of ethyl benzoate onto aluminum should be opposite to the direction of electron flow for the transfer reaction, which makes the transfer reaction more difficult to take place.

Syndiotactic propagation of propylene is know to be catalyzed by homogeneous vanadium catalyst (18). In the polypropylene samples prepared with the homogeneous catalysts, the relative population of iso-, hetero- and syndiotactic triads is in accordance with that predicted from the first order Markov model (25, 26). There is no chiral structure around the homogeneous vanadium species. The stereochemistry of the entering monomer is controlled by the chirality of the growing chain end, in contrast with the isotactic propagation.

Inoue et al. (27) found that a porphyrin-Zn alkyl catalyst polymerized methyloxirane to form a polymer having syndio-rich tacticity. The relative population of the triad tacticities suggests that the stereochemistry of the placement of incoming monomer is controlled by the chirality of the terminal and penultimate units in the growing chain. There is no chirality around the Zn-porphyrin complex. Achiral zinc complex forms syndio-rich poly(methyloxirane), while chiral zinc complex, as stated above, forms isotactic-rich poly(methyloxirane). The situation is just the same as that for propylene polymerizations. Achiral vanadium catalyst produces syndiotactic polypropylene, while chiral titanium catalyst produces isotactic polypropylene.

Literature Cited

1. Imai, N.; Narita, T.; Tsuruta, T. Tetrahedron Lett. 1971, 38, 3517.
2. Narita, T.; Imai, N.; Tsuruta, T. Bull. Chem. Soc. Japan 1973, 46, 1242.
3. Narita, T., Nitadori, Y.; Tsuruta, T. Polymer J. 1977, 9, 191.

4. Noren, G. K. J. Org. Chem. 1975, 40, 967.
5. Narita, T.; Yamaguchi, T; Tsuruta, T. Bull. Chem. Soc. Japan 1973, 46, 3825.
6. Tsuruta, T.; Narita, T.; Nitadori, Y.; Irie, T, Makromol. Chem. 1976, 177, 3255.
7. Nitadori, Y.; Tsuruta, T. Makromol. Chem. 1978, 179, 2069.
8. Maeda, M.; Inoue, S. Makromol. Chem., Rapid Commun. 1981, 2, 537.
9. Kimura, M.; Egashira, T.; Nishimura, T.; Maeda, M.; Inoue, S. Makromol. Chem. 1982, 183, in press.
10. Sekiguchi, H. Pure & Appl. Chem. 1981, 53, 1689.
11. Ikeda, M.; Hirano, T.; Nakayama, S.; Tsuruta, T. Makromol. Chem. 1974, 175, 2775.
12. Tsuruta, T.; Tsushima, R. Makromol. Chem. 1976, 177, 337.
13. Aoi, H.; Ishimori, M.; Yoshikawa, S.; Tsuruta, T. J. Organometal. Chem. 1975, 85, 241.
14. Takeichi, T.; Arihara, M.; Ishimori, M.; Tsuruta, T. Tetrahedron 1980, 36, 3391.
15. Aida, T.; Inoue, S. Macromolecules 1981, 14, 1162.
16. Aida, T.; Inoue, S. Macromolecules 1981, 14, 1166.
17. Tsuruta, T. Pure & Appl. Chem. 1981, 53, 1745.
18. Tsuruta, T. J. Polymer Sci. 1972, D6, 179.
19. Ishimori, M.; Tomoshige, T.; Tsuruta, T. Makromol. Chem. 1968, 120, 161.
20. Ishimori, M.; Hagiwara, T.; Tsuruta, T.; Kai, Y., Yasuoka, N.; Kasai, N. Bull. Chem. Soc. Japan 1976, 49, 1165.
21. Tsuruta, T. J. Polymer Sci. Polymer Symposium 1980, 67, 73.
22. Hagiwara, T.; Ishimori, M.; Tsuruta, T. Makromol. Chem. 1981, 182, 501.
23. Tsuruta, T. Makromol. Chem. Suppl. 1981, 5, 230.
24. Kashiwa, N. Paper presented at "International Symposium on Transition Metal Catalyzed Polymerizations: Unsolved Problems" (August, 1981, Michigan Molecular Institute).
25. Doi, Y. Macromolecules 1979, 12, 249; 1012.
26. Doi, Y.; Ueki, S.; Keii, T. Macromolecules 1979, 12, 814.
27. Takeda, N.; Inoue, S. Makromol. Chem. 1978, 179, 1377.

RECEIVED September 10, 1982

4

Muonium as a Hydrogen-like Probe to Study Monomer Initiation Kinetics

J. M. STADLBAUER, B. W. NG, Y. C. JEAN,[1] Y. ITO,[2] and D. C. WALKER

University of British Columbia, Department of Chemistry and TRIUMF, Vancouver, B.C. V6T 1Y6 Canada

The radioactive muonium atom(Mu) has the same ionization potential and Bohr radius as the hydrogen atom, but only 1/9 the mass. At the end of its 2.2μsec intrinsic life-time the positive muon (which acts as the Mu nucleus) decays into an energetic positron. This decay can be observed using fast single particle counting methods. Because of this ease of detection and its hydrogen-like properties, Mu makes an excellent probe to directly study hydrogen atom reactions: for example, hydrogen atom initiation of monomer polymerization. We have examined the addition reaction kinetics of Mu with acrylamide, acrylic acid, acrylonitrile, methylmethacrylate and styrene, all in aqueous solution. Their second order rate constants were found to be, respectively, 1.9, 1.6, 1.1, 1.0 and 0.11 x $10^{10}M^{-1}s^{-1}$. As proof that Mu does add across the vinyl bond we have observed the muonium containing free radicals in the pure liquid monomers and obtained their hyperfine coupling constants.

First of all what is muonium, what is its source, how do we observe it, and why is it useful? Muonium (Mu) is an atom comprised of a positive muon nucleus, (μ^+), and a bound electron. This bound electron can have its spin parallel or antiparallel to the muon nuclear spin resulting in 'triplet' and 'singlet' muonium atoms, respectively. The atom has a mass 1/9 that of the hydrogen atom, H; but because the reduced masses are

[1] Current address: University of Missouri–Kansas City, Department of Physics, Kansas City, MO 64110.

[2] Current address: University of Tokyo, Research Center for Nuclear Science and Technology, Tokyo, Japan.

0097-6156/83/0212-0035$06.00/0

nearly the same, Mu has essentially the same Bohr radius and ionization potential as hydrogen.

Muonium is formed when a positive muon thermalizes in a target and picks up an electron from the stopping medium into a bound state. Muons, both high energy (28 MeV) and low energy (4.1 MeV), are the product of positive pion decay, in flight or at rest, respectively.

$$\pi^+ \longrightarrow \mu^+ + \nu_\mu . \qquad (1$$

The pion itself is generally produced through the reaction

$$^{9}Be + {}^{1}p \longrightarrow {}^{10}Be + \pi^+. \qquad (2$$

Intense high energy proton beams required for these studies are currently available at TRIUMF (Vancouver, Canada), LAMPF (Los Alamos, U.S.A.), SIN (Villigen, Switzerland), and KEK (Tsukuba, Japan).

Under a transverse magnetic field the free muon and other diamagnetic muonic species precess with a Larmor frequency of 0.0136 MHz/G, while triplet muonium precesses at 1.39 MHz/G. The factor of 102 between these precession frequencies makes it easy to distinguish between the species present. The intrinsic lifetime of the muon, 2.2×10^{-6}s, allows enough time to study the chemical reactions of muonium in many chemical environments (1-4). When the muon decays it ejects a positron and two neutrinos. The positrons, with a maximum energy of 52 MeV, are easily detectable using nuclear physics single-particle fast counting systems. The experimental data are tabulated as time histograms that are then computer analyzed. A detailed discription of the MSR (Muonium Spin Rotation) method is available elsewhere (3).

Muonium has been observed in pure hydrocarbons (5), alcohols (6,7), and water (4). Because Mu reacts slowly with these pure liquids, giving observable reaction lifetimes of Mu up to 4μs, they can be used as solvents to study various solutes of interest. As the free triplet Mu atom reacts with the solute its observed precession frequency is damped and a decay constant, λ, can be obtained. The concentration dependence of the decay constant provides second order chemical rate constants for Mu addition, abstraction, spin conversion, and oxidation-reduction reactions. When analogous hydrogen atom rate constants are available the kinetic isotope effect can also be calculated.

Muonium is useful as a hydrogen-like probe to study the atomic and radical reactions of hydrogen because Mu's reaction decay constant is directly observable, while most hydrogen atom data come from measured relative rate constants. The more we know of the reactions of hydrogen, the simplest and most abundant element in the universe, the sooner we will be better able to understand more complex atoms and their reactions.

Experimental

Chemicals. Acrylonitrile (AN), methylmethacrylate (MMA), and styrene were supplied by Aldrich and purified before use. Acrylic acid (AA) and acrylamide (AM) were purchased from Eastman. The acrylic acid contained 200 ppm p-methoxyphenol as inhibitor. However, at the solute concentrations of 5.0 x 10^{-5}M and 7.4 x 10^{-5}M the inhibitor's concentration would be $\sim 10^{-8}$M and, therefore, could not contribute significantly to the observed muonium decay. The water solvent was triply distilled, initially from permanganate solution.

Apparatus. The schematic diagram in Figure 1 shows the general experimental set-up for a low energy 'surface' muon beam. Right and left positron detectors are made of plastic scintillator 'paddles' connected to RCA 8575 photomultiplier tubes by lengths of light pipe; the whole of which are wrapped to exclude outside light. Detectors can be set up for double or triple coincidence counting to help minimize background. Graphite degrader between the detectors helps maximize signal amplitudes. The Helmholtz coils, which provide the external magnetic field, are mounted transversely to the beam. A signal from the TM (thin scintillator-muon) counter starts the 1ns resolution clock. A signal from either right or left detectors in coincidence mode stops the clock if this signal appears between a set "gate" of 0 to 4μs. If the positron signal is not coincident or appears after 4μs it is not counted and the system starts over. Usually about 50,000 good events are counted per minute.

MSR: Muonium Spin Rotation Measurements. For the kinetic portion of this study a low energy (4.1 MeV) beam of muons from the M20 channel at TRIUMF was focused on a shallow Teflon cell with 0.007 cm Mylar as muon windows. These target cells contained approximately 80 ml of sample solution which was bubbled with high purity He to remove dissolved oxygen. In the case of these volatile acrylic solutes the bubbling gas was first passed through a prebubbler containing a solution of the same concentration as the sample. Muonium precession signals were observed at a field of 8G provided by Helmholtz coils transverse to the beam and centered on the target.

Figure 2 shows some time-histograms of both raw and fitted data. The data is computer fitted using MINUIT, a χ^2-minimization program, to a nine parameter function. Equation (3) shows that function with two parameters, background and muon decay, subtracted out.

$$A(t) = A_\mu \cos(\omega_\mu t + \phi_\mu) + A_M \exp(-\lambda t) \cos(\omega_M t - \phi_M) \qquad (3$$

where A_μ and A_M are, respectively, the muon and muonium signal amplitudes, ω_μ and ω_M are their frequencies, ϕ_μ and ϕ_M their

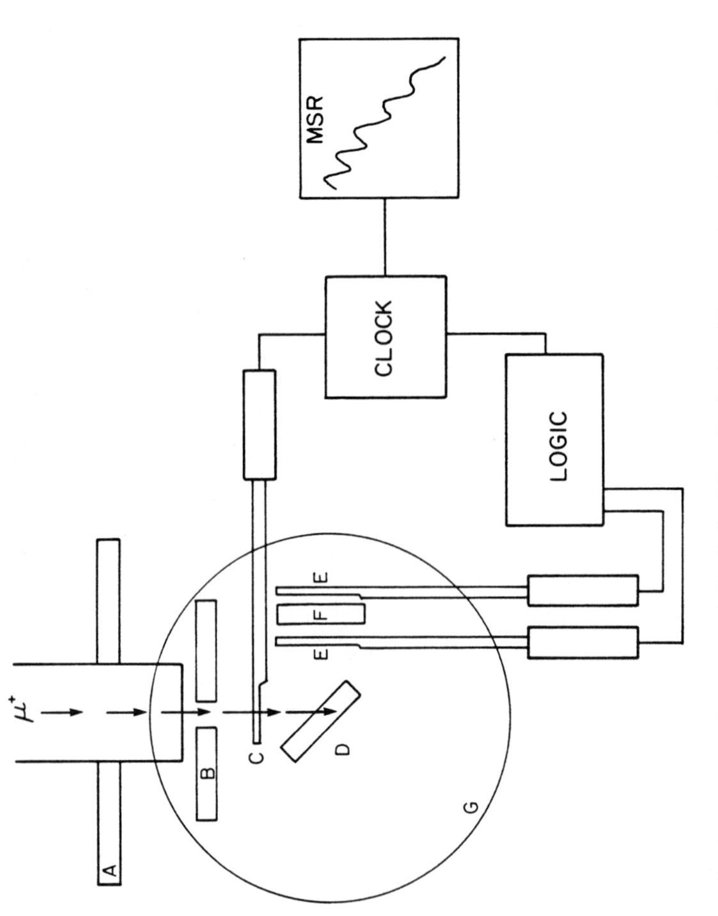

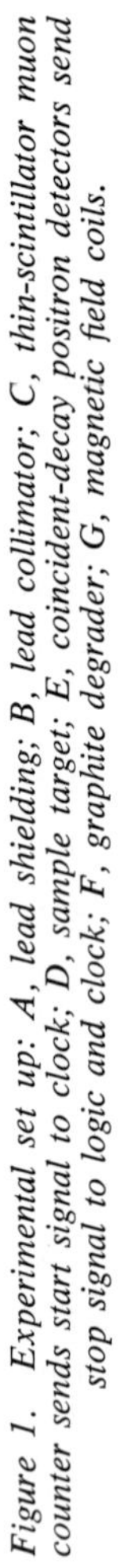
Figure 1. Experimental set up: A, lead shielding; B, lead collimator; C, thin-scintillator muon counter sends start signal to clock; D, sample target; E, coincident-decay positron detectors send stop signal to logic and clock; F, graphite degrader; G, magnetic field coils.

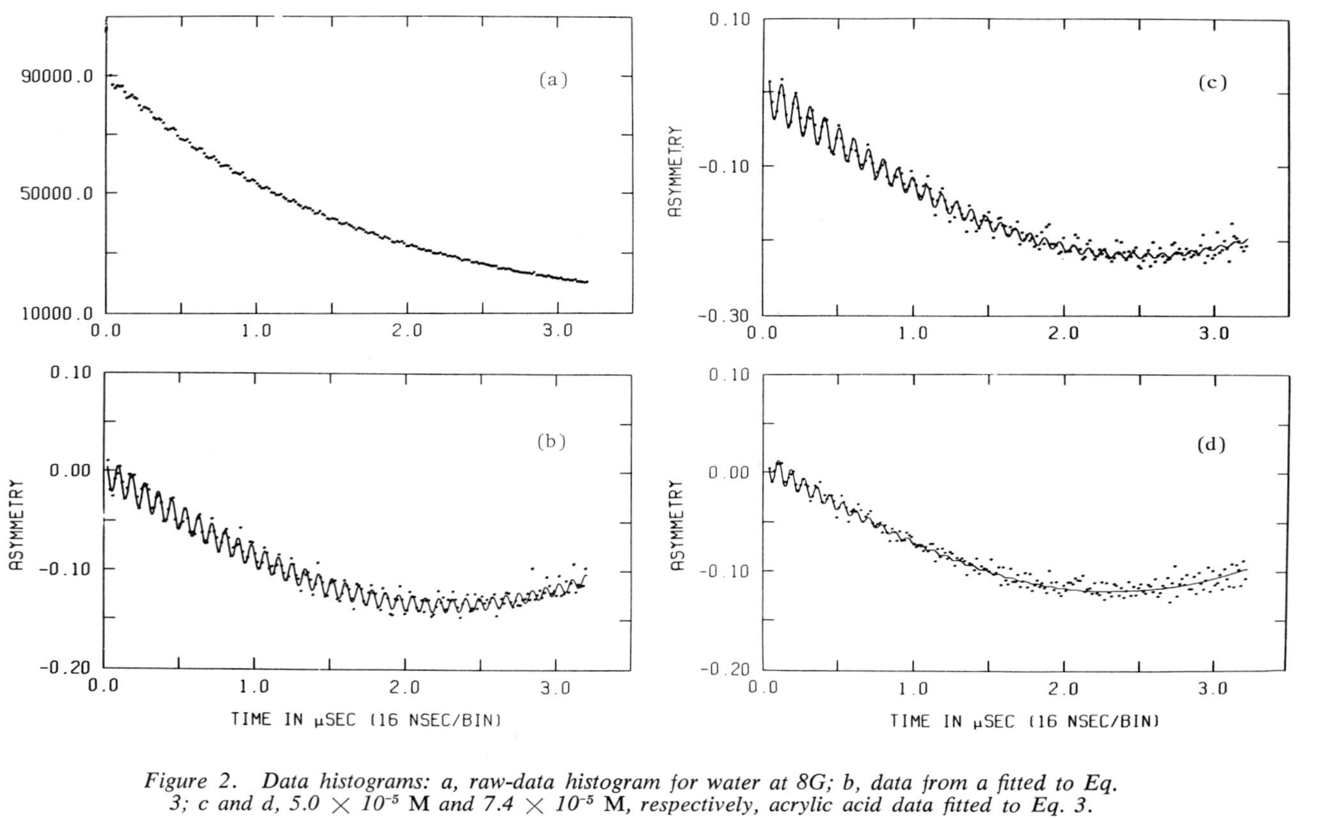

Figure 2. Data histograms: a, raw-data histogram for water at 8G; b, data from a fitted to Eq. 3; c and d, 5.0 × 10^{-5} M *and* 7.4 × 10^{-5} M, *respectively, acrylic acid data fitted to Eq. 3.*

initial phases, while λ is the muonium decay constant. Using equation (4), the obseved λ and the known solute concentration [S], the bimolecular rate constant, k_M, can be calculated for the reaction between Mu and the solute.

$$\lambda = \lambda_o + k_M[S] \qquad (4$$

The decay constant for the pure water solvent, λ_o, has a value of $(2.4 \pm 0.6) \times 10^5 s^{-1}$ (5-8). Two λs, right and left, are obtained per experiment and are plotted against solute concentration. The slope of the best line is taken as k_M.

MRSR: Muonium Radical Spin Rotation. In order to prove that Mu was indeed adding across the vinyl double bond of styrene and the liquid acrylics, high energy muons (24 MeV) were made to strike glass, round-bottomed flasks containing neat samples. Two methods of degassing were utilized: i) a freeze-pump-thaw cycle followed by vacuum sealing, and ii) by bubbling the samples on line using a special probe and ground glass fittings. Data were analyzed using a Fast Fourier Transform (FFT) program to obtain the radical frequencies and the hyperfine coupling constant, a_μ, as described elsewhere (9,10).

Results

Table I lists the decay constants, λ, obtained for the different concentrations of the monomers studied. These λs are the average of the left and right values obtained for each concentration (11). Though statistical errors range from 5% to 17%, experimental irreproducibilities in target geometry, field homogeneity, detector thresholds, muon beam asymmetry and background result in a more probable error of ±25% (8). This level of reproducibility is quite reasonable when compared to rate constants obtained by competitive rate techniques and direct physical methods.

Figure 3 shows the FFT spectra of styrene at 1500G and 2500G. The large low frequency peak is due to the muon while the two higher frequency peaks are due to the radical. Addition of these two radical frequencies yields the muon hyperfine coupling constant a_μ. [Peaks at 23 MHz are due to the radio frequency (RF) of the cyclotron. Peaks at multiples of 125 MHz are due to the counting system's clock frequency]. These spectra show that the coupling constant for styrene is a_μ = 213.4 MHz, and that it is independent of field. Figure 4 compares the FFT spectra of styrene and benzene at 3000 G. Table II lists the muon hyperfine coupling constants a_μ and a_μ-corrected for the muon's magnetic moment for direct comparison to a_p. The corresponding proton radical's constants, a_p, taken from the literature (12), are included in Table II.

Table I

Reaction Rate Data

Monomer	[S]/M	$\lambda/10^6 s^{-1}$	$k_M/10^{10} M^{-1} s^{-1}$
Styrene	$5.0x10^{-5}$ $8.0x10^{-5}$	0.38 0.41	0.11±0.01
Methylmethacrylate	$2.6x10^{-5}$ $5.0x10^{-5}$	0.58 0.80	0.95±0.19
Acrylonitrile	$2.0x10^{-5}$ $5.0x10^{-5}$	0.45 0.88	1.14±0.20
Acrylic acid	$5.0x10^{-5}$ $7.4x10^{-5}$	0.89 1.63	1.55±0.25
Acrylamide	$2.0x10^{-5}$ $5.0x10^{-5}$	0.83 1.23	1.90±0.13

Table II

Hyperfine Coupling Constant

Monomer	MuR a_μ (MHz)	MuR a_μ x0.3141 (MHz)	HR a_p (MHz)
Styrene	213.4	67.0	50.1
MMA	274.9	86.3	59.7
AN	280.4	88.1	64.4
AA	319.3	100.3	70.8
Benzene	514.6	161.6	133.7

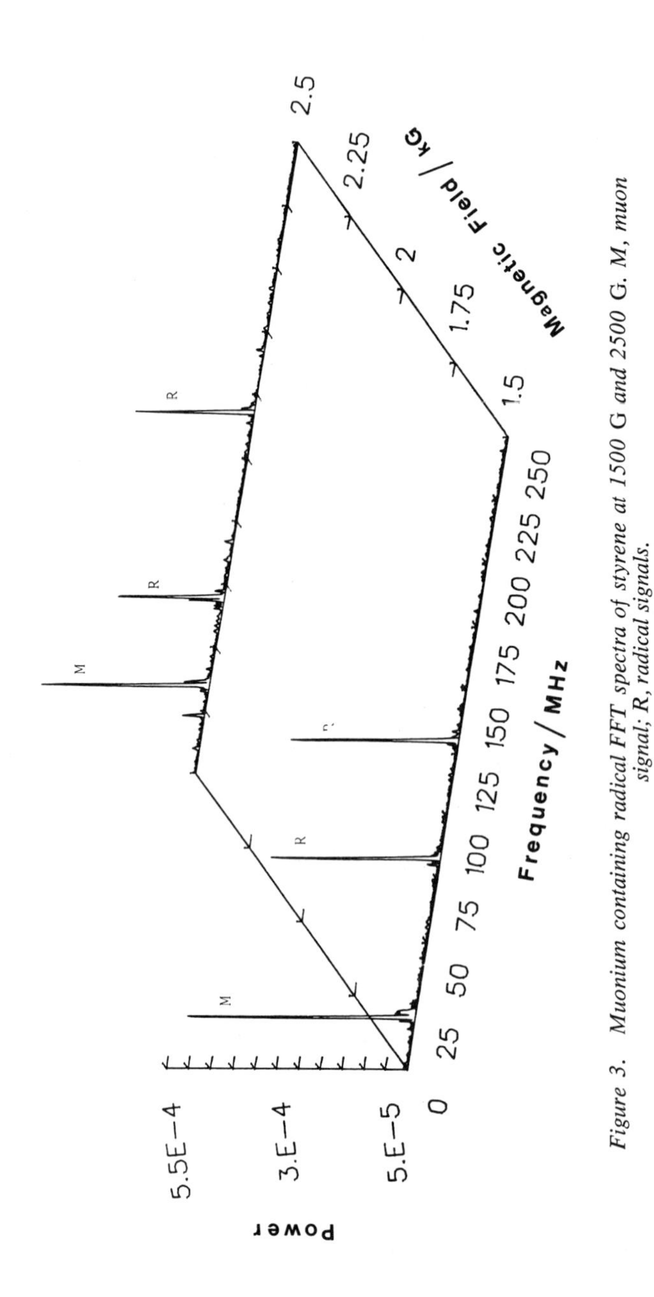

Figure 3. *Muonium containing radical FFT spectra of styrene at 1500* G *and 2500* G. *M, muon signal; R, radical signals.*

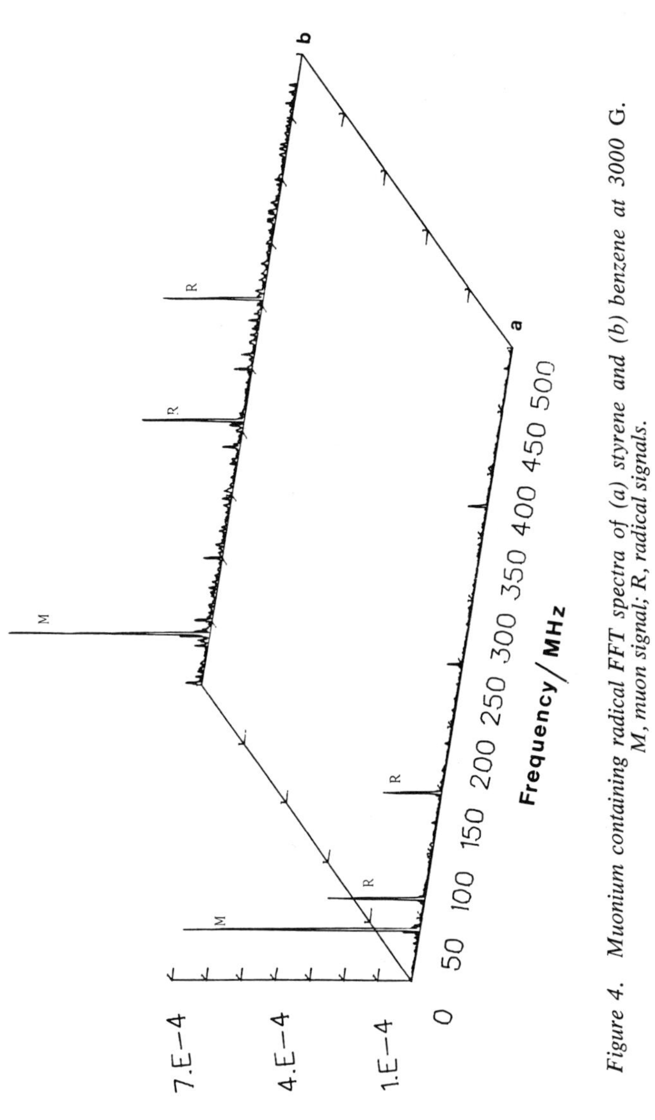

Figure 4. *Muonium containing radical FFT spectra of (a) styrene and (b) benzene at 3000 G. M, muon signal; R, radical signals.*

Discussion

The Fourier Transform spectra of the monomers studied indicate that Mu reacts to give only one free radical (at the time observed, $\sim 10^{-6}$s) because there is only one pair of radical frequencies in each system. The comparison with a_p substantiates the assumption that Mu is indeed adding across the vinyl bond of these monomers as shown in equation (5):

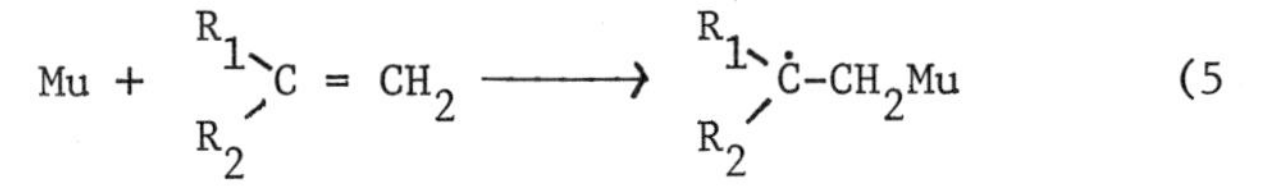

where R_1 and R_2 vary through H, C_6H_5, $CONH_2$, CO_2H, CH_3 and CO_2CH_3 depending on the monomer. Table II shows that the coupling constants for the acrylics vary between 200 to 300 MHz while the a_μ for Mu adding to benzene, giving the cyclohexadienyl radical, is 514 MHz. This value agrees with the published value for benzene (9). These data, plus the observed spectra for styrene and benzene in Figure 4, indicates that Mu is adding preferentially to the vinyl bond of styrene and not to the ring. This counters the suggestion by Swallow (13), based on band labelling studies, that H adds to styrene's vinyl bond 15% of the time and to the ring 85% of the time. However, as the MuR radicals are observed in the MRSR time window of 10^{-7} to 10^{-5}s, it is conceivable that the observed radicals are not the primary ones but rather the most stable ones after intramolecular rearrangements.

Table III compares the rate constants for Mu addition with those for H addition for the only two available in the literature

TABLE III

ISOTOPE EFFECTS

Compound	Rate constant/$M^{-1}s^{-1}$		
	k_M	k_H	k_M/k_H
Acrylonitrile	$(1.14 \pm 0.20) \times 10^{10}$	$(4.0 \pm 0.5) \times 10^{9}$	2.8
Acrylamide	$(1.90 \pm 0.13) \times 10^{10}$	1.8×10^{10}	1.1
Maleic acid	1.1×10^{10}	8.0×10^{9}	1.4

(14,15). The values for maleic acid (6,7,8) are included as a non-acrylic comparison. Acrylamide's isotope effect, k_M/k_H=1.1±0.4 compares nicely with that of maleic acid indicating that there is little mass effect in this type of reaction. As these rate constants are near the diffusion controlled limit of $2x10^{10}M^{-1}s^{-1}$ there is a question as to whether the Stokes-Einstein relationship holds for small particles like Mu and H, or whether the isotope effect should be 3, as with acrylonitrile, because the mean velocity of Mu (with mass 1/9 H) is three times faster than that of H. But since there is no obvious reason for the isotope effects for Mu adding to acrylamide and acrylonitrile to be different we reserve judgement on the question of isotope effects until a better understanding is attained.

As mentioned, these reported muonium rate constants, except for styrene, are very near the $\sim 2x10^{10}M^{-1}s^{-1}$ diffusion-controlled limit. Assuming that the specific rates listed in Table I are indeed different, we explain those differences through simple resonance and steric effects. Styrene, with a k_M=$1.1x10^9M^{-1}s^{-1}$, is the least reactive and most highly conjugated of the monomers studied. We believe that the large π-delocalization stabilizes the molecule and reduces its reactivity toward Mu. The rate constants for the other monomers studied increase from $9.5x10^9M^{-1}s^{-1}$ in the order MMA<AN<AA<AM. MMA has the least accessible double bond due to the 'blocking' effect of the α-methyl group. The other three are all straight chain three carbon conjugated compounds with their double bonds equally open to Mu attack. Mu (which acts like a nucleophile) would encounter less electron shielding and react faster with acrylamide or acrylic acid, both of which have carbonyl oxygens to withdraw electron density from target atoms, than it would with acrylonitrile. In these unbuffered solutions acrylamide is undissociated while acrylic acid is 64% dissociated [pK_a = 4.25 (16)] at the solute concentration of $5x10^{-5}M$. The acid anion might be expected to react somewhat slower than the acid because the anionic charge would be delocalized over the molecule and contribute some electron density to the shielding of the target carbon atoms.

Conclusion

We believe that the preceding resonance and steric arguments explain the differences in the reaction rates of the studied monomers, between themselves and the diffusion controlled limit. Because of the lack of H atom data for addition reactions at the diffusion controlled limit we wish to reserve judgement on the kinetic isotope effect. However, it appears that for these reported reactions there is little if any mass effect between H and Mu. Once the kinetic isotope effect is understood, Mu data observed directly by MSR methods could replace the more difficult

H-atom techniques. It is also possible that Mu initiation rates might fill the gaps in the data for initiation of polymerization by the simplest radical-atom hydrogen.

Acknowledgments

We wish to thank the Natural Sciences and Engineering Research Council (NSERC) of Canada for funding and Drs. T. Suzuki and P.W. Percival for their kind assistance.

Literature Cited

1. Hughes, V.W. Ann. Rev. Nucl. Sci. 1966, 16, 445.
2. Brewer, J.H. and Crowe, K.M. Ann. Rev. Nucl. Part. Sci. 1978, 28, 239.
3. Fleming, D.G.; Garner, D.M.; Vaz, L.C.; Walker, D.C.; Brewer, J.H.; and Crowe, K.M. Adv. in Chem. Ser. 1979, 175, 279.
4. Percival, P.W.; Ficher, H.; Camani, M.; Gygax, F.N.; Ruegg, W.; Schenck, A.; Schilling, H.; and Graf, H. Chem. Phys. Lett. 1976, 39, 333.
5. Ito, Y.; Ng, B.W.; Jean, Y.C.; and Walker, D.C. Can. J. Chem. 1980, 58, 2395.
6. Percival, P.W.; Roduner, E.; and Ficher, H. Adv. in Chem. Ser. 1979, 175, 335.
7. Percival, P.W.; Roduner, E.; Ficher, H.; Camani, M.; Gygax, F.N.; and Schenck, A. Chem. Phys. Lett. 1977, 47, 11.
8. Ng, B.W.; Jean, Y.C.; Ito, Y.; Suzuki, T.; Brewer, J.H.; Fleming, D.G.; and Walker, D.C. J. Phys. Chem. 1981, 85, 454.
9. (a) Roduner, E.; Percival, P.W.; Fleming, D.G.; Hochmann, J.; and Fischer, H. Chem. Phys. Lett 1978, 57, 37.
   (b) Roduner, E. and Fischer, H. Chem. Phys. 1981, 54, 261.
10. Roduner, E. Ph.D. Thesis, Universitat Zurich, 1979.
11. Stadlbauer, J.M.; Ng, B.W.; Walker, D.C.; Jean, Y.C.; Ito, Y. Can. J. Chem. 1981, 59, 3261.
12. Smith, P.; Pearson, J.T.; Wood, P.B.; and Smith, T.C. J. Chem. Phys. 1965, 43, 1535.
13. Swallow, A.J. Adv. in Chem. Ser. 1968, 82, 499.
14. Chambers, K.W.; Collinson, E.; and Dainton, F.S. Trans. Faraday Soc. 1970, 66, 142.
15. Buxton, G.V.; Ellis, P.G.; and McKillip, T.F.W. J. Chem. Soc. Faraday Trans I 1979, 75, 1050.
16. Madhaven, V.; Lichtin, N.N.; and Hayon, E. J. Org. Chem. 1976, 41, 2320.

RECEIVED September 10, 1982

COLLOID AND SURFACE CHEMISTRY

5

Electron-Transfer Behavior of the Metal Complexes Attached to Polymer Matrices

EISHUN TSUCHIDA and HIROYUKI NISHIDE

Waseda University, Department of Polymer Chemistry, Tokyo 160, Japan

Interfacial electron-transfer reactions between polymer-bonded metal complexes and the substrates in solution phase were studied to show colloid aspects of polymer catalysis. A polymer-bonded metal complex often shows a specifically catalytic behavior, because the electron-transfer reactivity is strongly affected by the polymer matrix that surrounds the complex. The electron-transfer reaction of the amphiphilic block copolymer-bonded Cu(II) complex with Fe(II)(phenanthroline)$_3$ proceeded due to a favorable entropic contribution, which indicated hydrophobic environmental effect of the copolymer. An electrochemical study of the electron-transfer reaction between a poly(xylylviologen) coated electrode and Fe(III) ion gave the diffusion constants of mass-transfer and electron-exchange and the rate constant of electron-transfer in the macromolecular domain. A polymer-Cu complex coated electrode was successfully applied as the interfacial catalyst for the oxidative polymerization of phenol.

The metal complex bound to a polymer often shows a specific catalytic behavior compared with that of the corresponding monomeric complex, because the reactivities of the complex are strongly affected by the polymer chain that surrounds the complex (1,2). The catalytic cycle is illustrated in Scheme 1, the example used being the Cu complex catalyzed oxidation of 2,6-dimethylphenol (3). In the first step, the substrate phenol coordinates to the Cu(II) complex and one electron transfers from the substrate to the Cu(II) ion. Then the activated substrate dissociates from the catalyst and the reduced Cu(I) catalyst is reoxidized to the original Cu(II) complex. Among these elementary reactions, the electron-transfer step is the most important process governing the catalytic behavior of a polymer-metal complex for the following reasons: (i) The electron-transfer step is often the slowest

0097-6156/83/0212-0049$06.00/0

elementary reaction in the catalysis. (ii) The electron-transfer step is an intracomplex process, and it is expected that the property of ligands in the complex catalyst as well as the property of the coordinated substrate directly affects the rate of the electron-transfer step. (iii) The electron-transfer reaction is strongly influenced by chemical environment around the complex, such as solvent effect and also the environmental effect of a polymer. (iv) The electron-transfer process requires rearrangement of the complex structure, which accompanies conformational change of the polymer-ligand. (v) Metal complexes are concentrated within the macromolecular domain, where an electron-exchange pathway is inferred. The present paper describes interfacial electron-transfer reactions between the metal complexes bound to polymer matrices and the substrates in solution phase to show colloid aspects of polymer catalysis.

Electron-transfer Reaction of the Metal Complexes Attached to Polystyrene Beads

The importance of the electron-transfer step is demonstrated, e.g. in the oxidative polymerization of 2,6-dimethylphenol catalyzed by the Cu complex attached to polyvinylpyridine beads. A metal complex catalyst is often insolubilized by using a cross-linked polymer matrix in order to handle it easier, but the insolubility causes a decrease in the catalytic activity. The Cu catalyst bound to divinylbenzene cross-linked poly(4-vinylpyridine) beads resulted in much lower catalytic activity, which was caused by the slower electron-transfer step and its higher activation energy (4). The structure of the Cu complex catalyst must be rearranged during the electron-transfer step from the square planar structure to the tetragonal structure (Scheme 2). Thus, the energy required to rearrange the complex structure is larger and electron-transfer step occurs with more difficulty for the polymer complex in which Cu ions are situated in the cross-linked polymer matrix and in which the polymer-ligand is lacking in flexibility. In order to improve this point, the metal complex catalysts have been attached to the polymer matrices with spacer groups (5, 6). Cupric ion was complexed with the pyridine derivative to polystyrene beads with oligo(ethylene oxide) (1) (7). The rate constant of the electron-transfer step was increased several times for the catalyst (1). This agreed with the result that 1 acted as an effective catalyst for the oxidative polymerization of dimethylphenol to yield high molecular weight poly(phenylene oxide) (7).

The second example which emphasized the electron-transfer step as a key process in polymer catalysis is the Ru(bipyridine)$_3$ complex sensitized photoreduction of methylviologen, which is of interest as a means to reduce protons leading to hydrogen evolution (Scheme 3). To suppress a spontaneous backward reaction which consumes the acquired energy, a heterogeneous reaction

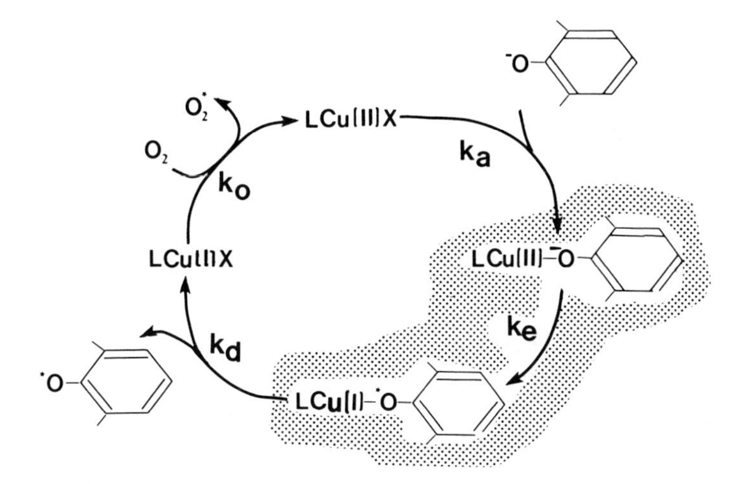

Scheme 1.

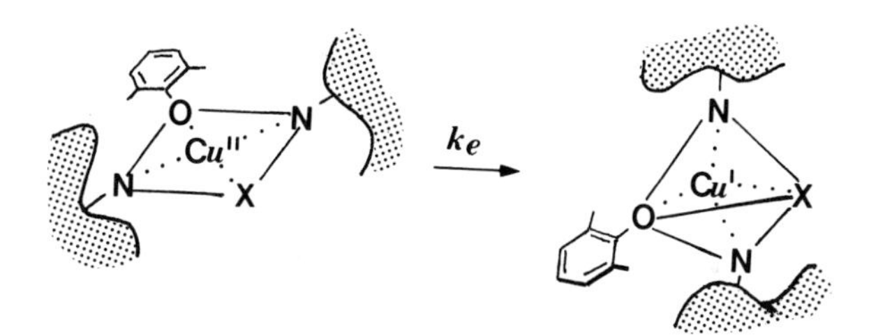

Scheme 2.

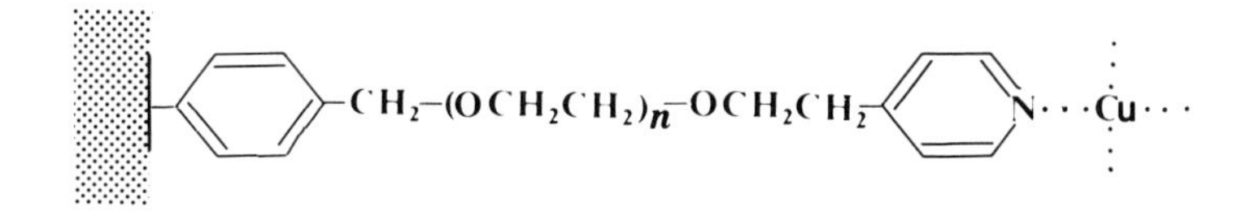

Structure 1.

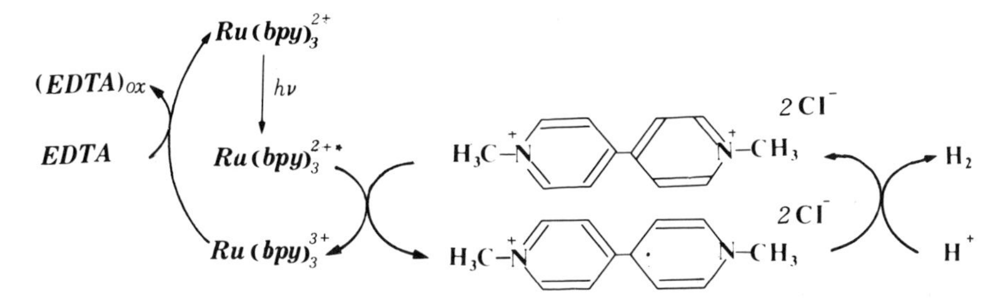

Scheme 3.

system using a polymer Ru complex was proposed (8). The aqueous solution of methylviologen and EDTA as an electron donor with the Ru complex was vigorously stirred under visible light irradiation. The cation radical of methylviologen was successfully accumulated in the presence of the Ru complex bound to polystyrene by a spacer group (2), while the Ru complex directly bound to polystyrene beads did not show any photoactivity (9). The emission of the Ru complex was effectively quenched with methylviologen by electron-transfer from the excited Ru complex to methylviologen even for the heterogeneous photoreaction system.

Electron-transfer Reaction of the Cu Complex Bound to Amphiphilic Block Copolymer

A water-soluble block copolymer composed of hydrophilic poly(ethylene oxide) and hydrophobic poly(styrene-co-vinyl-imidazole) was synthesized (3). Cu ion was complexed with the imidazole residue of the central hydrophobic residue. The complexation of about one Cu ion per block copolymer molecule resulted in a pale-blue aqueous and homogeneous solution.

Visible absorption and ESR spectra of the aqueous solution of the Cu/3 complex are shown together with those of the corresponding monomeric Cu/1-ethylimidazole complex in aqueous solution and in toluene (Figure 1). The visible absorption maximum and the hyperfine structure of the perpendicular ESR spectrum of the Cu/3 solution do not agree with those of the Cu/ethylimidazole complex in aqueous solution, but they are close to or agree with those in toluene. These results suggest that the Cu complex bound to 3 is situated in a rather hydrophobic environment like that of toluene even in an aqueous medium.

The spin-labeled analogous compound (4) was synthesized to estimate the mobility of the complex moeity bound to 3. The spin-label piperidine-oxyl in solution gives a narrow line ESR spectrum which is broadened in the solid state (Figure 2). A broadened spectrum was observed for 4 aqueous solution, which means that the motion of the complex is suppressed by the styryl residues of the polymer even in homogeneous aqueous solution. This result confirms that the Cu complex is in the hydrophobic environment of 3.

The reactivity of the Cu/3 complex was studied kinetically in the two electron-transfer reactions shown below. Cu(II) ion was reduced with ascorbic acid through an inner-sphere process in which ascorbic acid coordinates to the Cu complex and then transfers an electron to the Cu(II) ion. Cu(II) ion was also reduced with $Fe(II)(phenanthroline)_3$ via an outer-sphere mechanism in which electron-transfer occurs when a reductant and an oxidant move into sufficient proximity.

$$\text{Cu(II)} + \text{Ascorbic acid} \rightarrow \text{Cu(I)} + \text{Dehydroascorbic acid} \qquad (1)$$

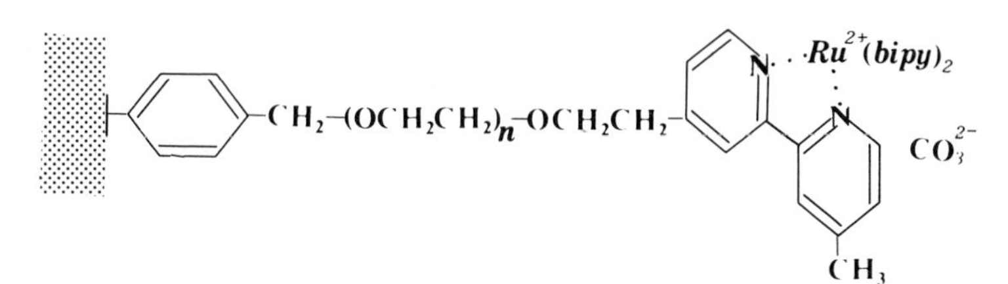

Structure 2.

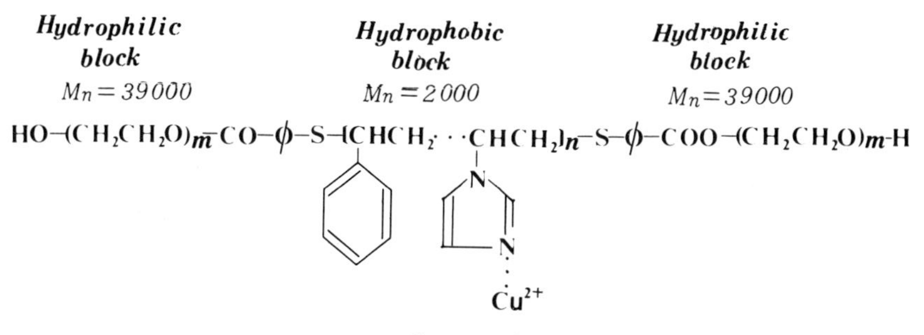

Structure 3.

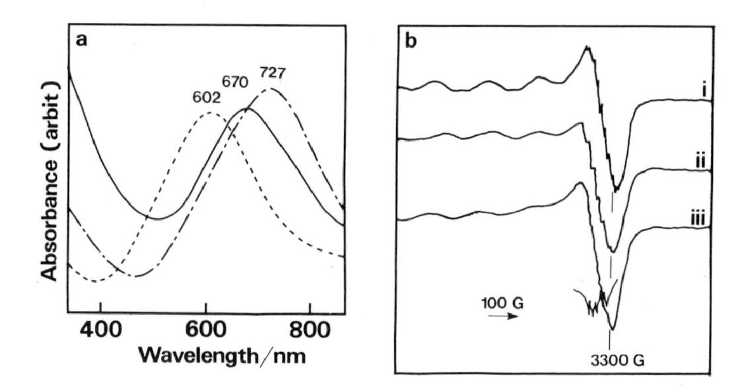

Figure 1. Visible adsorption spectra (a) and ESR spectra (b) of the Cu complex bound to the amphiphilic block copolymer (3). Key to a: —, Cu/3 in aqueous solution, [imidazole residue of 3]/[Cu] = 10; – – –, Cu/ethylimidazole in aqueous solution; — – —, Cu/ethylimidazole in toluene, [ethylimidazole]/[Cu] = 200, at 25 °C. Key to b: i, Cu/ethylimidazole in aqueous solution; ii, Cu/ethylimidazole in toluene; iii, Cu/3 in aqueous solution. The aqueous medium is pH 5.5 acetate buffer, at −196 °C.

$HO{-}(CH_2CH_2O)_m{-}CO{-}\phi{-}S{-}(CHCH_2\cdots CHCH_2)_n{-}S{-}\phi{-}COO{-}(CH_2CH_2O)_m{-}H$

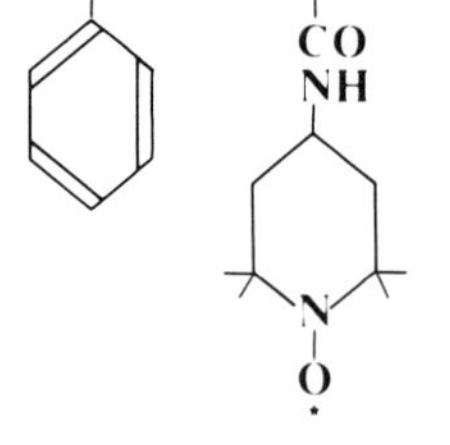

Structure 4.

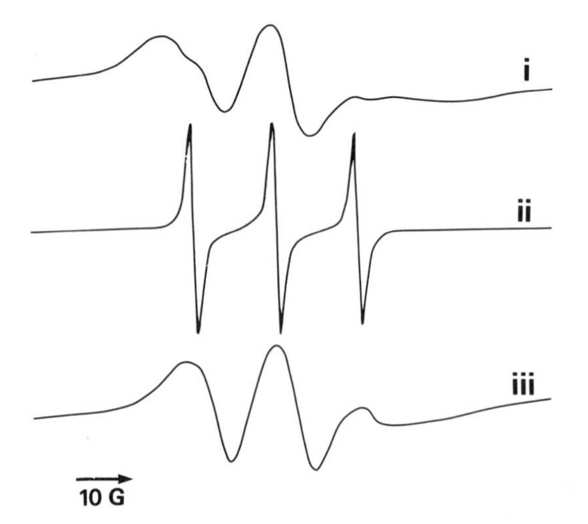

Figure 2. ESR spectra of the spin-labeled compound (4). i, 2,2,6,6-tetramethyl-4-aminopiperidine-1-oxyl in the solid state; ii, in toluene; iii, 4 in aqueous solution, at room temperature.

$$Cu(II) + Fe(II)(phenanthroline)_3 \rightarrow Cu(I) + Fe(III)(phenanthroline)_3 \quad (2)$$

The rate constant and activation parameters for the electron-transfer reactions are given in Table I. The reaction rate of the Cu/3 complex via the inner-sphere was smaller than that of the Cu/ethylimidazole complex. The coordination of the substrate to Cu(II) ion was enthalpically unfavored as compared to the homogeneous Cu complex. On the other hand, the outer-sphere reaction with $Fe(II)(phenanthroline)_3$ proceeded faster for the Cu/3 system than for the homogeneous Cu/ethylimidazole complex. 3 made a significant favorable entropic contribution to the outer-sphere electron-transfer reaction.

The reactivity ratios of the Cu/3 system and the monomeric Cu/ethylimidazole system were measured in aqueous-alcoholic solvents (Figure 3). The alcohol addition brought about a decrease in reactivity for the outer-sphere reaction, while the reactivity ratio for the inner-sphere reaction was practically independent of alcohol addition. A hydrophobic interaction between 3 and $Fe(phenanthroline)_3$ is assumed to account for the higher reactivity of the Cu/3 complex.

Plastocyanin is a protein with a single polypeptide chain with molecular weight of 10500 and a single Type 1 Cu ion, and its electron-transfer function is essential in photosynthesis. The X-ray crystal structure of plastocyanin has recently been established (10), which indicated that the core of the molecule is hydrophobic and notably aromatic because six of the seven phenylalanine residues are clustered there. Polar side chains are distributed on the exterior of plastocyanin molecule. Many hypotheses have been proposed to explain the electron-transfer pathway to and from the metal center of plastocyanin, such as a tunnelling mechanism along hydrophobic channels (11). High reactivity and entropic favorability have been reported for the electron-transfer reaction of plastocyanin with Fe(II) complex (12). The Cu complex bound to the amphiphilic block copolymer is interesting as a metal compound of plastocyanin, because both polymer and apoprotein environments are considered to produce a hydrophobic environment and a large effect on the electron-transfer reaction through its entropic contribution.

Electron-transfer Step in the Macromolecular Domain

It is expected that the electron-transfer reactivity in a macromolecular domain is strongly influenced by the primary structure of polymer matrices, distribution of redox sites, charge density and polarity of polymer domain, etc. In order to assess these factors, the electron-transfer reaction of poly(xylylviologen) (5) with Fe(III) ion was selected for investigation. The study gives parameters to represent the reactivities of the electron-transfer between the electrode and the viologen site,

Table I Rate constants and activation parameters for the electron-transfer reaction at 20°C.

Reductant	Cu(II) complex	Ligand ratio	$k \cdot 10^{-2}$ (l/mol·min)	$\Delta H^{\ddagger}$ (kcal/mol)	$\Delta S^{\ddagger}$ (e.u.)
Ascorbic Acid	Cu/Block Copolymer	10	1.8_4	8.1_1	$-20._4$
	Cu/Ethylimidazole	10	5.6_2	6.3_4	$-24._1$
	Cu/Ethylimidazole	200	$13._8$		
$Fe(phen)_3$	Cu/Block Copolymer	10	0.40_6	$15._8$	3.7_0
	Cu/Ethylimidazole	10	0.18_5	$11._8$	$-14._7$
	Cu/Ethylimidazole	200	0.21_0		

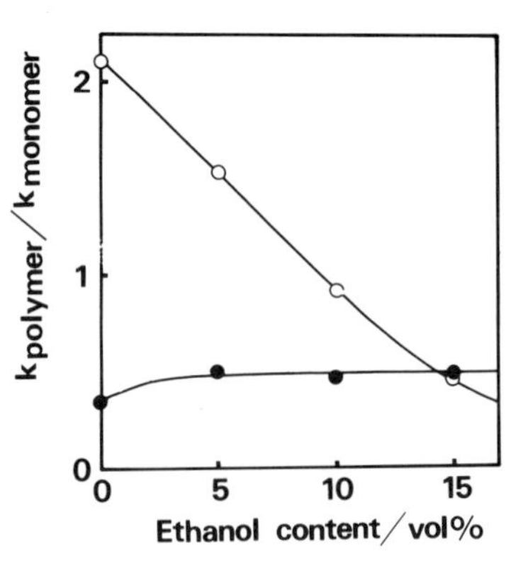

Figure 3. Effect of ethanol addition on the reduction rate of the Cu(II) complexes. Key: ○, reduction with Fe(II)-(phenanthroline)₃; ●, reduction with ascorbic acid.

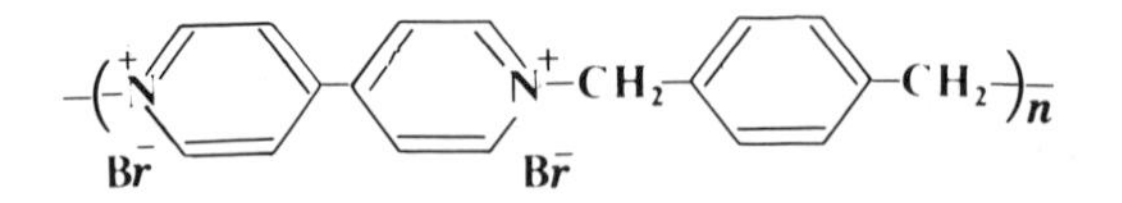

Structure 5.

the electron-exchange between the viologen sites in the polymer film, and the electron-transfer between the viologen site on the film surface and Fe(III) ion in the solution phase.

Poly(xylylviologen) or its complexes with polyanions, poly-(styrene sulfonic acid) or poly(glutamic acid), were coated on a graphite disk electrode. The disk current voltammogram was measured by rotating and scanning this ring-disk electrode in the aqueous solution of Fe(III) ion (Figure 4). Two limiting reduction currents were observed: i_1 at +0.1 volt and i_2 at -0.5 volt. Oxidation current of $Fe(CN)_6^{2-}$ was detected by the ring electrode set at 0.5 volt. These results indicate that i_1 and i_2 correspond to the mass-transfer and electron-transfer processes (Figure 5). That is, the current i_1 is caused by the reaction in which the substrate passes through the polymer domain and an electron directly transfers from the electrode to the substrate. The current i_2 corresponds to the reaction in which electron exchanges through the redox sites in the polymer domain and transfers to the substrate at the interface of the polymer layer and the solution phase. In the latter process, the viologen redox sites function as mediators of electron-transfer from the electrode to the substrate.

The mass-transfer and electron-transfer in the polymer domain is shown as a function of the thickness of the coated polymer film in Figure 6. The currents i_1 and i_2 at infinite rotation were calculated by using Koutecky-Levich equation (13): they represent the mass-transfer and electron-transfer in the polymer domain. The i_1 infinite value decreases with the film thickness, which means that contribution of the mass-transfer process to the redox reaction decreases with the film thickness. On the other hand, the i_2 infinite value increases with the film thickness, which reflects an expansion of the macromolecular domain, i.e. increase of the redox sites. At intermediate thickness of the polymer layer, the electron-transfer reaction proceeds most rapidly. Further increase of the thickness brings about the steep decrease in the reaction rate via the electron-transfer process, probably caused by the decrease in the electron-transfer efficiency.

Diffusion coefficient of the substrate (D_s) and diffusion coefficient of the electron-exchange (D_e) were calculated from cyclic and disk current voltammograms by using the Koutecky-Levich equation and Fick's first law (14, 15) (Table II). D_s in the polymer domains was estimated as 10^{-7} - 10^{-8} cm^2/sec, much smaller than in solution (10^{-5} cm^2/sec). D_s is affected by charge density of the polymer domain, e.g., the diffusion of cations is suppressed in the positively charged domain composed of cationic polyelectrolyte, while anions moves faster. A larger D_s value was observed, of course, for the porous film and not for the film with high density. On the other hand, D_e in the polymer domain was also very small, i.e. 10^{-10} - 10^{-11} cm^2/sec. This may be explained as follows. An electron-transfer reaction always alters the

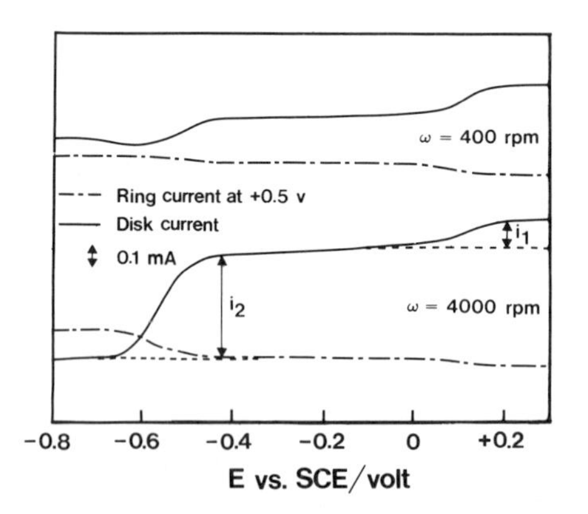

Figure 4. Voltammograms of the complex of poly(xylylviologen) with poly(styrene sulfonic acid) coated on electrode. 0.2 M *KCl aqueous solution containing 2 m*M *$K_3Fe(CN)_6$.*

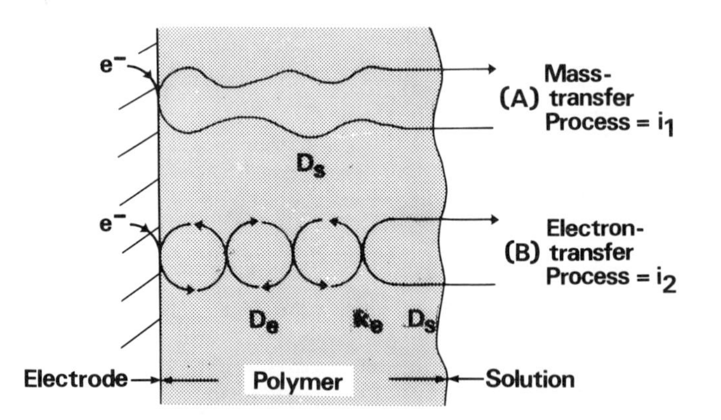

Figure 5. Redox pathways in the macromolecular domain coated on electrode.

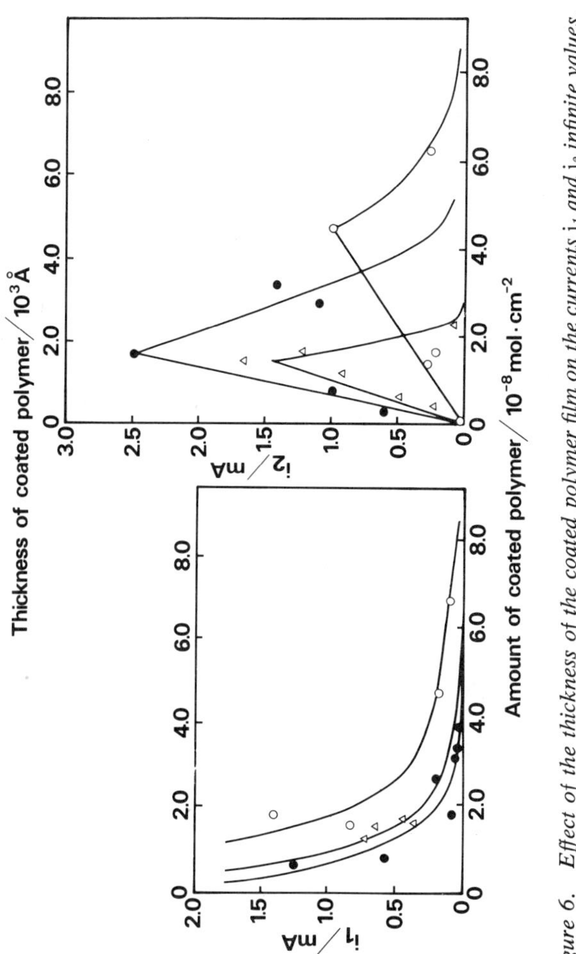

Figure 6. Effect of the thickness of the coated polymer film on the currents i_1 and i_2 infinite values. Key: ○, electrode coated with poly(xylylviologen) (5); △, electrode coated with the 5 complex of poly(styrene sulfonic acid); ●, electrode coated with the 5 complex of poly(glutamic acid), 0.2 M $NaClO_4$ aqueous solution containing 2 mM $K_3Fe(CN)_6$.

Table II Diffusion constants of substrate (D_s: cm^2/sec), of electron-exchange (D_e: cm^2/sec), and rate constants of electron-transfer (k_e: $l/mol \cdot sec$) in the macromolecular domain.

Polymer Matrix	Domain Charge	Domain Structure	Substrate	Mass-transfer $D_s \cdot 10^7$	Electron-transfer $D_e \cdot 10^{10}$	$k_e \cdot 10^{-3}$
5	2+~1+	loose	$[Fe(CN)_6]^{3-}$	1.32	> 10	0.41
			Fe^{3+}	0.31		
5-PSS	1+~0	tight	$[Fe(CN)_6]^{3-}$	0.68	0.66	2.56
			Fe^{3+}	0.50		
5-PGA	1+~0	medium	$[Fe(CN)_6]^{3-}$	0.71	10	1.58

PSS: poly(styrene sulfonic acid), PGA: poly(glutamic acid), 0.2 M $NaClO_4$ aqueous solution containing 2 mM Fe(III) ion.

charge of the redox site, which accompanies the movement of a counter ion. The diffusion of this counter ion is assumed to be rate determining in the polymer domain. Thus, the collision frequency between the redox sites situated in the polymer domain is not so great as expected.

These results suggest some factors for the preparation of electrodes with electron-transfer ability. (i) There is an optimum thickness of the coated polymer film for the electron-transfer reaction. (ii) The polymer matrix should be flexible. Otherwise, the matrix retards the diffusion of a counter ion and suppresses the effective collision between redox sites. (iii) A hydrophilic but uncharged polymer domain is suitable for the mass-transfer process in catalysis. A series of polymer complex coated electrodes were prepared as interfacial catalysts (16), one example is given below.

It is well known that 2,6-dimethylphenol is oxidatively polymerized to yield poly(2,6-dimethyl phenylene oxide) (Scheme 4) (17) This polymerization proceeds not only in the presence of Cu-pyridine catalyst under aerobic conditions but also electrolytically (18). Figure 7 shows the anodic oxidation of dimethylphenol by supplying 200 mA/sec with platinum counter electrodes and calomel reference electrode under anaerobic conditions. The phenol was polymerized with an equivalent hydrogen evolution to yield poly(phenylene oxide) with molecular weight over 10^4. A small amount of biphenoquinone was formed by the side reaction via C-C coupling, whose yield increased ca. 10 fold by stirring the electrolytic solution, i.e. the C-O coupling to produce poly(phenylene oxide) occurs at the surface or within the diffusion layer of electrode.

There remain some problems to be solved for this electrolytic oxidation of phenols. (i) Considerable magnitude of overvoltage and low current efficiency. (ii) Poly(phenylene oxide) formed deposits on the electrode surface as a thin, insulating film passivating of the electrode. (iii) The side reaction which forms biphenoquinone.

The electrolytic oxidation was found to proceed much faster in the presence of Cu-pyridine as a redox mediator in the electrolytic cell divided with a membrane. The electrode coated with Cu/poly(4-vinylpyridine) was also effective for the oxidative polymerization, and what was more, without a partition membrane (Figure 8). Polymer-Cu complex film coated on the electrode prevented formation of the insulating film of the product polymer on the electrode surface and decreased the electrolytic potential. The oxidation using the electrode coated with a macromolecular Cu complex provides a facile method for forming poly(phenylene oxide)s.

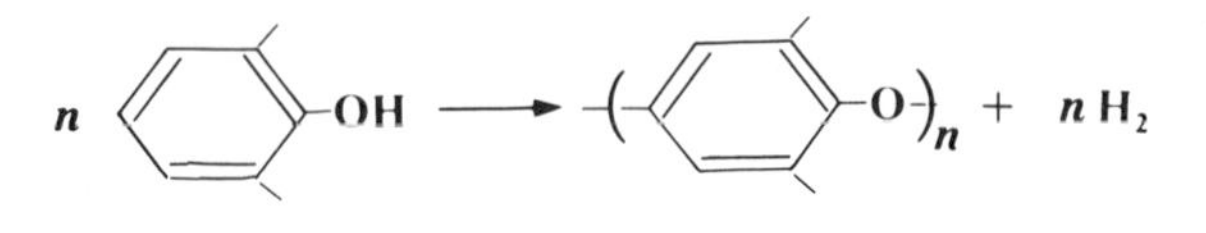

Scheme 4.

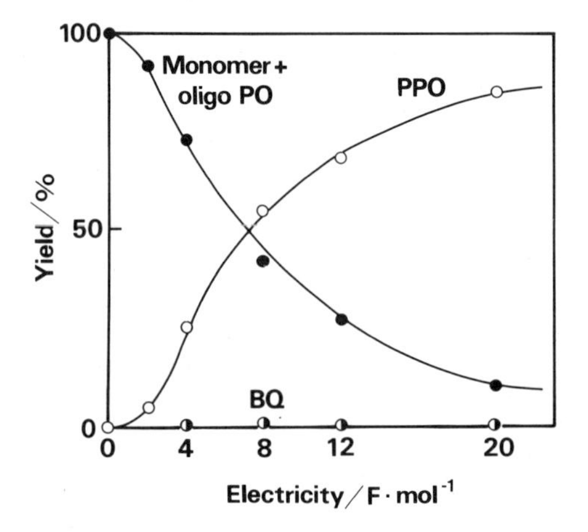

Figure 7. Electrolytic oxidation of 2,6-dimethylphenol. Key: ○, *poly(2,6-dimethyl phenylene oxide);* ●, *oligo(2,6-dimethyl phenylene oxide);* ◑, *2,2′6,6′-tetramethylbiphenoquinone, 0.1* M *dimethylphenol in 20% methanol-dichloromethane containing 0.2* M *tetraethyl ammonium bromide, current density 10 mA/cm²; platinum counter electrodes with saturated calomel reference under anaerobic condition at room temperature.*

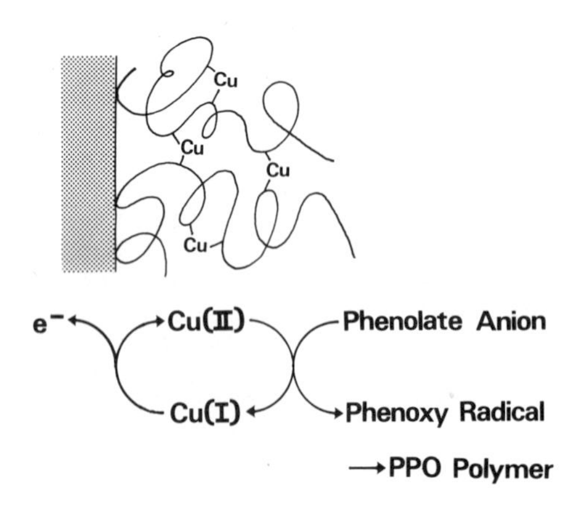

Figure 8. Oxidative polymerization using electrode coated with macromolecular Cu complex.

Literature Cited

1. Tsuchida, E.; Nishide, H. Adv. Polymer Sci. 1977, 24, 1-87.
2. Kaneko, M.; Tsuchida, E. J. Polymer Sci. Rev. 1981, 16, 397-522.
3. Tsuchida, E.; Nishide, H.; Nishikawa, H. J. Polymer Sci. Symp. 1974, 47, 47-54.
4. Nishide, H.; Suzuki, Y.; Tsuchida, E. Eur. Polymer J. 1981, 17, 573-577.
5. Warshawsky, A.; Kalir, R.; Deshe, A.; Berkovitz, H.; Patchornik, A. J. Am. Chem. Soc. 1979, 101, 4249-4256.
6. Nishide, H.; Shimidzu, N.; Tsuchida, E. J. App. Polymer Sci. in press.
7. Nishide, H.; Suzuki, Y.; Tsuchida, E. Makromol. Chem. in press.
8. Kaneko, M.; Motoyoshi, J.; Yamada, A. Nature 1981, 285, 468-470.
9. Tsuchida, E.; Nishide, H.; Shimidzu, N.; Yamada, A.; Kaneko, M.; Kurimura, Y. Makromol. Chem. Rapid Commun. 1981 2, 621-626.
10. Colman, P.M.; Freeman, H.C.; Guss, J.M.; Murata, M.; Norris, J.A.M.; Venkatappa, M.P. Nature 1978, 272, 319-324.
11. Moore, G.R.; Williams, R.J.P. Coord. Chem. Rev. 1976, 18, 125.
12. Wherland, S.; Holwerda, R.A.; Rosenberg, R.C.; Gray, H.B. J. Am. Chem. Soc. 1975, 97, 5260-5262
13. Koutecky, J.; Levich, V.G. Zh. Fiz. Khim. 1956, 32, 1565-1567.
14. Levich, V.G. "Physicochemical Hydrodynamic"; Prentice-Hall: Englewood Cliffs, N.J. 1962, pp 345-357.
15. Oyama, N.; Anson. F.C. Anal. Chem. 1980, 52, 1192-1200.
16. Oyama, N.; Oki, N.; Ohno, H.; Ohnuki, Y.; Matsuda, H.; Tsuchida, E. J. Electroanal. Chem. in press.
17. Hay, A.S. Adv. Polymer Sci. 1967, 4, 496-542.
18. U.S. Patent 1967, 3 335075.

RECEIVED August 24, 1982

6

Esterolytic Reactions of Active Esters Using Heterogeneous Polymeric Catalysts Containing Imidazole Groups

C. G. OVERBERGER and BYONG-DO KWON

The University of Michigan, Department of Chemistry and the Macromolecular Research Center, Ann Arbor, MI 48109

The preparation of polymeric catalysts and substrates containing imidazole groups and nitrophenyl esters, respectively, grafted onto crosslinked polystyrene beads has been described and the effects of the acyl chain length in the substrate in the aqueous alcohol solvent systems on the rate of hydrolysis of 3-nitro-4-acyloxybenzoic acid substrates catalyzed by insoluble catalysts were determined.

Over the last decade we have studied extensively the esterolytic activity of polymers that contain pendant imidazole groups. (1-12) In the poly-4(5)-vinylimidazole catalyzed reaction, three major factors which contribute to large rate enhancements have been defined; cooperative effects(13,14), electrostatic effects (15,16), and hydrophobic effects.(17,18)

The hydrophobic effect can be described in a broad sense as a type of non-specific apolar bonding between catalyst and substrate. These hydrophobic interactions, especially in an aqueous environment, have been predominant in determining the efficiency of the catalysts.(8,10,12) Toward this end, considerable effort on our part has been directed toward the preparation of polymeric catalysts that contain pendant imidazole groups and apolar bonding sites that are soluble in highly aqueous solvent systems. The efficiency of our recently investigated catalysts was enhanced by apolar bonding in an aqueous environment between substrate and the apolar polymer backbone of the catalysts,(8,12,13) a pendant apolar group associated with the catalyst,(10,18) or a long-chain N-acyl-imidazole intermediate that occurs during the solvolysis reaction.(17) Other workers have also used the hydrophobic effect in imidazole-catalyzed esterolytic reactions. Recently, Nango and Klotz(20,21) reported esterolytic catalysis by polyethylenimine derivatives that contain pendant imidazole groups, diethylamino groups, or functionalized triazine residues which increased catalytic effectiveness due to cooperative effects and increased

0097-6156/83/0212-0065$06.00/0

apolar binding. Mirejovsky(22) investigated the optimal environment for imidazole-bound polyethylenimine catalysts that contained pendant apolar lauryl groups as binding sites. In an effort to design a more effective macromolecular catalyst, Shinkai and Kunitake(23) elucidated the correlation between binding capacity and catalytic activity in phenylimidazole-containing copolymers.

More recently, we turned our attention toward insoluble polymer catalysts to provide additional insight into this area. This article describes the synthesis of insoluble polymeric catalysts and substrates and a preliminary study on their esterolytic reactivity towards various long-chain *p*-nitrophenyl esters.

Experimental

Materials. Crosslinked polystyrene beads were purchased from Bio-Rad Laboratories and were dried under vacuo at 40°C. Merrifield resin (chloromethyl group content 0.9 meq/g) was purchased from Pierce Chemical Co. A series of 3-nitro-4-acyloxybenzoic acids(7) were prepared by methods described in the literature. The long-chain *p*-nitrophenyl ester substrates were purchased from Sigma Chemical Co. *p*-Nitrophenyl acetate was a product of Pierce Chemical Co. and sublimed in vacuo before being used. Polystyrene (average MW 22,000) was purchased from Aldrich Chemical Co.

General Procedure. The elemental analyses were determined by Spang Laboratories, Eagle Harbor, Michigan, or Galbraith Laboratories, Inc., Knoxville, Tennessee.

The NMR spectra on low molecular weight compounds were recorded on a Varian T-60A spectrometer. NMR of polymers were recorded on a JEOL JNM-FX902 Fourier Transform NMR Spectrometer equipped with a FAFT50 FG/BG disc unit, and WM-360 FT NMR spectrometer equipped with ASPEC 2000 computer system manufactured by Bruker Instruments, Inc.

Rates of hydrolysis were measured on a Varian Cary 219 UV spectrophotometer equipped with a Haake constant temperature bath. A stirred batch tank reactor was placed in a water bath maintained at 26.0 ± 0.1°C and the solution was transported to a micro flow UV cell located in the spectrophotometer via silicone rubber tubing using a precision flow peristaltic pump (MHRE 22) manufactured by the New Brunswick Scientific Company.

Preparation of Polymer Catalysts (I-VI). Crosslinked polystyrene beads with 1% divinylbenzene mesh size 200-400 (SX1) purchased from Bio-Rad Laboratories, Inc., were chloromethylated by the method of Pepper, Paisley, and Young(24). The resin contained 4.59 meq/g of chloromethyl groups according to the modified Volhard method.(25) The chloromethylated resin was allowed to swell in dioxane-ethanol and reacted with histamine; pyridine or triethylamine was used as an acid captor. The reaction mixture was stirred and kept at 80°C for a period of 7 days. When the reaction was completed, the resin(I) was washed with dioxane:H_2O (1:1/v:v) dioxane:methanol (1:1/v:v), dioxane, successively. The ring substitution ratio was 44.4%, histamine and 17.0% remaining $-CH_2Cl$ units. Polystyrene bead crosslinked with 2% divinylbenzene mesh size 200-400 was chloromethylated followed by substitution with histamine using the same method described above.(II) The aromatic rings are substituted by histamine (41.7%) and 9.93% with $-CH_2Cl$ groups.

Parts of I and II were treated with potassium acetate in dioxane for 48 hr at 120°C, then with diluted 2N NaOH;dioxane (1:2/v:v) at room temperature overnight to remove the remaining chloride residue (III,IV).

Merrifield resin purchased from Pierce Chemical Co. which contains 0.9 meq/g of chloromethyl groups was treated with histamine free base in the presence of triethylamine.(V) The ring substitution ratio was 2.8% by histamine, 4.5% by chloromethyl group. Some Merrifield resin and potassium phthalate in dry DMF was heated at 120°C overnight. When the reaction was completed, the resin was filtered, washed and treated with $NH_2NH_2 \cdot H_2O$ followed by 5% NaOH to give an orange-colored resin. No chlorine was detected in the resin.

The second polymer derivative prepared for use as a polymeric catalyst is of the type

$$XPS{-}\overset{O}{\overset{\|}{C}}(CH_2)_n{-}N\langle\text{imidazole}\rangle N. \qquad \text{(VI)}$$

One of these was synthesized by the Friedel-Crafts acylation of ω-chloroacid chloride(26) (0.1 mole) and $AlCl_3$ (0.15 mole) with Bio-beads SX1, SX2, SX12, and SM2 (0.1 mole) in 200 ml of nitrobenzene for 2 hours at 5°C. The resin was washed with acetic acid, 6N HCl:dioxane (1:1), H_2O-dioxane (1:1), dioxane, CH_2Cl_2 and MeOH. The resin was dried under vacuo at 45°C and treated with the sodium derivative of imidazole in DMF.

Structural characterization of these catalysts is under investigation using infrared, elemental analysis and NMR spectroscopy using equipment stated earlier.

A summary of the polymeric catalysts is in Table I.

Preparation of Polymeric Substrates. Crosslinked polystyrene containing carboxyl groups was prepared by the Letsinger method. (27) Diphenylcarbamyl chloride (18.5 g, 80 mmole) in 40 ml nitrobenzene was added over a period of 15 min to a well-stirred mixture of 8.3 g (80 mmole) Bio-Bead SX2 and 15.0 g (0.113 mole) of aluminum chloride in 350 ml of dry nitrobenzene. The dark mixture was then warmed at 80°C for 2.5 hr, cooled and treated with 200 ml of water. The resin was separated and washed in succession with dilute hydrochloric acid, methanol, and ether. For hydrolysis, the carboxamido polymer was heated at 103-138°C for 32 hr with a

Table I. Catalysts

Catalyst	Backbone	Active Site	Degree of Substitution		Catalyst Concentration
I	Bio-Beads SX1 (Bio Rad)	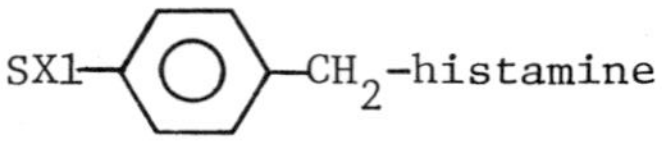 SX1–C_6H_4–CH_2-histamine	Histamine	44.4%	2.7×10^{-3} mole/g
			$-CH_2Cl$	17.0%	
II			Histamine	44.2%	2.7×10^{-3} mole/g
			$-CH_2OH$	17.0%	
III	Bio-Beads SX2 (Bio Rad)	SX1-CH_2-histamine	Histamine	44.7%	3.01×10^{-3} mole/g
			$-CH_2Cl$	9.93%	
IV			Histamine	44.7%	
			$-CH_2OH$	9.95%	
V	Merrifield Resin (Pierce)		Histamine	2.8%	2.518×10^{-4} mole/g
			$-CH_2Cl$	4.5%	
VI			Histamine	10.6%	9.76×10^{-4} mole/g
B_nXN	Bio-Beads SX1, SX2 SX12, SM_2	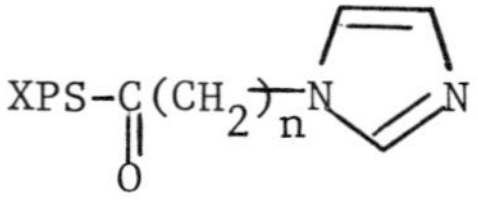 XPS–C(=O)$(CH_2)_n$–N(imidazole)	XPS = crosslinked polystyrene bead		
C_nXN		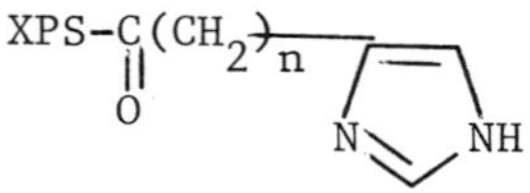 XPS–C(=O)$(CH_2)_n$–(imidazole, NH)			

mixture of 335 ml of acetic acid, 250 ml of sulfuric acid, and 150 ml of water. Filtration and washing with water and ether afforded polymer-containing carboxyl groups as a pale greenish material showing the characteristic carboxyl absorption in the infrared at 1730 cm^{-1} and 1680 cm^{-1}. Titration was effected by suspending 0.500 g of polymer in 25 ml of 95% ethanol, adding 25 ml 0.1N sodium hydroxide, warming the mixture to reflux, cooling, and back titrating with hydrochloric acid: 1.4 g mequivalent of carboxyl per gram of polymer was found. Warming a portion of carboxyl polymer with excess thionyl chloride in benzene for 5 hr afforded the acid chloride and

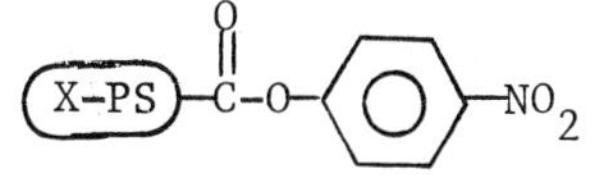

on reaction with p-nitrophenyl and pyridine. A strong band at 1530 cm^{-1} and a weak C-N stretching (1345 cm^{-1}) indicates the nitrobenzene moiety clearly.

The second polymeric substrate with a longer chain length was prepared by the Friedel-Crafts alkylation of Bio-Bead SX1 and $AlCl_3$ with 4-chlorobutyric acid in nitrobenzene for 5 hours at 80°C. The resin was washed with AcOH, 6N HCl-dioxane (1:1), H_2O:dioxane (1:1), dioxane and CH_2Cl_2. Titration was effected as stated earlier; 0.9 mequiv of carboxyl per gram of polymer was determined.

Kinetic Measurement. A preliminary kinetic study has been carried out as follows: Solvent and catalyst were added to a 50 ml 3-neck round-bottom flask equipped with mechanical stirrer, and the flask was placed in a constant temperature bath at 26°C ± 0.02. Time was recorded as substrate was added in the flask with stirring. An aliquot of reaction mixture was taken out using a syringe equipped with a millipore filter and the absorption was measured.

A typical example of reaction was the following: (catalyst) = 5.185 x 15^{3}M (substrate) = 3.085 x 10^{-4}M pH = 9.1 Tris 0.02M μ = 0.02 T = 26°C ratio = [catalyst]/[substrate] = 16.8. Absorbance versus time curves were obtained. A blank on each kinetic run was prepared and the blank value was subtracted from the catalytic value. All data obtained under conditions of [catalyst] >>[substrate] were, unless otherwise noted, treated as pseudo first-order kinetics by plotting $\ln(A_\infty - A_t)$ versus time. A_{max} was used instead of A_∞. The slope of the straight line was taken as k_{obs} and in the case of accelerative behavior the slope at 75% of the maximum absorption was taken; k_{cat} was calculated from k_{obs}: $k_{cat} = k_{obs}/[catalyst]$.

Recently a continuous flow type measurement was set up as stated earlier in the Materials Section.

Results and Discussion

Kinetic Studies. The previous work of Overberger(1,7) provided a rationale to this research. These investigators utilized the hydrophobic backbone of poly-4(5)-vinylimidazole in conjunction with hydrophobic substrates (Sn^-) to demonstrate enhanced rates of esterolysis. In particular, a buildup of long-chain acylated imidazole residues, which increased hydrolysis rates dramatically, inspired the concept of using hydrophobic polymer side chains.

In this study, insoluble polymers containing imidazole groups were tested for esterolytic activity with several substrates of differing hydrophobic chain lengths. The substrates used were 3-nitro-4-acyloxybenzoic acid (Sn^-). The concentration of substrate used does not allow the formation of substrate micelles, the concentration being below the critical micelle concentration for these esters (28,29).

A ten- to twenty-fold concentration excess of polymeric imidazole residues over that of substrate molecules was usually employed. This allowed a pseudo first-order presentation of the kinetic data. In many cases, curvature in the plots of $\ln(A_{max}-A)$ versus time was found. Observation of complex kinetics in hydrolysis of functional groups on polymer chains is not uncommon. (30,31) Letsinger and Klaus(32) have observed some phenomena in the study of synthetic polymeric catalysts and substrates. In treating the data, they used the empirical relation

$$A-A_o/A_\infty-A = k't$$

A_o, A_∞, and A represented the absorbancies at initial time, at completion of the reaction, and at the time of measurement, respectively. k' was considered to be a pseudo first-order rate constant defined by the expression $-dc/dt = k'C(C/C_o)$, where C/C_o represents the relative reactivity of the ester groups on a polymer chain as a function of extent of reaction. A reasonably good fit to this was obtained for reactions conducted in the absence of catalysts as well as in those catalyzed by N-methylimidazole and poly-(N-vinylimidazole).

An empirical equation was applied to the data to obtain nonlinear plots. Therefore, in the case of accelerative behavior, the slope at 75% of the maximum absorption was taken. A summary of the second-order rate constant are provided in Table II.

NMR Studies. Previous ^{13}C NMR investigation of some polymers at temperatures well above the Tg and in solvent-swollen gels have shown that high resolution spectra may be obtained in good solvents without use of cross-polarization, magic angle spinning, or dipolar decoupling.(33-35) The only published ^{13}C NMR spectra of crosslinked polystyrenes show high resolution in swelling solvents and the technique has been used to estimate the degree

Table II. Apparent Second-Order Rate Constants (k_{cat}) for the Hydrolysis of Sn^-

Catalysts	k_{cat} $\ell M^{-1} min^{-1}$		
	S_2^-	S_5^-	S_{12}^-
I	28.92	19.37	24.75
II	15.80	10.73	23.05
III	27.03	- - -	- - -
IV	12.56	- - -	- - -
V	18.4	- - -	- - -

Solvent EtOH = H_2O = 50:50

pH = 9.1 Tris μ = 0.02

Table III. Comparison of Nuclear Overhauser Effects

	With NOE	No NOE	r
Linear polystyrene[a]	0.381	0.289	1.32
Funetionalized Linear Polystyrene[b]	0.589	0.493	2.24
Catalyst A[c]	0.788	0.621	1.27

a: Average MW 22,000; 5000 scans.

b: L-PS-C(=O)-CH_2CH_2Cl; 4000 scans.

c: X-PS-C(=O)-$CH_2CH_2CH_2$-N(imidazole)N : 10,000 scans.

In order to get analytical and structural information from cross-linked polystyrenes by solvent swelling, extensive H^1- and C^{13} NMR study is under investigation.

of functionalization of resins for solid phase peptide synthesis (36) by Horowitz, Horowitz, and Pinnell. They reported NMR analytical results on solvent-swollen, crosslinked polystyrenes which demonstrated a new approach for obtaining analytical and structural information on functionalized polystyrenes that are important as the backbone for many solid-state syntheses and reagents. They obtained the extent of $-CH_2Cl$ and $-CH_2OH$ in chloromethylated polystyrene resin using proton noise decoupled spectra. In principle, NMR spectra with a Nuclear Overhauser Effect (NOE) cannot be used quantitatively, because the Nuclear Overhauser Effect cannot be the same for all pendant groups.

Therefore, a preliminary NMR study of the Nuclear Overhauser Effect on linear polymers and crosslinked polymer was made. The NMR spectra were taken on resin slurries in deuterochloroform. The slurries were prepared by introducing a quantity of resin into an NMR tube; a portion of solvent was pipetted in and the resin was allowed to swell while the tube was agitated. Additional solvent was added with further agitation (sometimes an ultrasonic device helped the agitation) as the swelling continued so that the slurry remained mobile. A vortex plug was used to contain the swollen resin. Complete proton noise decoupled spectrum were obtained under the conditions: 45° pulse at 0.9 sec repetition; $CDCL_3$ solvent; internal deuterium lock. Gated decoupling spectra without Nuclear Overhauser Effect were obtained under the conditions: 90° pulse at 4 sec repetition; $CDCL_3$ solvent internal deuterium lock.

The ratio of height of peak in the aromatic region was compared to that of peak at aliphatic region (Table III).

Acknowledgments

The authors are grateful for financial support from the Department of Chemistry and the Macromolecular Research Center, The University of Michigan, Ann Arbor, Michigan 48109 USA.

Literature Cited

1. Overberger, C. G; Salamone, J. C. Acc. Chem. Res., 1969, 2, 217.
2. Overberger, C. G.; Morimoto, M.; Cho, I.; Salamone, J. C. J. Am. Chem. Soc., 1971, 93, 3228.
3. Overberger, C. G.; Morimoto, M. J. Am. Chem. Soc., 1971, 93, 3222.
4. Overberger, C. G.; Okamoto, Y. Macromolecules, 1972, 5, 363.
5. Overberger, C. G.; Okamoto, Y. J. Polym. Sci., Polym. Chem. Ed., 1972, 10, 3387.
6. Overberger, C. G.; Glowaky, R. C. J. Am. Chem. Soc., 1973, 95, 6014.
7. Overberger, C. G.; Glowaky, R. C.; Vandewyer, P.-H. J. Am. Chem. Soc., 1973, 95, 6008.

8. Overberger, C. G.; Smith, T. W. Macromolecules, 1975, 8, 401, 407, 416.
9. Overberger, C. G.; Kawakami, Y. J. Polym. Sci., Polym. Chem. Ed., 1978, 16, 1237, 1248.
10. Overberger, C. G.; Guterl, A. C., Jr. J. Polym. Sci., Polym. Symp., 1978, 62, 13.
11. Overberger, C. G.; Smith, T. W.; Dixon, K. W. J. Polym. Sci. Polym. Symp., 1975, 50, 1.
12. Overberger, C. G.; Dixon, K. W. J. Polym. Sci., Polym. Chem. Ed., 1977, 15, 1863.
13. Overberger, C. G.; Salamone, J. C.; Cho, I.; Maki, H. Ann. N.Y. Acad. Sci., 1969, 155, 431.
14. Overberger, C. G.; St. Pierre, T.; Vorchheimer, N.; Lee, J.; Yaroslavsky, S. J. Am. Chem. Soc., 1965, 87, 296.
15. Overberger, C. G.; Corett, R.; Salamone, J. C.; Yaroslavsky, S. Macromolecules, 1968, 1, 331.
16. Overberger, C. G.; Salamone, J. C.; Yaroslavsky, S. J. Am. Chem. Soc., 1967, 89, 6231.
17. Overberger, C. G.; Glowaky, R. C.; Vandewyer. P.-H. J. Am. Chem. Soc., 1973, 95, 6008.
18. Pacansky, T. J. Ph.D. Thesis, The University of Michigan, Ann Arbor, Michigan, 1972.
19. Overberger, C. G.; Mitra, S. Pure Appl. Chem., 1979, 51, 1391.
20. Nango, M.; Klotz, I. M. J. Polym. Sci., Polym. Chem. Ed., 1978, 16, 1265.
21. Nango, M.; Gamson, E. P.; Klotz, I. M. J. Polym. Sci., Polym. Chem. Ed., 1979, 17, 1557.
22. Mirejovsky, D. J. Org. Chem., 1979, 44, 4881.
23. Shinkai, S.; Kunitake, T. J.Am. Chem. Soc., 1971, 93, 4256.
24. Pepper, K. W.; Paisley, H. M.; Young, M. A. J. Chem. Soc., 1953, 4097.
25. Stewart, J. M.; Young, J. D. "Solid Phase Peptide Synthesis," W. H. Freeman, San Francisco, 1969, p. 54.
26. Tilak, M. A.; Hollinden, C. S. Tetrahedron Lett., 1968, 1297.
27. Letsinger, R. L.; Kornet, M. J.; Mahedevon; Jerina, D. M. J. Am. Chem. Soc., 1964, 86, 5163.
28. Glowaky, R. C. Ph.D. Thesis, The University of Michigan, 1972.
29. Smith, T. W. Ph.D. Thesis, The University of Michigan, 1973.
30. Moens, J.; Smets, G. J. Polym. Sci., 1957, 23, 931.
31. Gaetjens, E.; Morawetz, H. J. Am. Chem. Soc., 1961, 83, 1738.
32. Letsinger, R. L.; Klaus, I. J. Am. Chem. Soc., 1964, 86, 3884.
33. Yokota, K.; Abe, A.; Hosaka, S.; Sakai, I.; Saito, H. Macromolecules, 1978, 11, 95.
34. Schaefer, J. Macromolecules, 1971, 4, 110.

35. Torchia, D. A.; VanderHart, D. L. "Topics in Carbon-13 NMR Spectroscopy," Vol. 3, Levy, G. C., Ed., John Wiley and Sons, Inc., New York, N.Y., 1979, pp. 325-360.
36. Horowitz, D.; Horowitz, R.; Pinnell, R. P. Anal. Chem., 1980, 52, 1529.

RECEIVED August 24, 1982

Catalytic Effects of Micellar Poly(3-alkyl-1-vinylimidazolium) Salts on the Hydrolysis of Phenyl Esters

S. C. ISRAEL, KONSTANTINOS I. PAPATHOMAS,[1] and J. C. SALAMONE

University of Lowell, Department of Chemistry, Polymer Science Program, Lowell, MA 01854

The catalytic effects of poly(3-methyl-1-vinylimidazolium iodide) poly(3-n-hexadecyl-1-vinylimidazolium iodide), and their monomeric analogs on the alkaline hydrolysis of a series of neutral and anionic nitro-phenyl esters were studied in order to elucidate the contribution of hydrophobic and electrostatic interactions to their esterolytic efficiency. The base hydrolyses of these substrates were enhanced by the addition of cationic polyelectrolyte and polysoap. Saturation type kinetics were observed and the values of rate constants for the uncatalyzed and catalyzed reactions are given as well as the values for the association constant. The results suggest that the hydrophobic interactions play a major role in the hydrolysis enhancements of the neutral and charged substrates, and that these polymeric micelles can provide both the characteristics of polyelectrolytes and micelles and thus give higher rate enchancements.

The interactions of micelles with small molecule reactions and the catalytic effects that result from these interactions have been of great interest for a number of years (1,2). Much of the attention has been focused on the properties of polymeric micelles (2,3) and their effectiveness as catalytic agents. In addition to the effect of polymer micelles it has been

[1] Current address: IBM Endicott, Endicott, NY 13760.

0097-6156/83/0212-0075$06.00/0

shown that synthetic polyelectrolytes may also be effective catalysts for the same small molecule reactions (4,5,6). These micellar and synthetic polymeric systems are of particular interest because they exhibit many features observed in enzyme reactions (1-6). These features include reactivities higher than those of the corresponding monomeric systems, substrate specificity, competitive inhibition, bifunctional interactions of functional groups and substrate, and saturation by low and high molecular weight substrates (1-6).

Many hydrolytic studies have been reported utilizing both micellar and synthetic polymeric systems. Cordes (7) and co-workers, and Ocubo and Ise (8) have reported on systems which incorporate both the novel features of micelles and polyelectrolytes, i.e., catalyst systems which have binding sites available for both strong hydrophobic and electrostatic interaction with suitable substrates. Such hydrophobic polyelectrolyte systems have been prepared and have been termed "polysoaps" (9,10). The catalytic properties of these polysoaps has remained largely unexplored.

We have initiated a series of investigations to study the catalytic effects of a class of cationic polyelectrolytes ranging in solution behavior from "normal" polyions to polysoaps upon the alkaline hydrolysis of neutral and anionic phenyl esters of varying chain lengths. Employing these catalysts of varying hydrophylic-hydrophobic character in reactions of neutral and anionic substrates of varied hydrophilic-hydrophobic character, it should be possible to elucidate the contributions of both the hydrophobic interactions and electrostatic interactions on the rate of reaction.

Experimental

A. Synthesis of Substrates. p-Nitrophenyl acetate (PNPA) was synthesized from p-nitrophenol and acetic anhydride according to the general method of Bender and Nakamura (11). The ester was recrystallized four times from petroleum ether giving a melting point of 78-79°C (lit. 81-82°C).

p-Nitrophenyl laurate (PNPL) was prepared by a modification of the procedure reported by Zahn and Schade (12): 0.125 mole of p-nitrophenol, 0.10 mole of lauroyl chloride, and 0.10 mole of pyridine were dissolved in 80 ml of dry toluene and refluxed for 1 hr. The solution was neutralized with saturated $NaHCO_3$

and successively washed with water, 5% NaOH, 0.1N HCl, and finally with water. The toluene solution was dried over anhydrous $MgSO_4$, filtered and then evaporated. The obtained light yellow waxy solid was recrystallized four times from absolute ethanol, gave a melting point of 46-47°C (lit. 46°C).

Sodium 3-nitro-4-hydroxybenzenesulfonate was prepared from 2-nitrophenol and chlorosulfonic acid by the method of Gnehm and Krecht (13). The sulfonic acid was converted to the sodium salt by precipitation from saturated sodium chloride solution.

Sodium 3-nitro-4-acetoxybenzenesulfonate (NABS) was prepared by refluxing 1g (0.0042 mole) of sodium 3-nitro-4-hydroxybenzenesulfonate and 25 ml of acetic anhydride for 20 hrs. The mixture was filtered and the precipitate recrystallized four times from acetic acid. A pale yellow solid was obtained which did not melt below 300°C.

Sodium 3-nitro-4-dodecanoyloxybenzenesulfonate (NDBS) was prepared by the general procedure reported by Bruice (14). Equimolar quantities of trifluoroacetic anhydride and dodecanoic acid were allowed to react at room temperature for 30 min; a 0.5M quantity of phenol was added with stirring and the mixture heated at 80°C for 5 hrs. The cooled reaction mixture was washed with ether to remove the excess anhydride and the residue recrystallized four times from methanol-ether.

B. Synthesis of Catalysts. 3-Methyl-1-vinylimidazolium iodide (VII-C1) and 3-*n*-hexadecyl-1-vinylimidazolium iodide (VII-C16) were obtained by the quaternization of 1-vinylimidazole with the corresponding *n*-alkyl-iodides, according to the method previously reported (15).

Poly(3-methyl-1-vinylimidazolium iodide) (PVII-C1) and poly(3-*n*-hexadecyl-1-vinylimidazolium iodide) (PVII-C16) were homopolymerized from their corresponding vinylimidazolium iodide monomers according to our previously reported method (15).

C. Kinetic Measurements. Catalyst solutions were prepared in 20 mole% ethanol-water buffered to 8.21 pH. Sufficient potassium chloride was added to adjust the ionic strength to 0.002M. To 20 ml of the buffered catalyst solution, in a volumetric flask thermostated at 30.0°C, 3 ml of an ethanol solution of the substrate was added.

The rate of hydrolysis of substrate was followed in a Beckman Model 25 Spectrophotometer equipped with a sipper cell. The absorbance of the nitrophenolate anion was measured as a function of time at 406 nm, utilizing the sipper cell. Absorbance at infinite time was taken after at least ten half-lifes. The reaction mixtures and spectrophotometer cell were thermostated to constant temperature.

Results and Discussion

The main thrust of this work is the elucidation of the contribution of electrostatic and hydrophobic interactions to the enzyme like catalytic behavior. The catalysts of choice for this study are shown below:

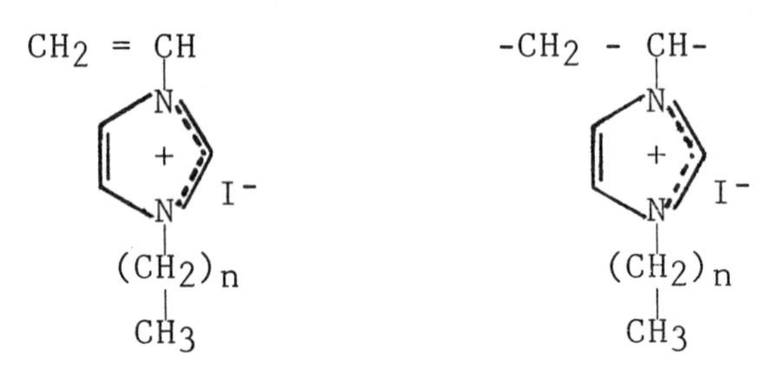

n = 0,15

The short side-chain monomer, 3-methyl-1-vinylimidazolium iodide (VII-C1), is hydrophilic and would not be expected to exhibit catalytic behavior. The long side-chain monomer, 3-n-hexadecyl-1-vinylimidazolium iodide (VII-C16) does form micelles in solution and would be expected to participate in hydrophobic interaction with the appropriate substrate. The short side-chain polymer, poly(3-methyl-1-vinylimidazolium iodide) (PVII-C1), exhibits polyelectrolyte behavior and would be expected to exhibit electrostatic interaction only, whereas the long side-chain polymer, poly(3-n-hexadecyl-1-vinylimidazolium iodide) (PVII-C16), is a polyelectrolyte and forms micelles in aqueous media; (16,17) and thus would be expected to participate in both hydrophobic and electrostatic interaction with the substrates.

The substrate reactions under investigation involve the base hydrolysis of the neutral esters as shown in Reaction (1) and the anionic esters as shown in Reaction (2).

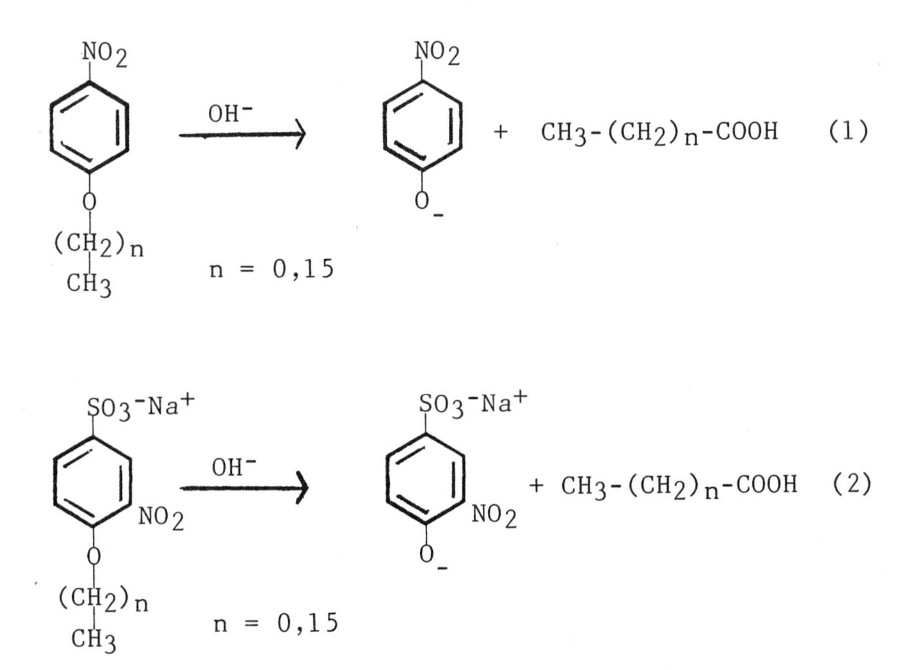

By varying the number of methylene units on the neutral and charged substrate and by changing the length of the pendent group on the catalysts, it is possible to control the hydrophilic-hydrophobic nature of both substrate and catalyst.

The rate constants for these reactions were evaluated from the plots of $1/(k_{obs}-k_1)$ vs. $1/[C]$ according to the following kinetic scheme:

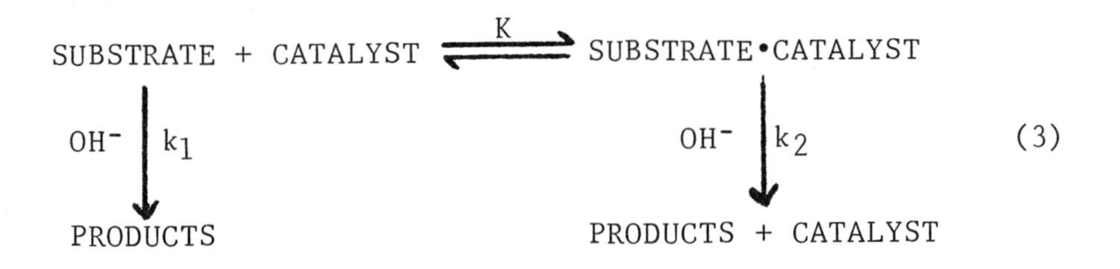

$$\frac{1}{k_{obs} - k_1} = \frac{1}{k_2 - k_1} + \frac{1}{K(k_2 - k_1)} \cdot \frac{1}{[C]} \qquad (4)$$

Where [C] denotes the catalyst concentration, k_{obs} is the measured pseudo first-order rate constant; k_1, the rate constant for the hydrolysis reaction of the substrate; and k_2, the rate constant for the hydrolysis reaction of the catalyst-substrate complex.

Figure 1 shows the first-order rate plots for the neutral substrates and the plots for the anionic substrates are shown in Figure 2. The kinetic parameters may be evaluated from Equation (4). A plot of $1/(k_{obs}-k_1)$ vs. $1/[C]$ yields a straight line for each of the systems with the values of k_1, k_2, and the association constants K determined from the slope and intercept of these plots. These values are given in Table I for the neutral and anionic substrates.

In the case of the neutral esters, it can be seen from the data in Table I that the alkaline hydrolysis reactions of PNPA yield no catalytic enhancement of the rate, and we detected no catalyst-substrate binding with either the micellar monomers or polymers. It can be seen that the reactions of PNPL were enhanced by the addition of cationic polyelectrolyte and polysoap and the magnitudes of the enhancement were in the order of PVII-C16 $\rangle$ PVII-C1. Saturation type kinetics were observed [Figure (3)] strongly supporting the formation of a complex between polymer and ester (18). This complexation phenomenon implies the validity of the proposed kinetic scheme [Equation (3)]. In these systems, both K and k_2 increased according to the hydrophobicity of the polyion or the ester. Values of k_2 are seen to be greater than k_1 values. Thus the esters bound to polyions are more reactive than the esters themselves. Values of k_2 increased in the order PVII-C16 $\rangle$ PVII-C1. This indicates that the reactivity of the complex increases as the hydrophobic character of the complex increases. This is in agreement with the findings of several researchers (7,8,19) who have reported that an increase in the reaction rate and the binding constant can be correlated with an increase in hydrophobic character of the catalyst.

In the case of the PNPA, it was shown that the presence of VII-C16 had no effect on the hydrolysis, but in the case of PNPL, a plot of k_{obs} vs. [VII-C16] as shown in Figure 4 does not yield a straight line and indicates that the VII-C16 monomer inhibits the

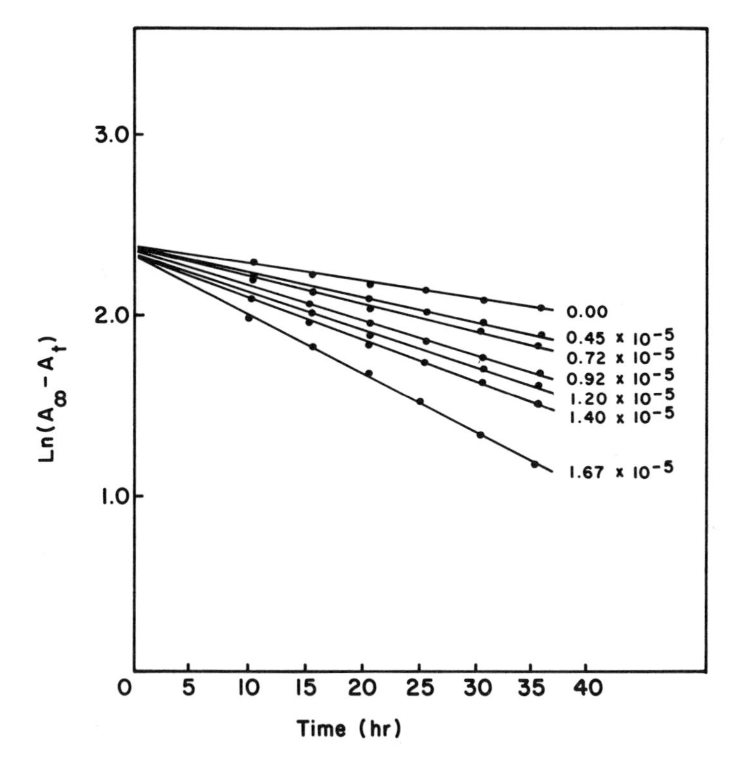

*Figure 1. Variation of ln(*A∞ − At*) with time for the hydrolysis of PNPL in the presence of PVII-C16. [PNPL] = 1.66 × 10⁻⁵* M*, pH = 8.21, 20% EtOH-H₂O.*

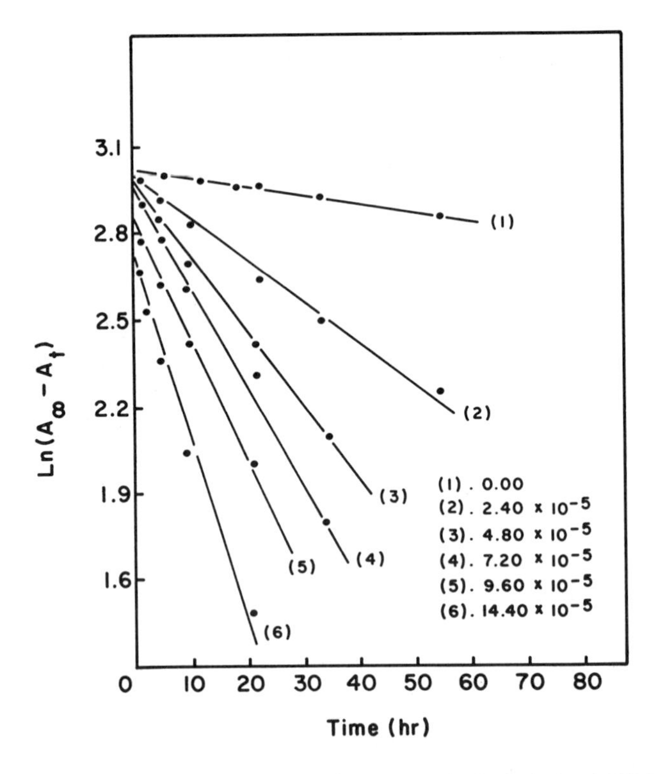

Figure 2. Variation of ln(A∞ − At) with time for the hydrolysis of NDBS in the presence of PVII–C16. [NDBS] = 1.15 × 10^{-4} M, pH = 8.21, 20% EtOH-H_2O.

TABLE I

Kinetic parameters for the hydrolysis in the presence of polyions and micelles at 30°C and 20% $EtOH-H_2O$.

Ester	Catalyst	pH	K	$k_1(min^{-1})$	$k_2(min^{-1})$
PNPA	$VII-C_{16}$	8.21	--	1.71×10^{-2}	--
	$PVII-C_{16}$	8.21	--	1.71×10^{-2}	--
PNPL	$PVII-C_1$	8.21	1080	0.98×10^{-4}	1.64×10^{-4}
	$PVII-C_{16}$	8.21	5530	0.98×10^{-4}	2.99×10^{-4}
NABS	$PVII-C_1$	8.21	815	4.21×10^{-4}	2.73×10^{-3}
	$PVII-C_{16}$	8.21	874	4.21×10^{-4}	3.06×10^{-3}
NDBS	$VII-C_1$	8.21	--	--	--
	$VII-C_{16}$	8.21	428	4.32×10^{-5}	3.07×10^{-4}
	$PVII-C_1$	8.21	867	4.32×10^{-5}	4.22×10^{-4}
	$PVII-C_{16}$	8.21	6380	4.32×10^{-5}	1.95×10^{-3}

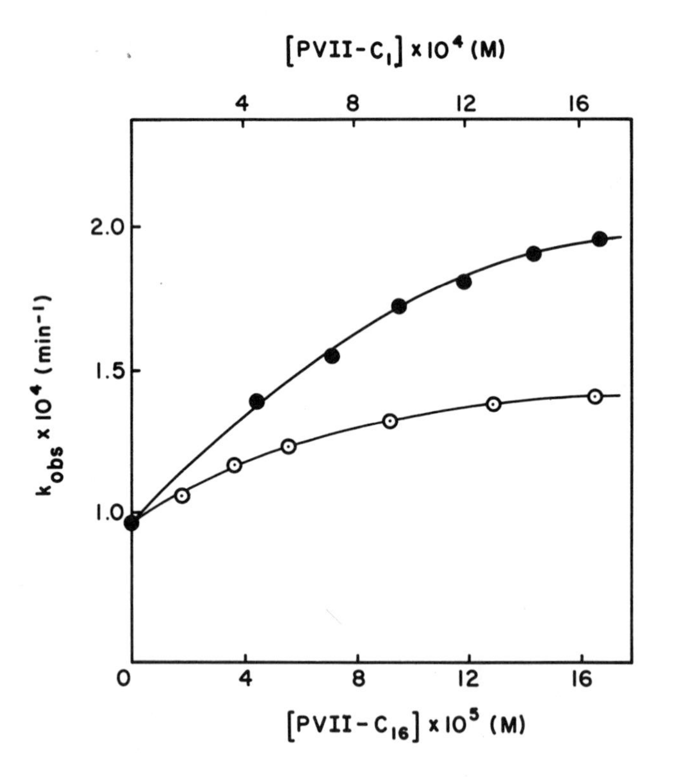

Figure 3. Variation of observed rate constants for the hydrolysis of PNPL with varying concentrations of PVII–C1 (⊙) and PVII–C16 (●) at 30 °C. [PNPL] = 1.66×10^{-5} M, pH = 8.21, 20% EtOH-H_2O.

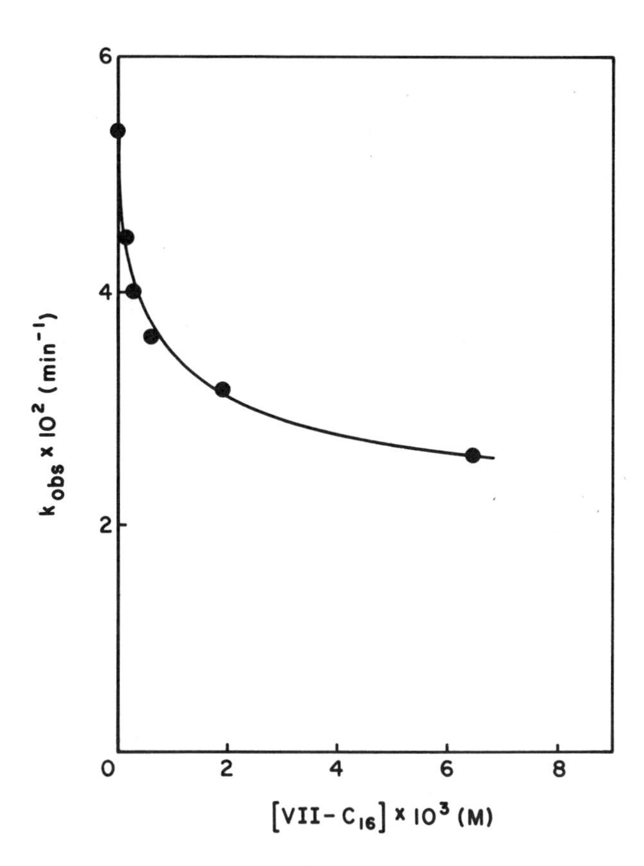

Figure 4. Variation of k_{obs} with concentration of VII–C16 for the hydrolysis of PNPL at 30 °C. [PNPL] = 4.49 × 10^{-5} M, pH = 10.4, 20% EtOH-H_2O.

hydrolysis of the PNPL and the rate measurements do not obey the kinetic scheme presented in Equation (4). The mechanism of this inhibition is not understood at the present time.

In the case of the anionic esters, the alkaline hydrolysis of NDBS and NABS are enhanced by the addition of cationic polyelectrolyte and polysoap. The most hydrophobic catalyst (PVII-C16) gave the largest enhancement for both substrates. Saturation behavior was observed for both anionic substrate reactions, indicating the formation of the complex between polymer and substrate (20). In the case of the NABS, the increase in the hydrophobic character of the catalysts yields a very small increase in both the binding constant (K) and k_2 and thus it can be concluded that the hydrophobic interaction plays a very minor role. The rate enhancement due to the electrostatic interactions in this system is over 6 fold. For the long side-chain ester NDBS we can see the effect of both the electrostatic and hydrophobic interactions. The short side-chain polymer PVII-C1 produced over a 9 fold increase in the rate due primarily to the electrostatic interaction. By far the most efficient catalyst studied is the long side-chain polymer, PVII-C16, which combines both electrostatic and hydrophobic interactions to yield over a 45 fold increase in the reaction rate.

Literature Cited

1. Cordes, E.H. and Dunlap, R.B., Accounts Chem. Res., 1969, 2, 1.
2. Cordes, E.H. and Gitler, C., Prog. Bioorg. Phys. Org. Chem., 1970, 8, 271.
3. Fendler, E.J. and Fendler, J.H., Advan. Phys. Org. Chem., 1970, 8, 271.
4. Morawetz, H., Advan. Catal., 1969, 20, 341.
5. Morawetz, H., Accounts Chem. Res., 1970, 3, 354.
6. Overberger, C.G. and Salamone, J.C., Accounts Chem. Res., 1969, 2, 217.
7. Rudolfo, T., Hamilton, J.A., Cordes, E.H., J. Org. Chem., 1974, 39, 2281.
8. Okubo, T. and Ise, N., J. Org. Chem., 1973, 38, 3120.
9. Strauss, U.P, Gershfeld, N.L. and Crook, E.V., J. Phys. Chem., 1956, 60, 577.
10. Freedman, H.H., Mason, J.P. and Medalia, A.I., J. Org. Chem., 1958, 23, 26.

11. Bender, M.L. and Nakamura, J. Amer. Chem. Soc., 1962, 84, 2577.
12. Zahn, H. and Schade, F., Chem. Ber., 1963, 96, 1747.
13. Gnehm, R. and Krecht, O., J. Prakt. Chem., 1906, 73, 521.
14. Bruice, T.C., Katzhendler, J. and Fedor, L.R., J. Amer. Chem. Soc, 1968, 90, 1333.
15. Salamone, J.C., Israel, S.C., Taylor, P., Snider, B., Polymer, 1973, 14, 639.
16. Salamone, J.C., Israel, S.C., Taylor, P., Snider, B., Polym. Prepr., 1973, 14(2), 778.
17. Salamone, J.C., Israel, S.C., Taylor, P., Snider, B., J. Polym. Sci.,Polym. Symp. Ed., 1974, 45, 65.
18. Israel, S.C., Papathomas, K.I. and Salamone, J.C., Polym. Prepr., 1981, 22(1), 221.
19. Fendler, J.H. and Fendler, E.J., "Catalysis in Micellar and Macromolecular Systems", Academic Press, New York, 1975.
20. Israel, S.C., Papathomas, K.I. and Salamone, J.C., Polym. Prepr., 22(2), 377.

RECEIVED August 24, 1982

8

Epitaxial Polymerization as a Tool for Molecular Engineering

JEROME B. LANDO, ERIC BAER, SCOTT E. RICKERT, HEMI NAE, and STEPHEN CHING

Case Western Reserve University, Department of Macromolecular Science, Cleveland, OH 44106

Epitaxial crystallization is the oriented overgrowth of a substance on a crystalline substrate. The interaction between the two is usually highly specific and has profound effects on the morphology and structure of the crystallizing material. Epitaxial polymerization is a new phenomenon in the field of macromolecular science. It combines the epitaxial crystallization of a monomer on a crystalline substrate, followed by solid-state polymerization, which is controlled by the epitaxial crystallization of the monomer. A study of the vapor phase epitaxial polymerization of disulfurnitride to polythiazyl, $(SN)_x$, resulted in three new crystalline phases of $(SN)_x$ and a new appreciation of the catastrophic effect of water on this polymer. The epitaxial polymerization of hexachlorocyclotriphosphazene to poly(dichlorophosphazene), $(NPCl_2)_x$ has been investigated. Deposition from both the vapor phase and solution has been studied. The polymer structure and morphology depend upon the monomer epitaxial crystals. The actual monomer morphology and structure have been found to be dependent on the geometry of the substrate. The application of this method to the epitaxial polymerization of diacetylenes will also be discussed.

A topochemical effect in a solid state reaction is any effect on the structure and properties of the product or the kinetics of the reaction that can be directly attributed to the geometric arrangement of the reacting groups or the distance between those groups. The degree of topochemical control in a solid state reaction can vary greatly depending upon the particular system investigated.(1) Reactions to be discussed here, in which there is a crystallographic correlation between the reactant and the resulting product, can occur in solid solution or with the nucleation and growth of a product phase.

Systematic investigation of solid state polymerization reactions began with the discovery that crystalline acrylamide polymerizes when exposed to ionizing radiation. (2,3) Since that

0097-6156/83/0212-0089$06.00/0

time investigators have been intrigued by the possibility of producing polymers with unusual structures, morphologies and properties through solid state polymerization. Although these goals have been fulfilled to some extent in the past twenty-five years, a pervasive problem has been a lack of control over monomer morphology and structure. In simple terms we are "stuck" with the structures that nature gives us. In the following paper, a method will be discussed that allows variation of monomer structure and morphology. This involves a technique we have termed epitaxial polymerization.[4] Monomers are crystallized on crystalline substrates. Lattice matching allows the variation of monomer structure and morphology, yielding different polymer structures and morphologies upon polymerization. Three types of monomers will be discussed in this paper, cyclic sulfur-nitrogen compounds, cyclic phosphazenes and diacetylenes.

Experimental

Tetrasulfur Tetramide. Tetrasulfur tetramide (S_4N_4) was sublimed at 100°C and 10^{-5} torr. The vapor was passed through Ag_2S wool at 210°C forming S_2N_2. The hot S_2N_2 vapor condensed and crystallized on alkali halide single crystals at -78°C.[4,5] Polymerization occurred during heating to room temperature.

Polyphosphazene. Hexachlorocyclotriphosphazene ($N_3P_3Cl_6$) was obtained as purified crystals from the Army Research Laboratories. This trimer was deposited as thin films on alkali-halide substrates from the vapor, melt, and solution. Vapor phase deposition was accomplished by subliming $N_3P_3Cl_6$ at 80°C in 10^{-5} torr vacuum. It was found that deposition times of around 20 minutes yielded incomplete films suitable for electron microscopy. Prior to sublimation the alkali halide crystal was always annealed at room temperature in 10^{-5} torr vacuum in order to ensure removal of adsorbed water from the crystal surface.

Crystallization from solution involved the preparation of a 20 wt % solution of the trimer in decane. The precipitation temperature for this solution was 23°C, and all epitaxial depositions were done at 29°C, or 6°C above this cloud point. Deposition times on in situ cleaved salt crystal (100) surfaces were near 10 minutes for film thicknesses suitable for election microscopy.

Another method used involved the casting of a thin film of $N_3P_3Cl_6$ from decane on a freshly cleaved alkali halide surface at room temperature, followed by heating to 130°C. The molten trimer film was then slow cooled to 110°C (4°C below the bulk melting temperature) and held at that temperature for 1 hour.

All epitaxial films were either prepared directly for microscopic examination, or polymerized using a post-polymerization technique for the first time on phosphazene monomer. The trimer films were irradiated with 2.5 Mrad of γ-radiation from a Co^{60}

source (24 hours exposure time) at 37°C. The now-activated trimer films were reacted by annealing in an inert atmosphere at 140°C for two hours.

Polydiacetylenes. Dimethanol-diacetylene ($HOCH_2C{\equiv}C{-}C{\equiv}CCH_2OH$) (DMDA) was used as received from Farehan Chemical Co. Diphenyl urethane-diacetylene (⌬-NH-COO-$(CH_2)_4$-C≡C-C≡C-$(CH_2)_4$-OOC-NH-⌬) (TCDU) wasused as received from Allied Chemical Co. Both monomers were successfully deposited as epitaxial films on various alkali halides from vapor and solution phases. The following conditions were used for these crystallizations: vapor phase DMDA = 40°C at 10^{-5} torr, vapor phase TCDU = 150°C at 10^{-5} torr, solution (1 wt % in toluene) phase DMDA = 70°C for 30 minutes, solution (.4 wt % in ethyl acetate) phase TCDU = 30°C for 5 minutes.

Polymerization was accomplished by a 2.5 Mrad exposure to γ-irradiation (24 hours) at 37°C.

Alkali Halide Substrates. (100) freshly cleaved surfaces of KCl, KI, KBr, NaCl, KF, and RbI were used as epitaxial substrates for these materials.

Film Preparation for Electron Microscopy. $(SN)_x$, $N_3P_3Cl_6$, $(NPCl_2)_x$, DMDA, $(DMDA)_x$, TCDU, and $(TCDU)_x$ were all prepared for examination by use of a collodion stripping technique or by substrate dissolution, after Pt, C, or Au coating to enhance film strength and contrast. A Jeol JEM 100B transmission electron microscope was used in both bright field and diffraction modes.

Results and Discussion

Poly(Sulfur Nitride). S_2N_2, deposited on KCl, KI, NaCl and KBr all produced polymer oriented along the <110> direction of the substrate.(5) However, no orientation was observed using the highly hydroscopic KF and RbI substrates. Fourier transform infrared studies of $(SN)_x$ produced on these substrates showed a high degree of degradation by water.(6) The $(SN)_x$ crystals produced on KCl, KI, NaF, NaCl and KBr are rectangular platelets. The sharpness of the edges on KI indicate rapid reaction while warming, whereas on some of the other substrates rounded edges were observed, indicating sublimation of S_2N_2 before complete polymerization. This type of difference can be attributed to a differing catalytic effect on polymerization of these substrates.

It should be noted that none of the $(SN)_x$ produced by direct epitaxial polymerization was fibrous in nature as is the conventional material in the normal α form.(7-10) Not only is the α form produced on KCl not fibrous, it contains none of the extensive crystal twinning observed in the conventional polymerization. Thus greater crystal perfection is obtained by epitaxial polymerization. The conventional fibrous α$(SN)_x$ can be produced if a second deposition of S_2N_2 is made on $(SN)_x$ crystals produced by epitaxial polymerization. This is shown in figure 1. Note that an epitaxial effect is retained - the fibers are oriented.

The most striking changes are differences in structure of $(SN)_x$ crystals on these substrates as well as a difference in morphology from conventionally obtained $(SN)_x$. The different crystal structures are shown in Table I. Examination of the (110) spacings of the salts indicated that lattice matching was responsible for the appearance of three new phases of $(SN)_x$, the γ, δ and ε phases.

Polyphosphazenes. Single crystals of $N_3P_3Cl_6$ were grown on a variety of alkali halide crystals. The resultant morphology of the trimer and polymer was identical and of 2 distinct types: pyramidal or square shaped crystals, and rodlike crystals. All crystals were oriented in the substrate's <110> directions independent of method of crystallization. Representative films from vapor and solution depositions are shown in figures 2 and 3.

Electron diffraction from these crystals indicated a high degree of crystallinity in both the trimer and polymer, and indexing of the patterns required the use of new unit cells different from the usual structures[11] (Figure 4). X-ray diffractometry was used to provide structural information about the third dimension in these epitaxially crystallized films. It should be noted that the characteristic spacing determined by X-ray diffractometry is the b-axis, only if the unit cell is orthorombic (Table II).

Polydiacetylenes. Single crystals of poly(TCDU) and poly(DMDA) were deposited on a variety of alkali halide crystals. The resultant morphology of both the monomers and polymers were identical and were elongated platelets oriented in the substrates <110> directions (Figures 5 and 6).

Electron diffraction from poly $(TCDU)_x$ is shown in Figure 7, and corresponds to the polymer phase 2, shown in Table III. This phase is not usually observed in bulk crystals polymerized under routine conditions.[12,13]

Diffraction from $poly(DMDA)_x$ could not be indexed as the usual polymer phase,[14] but required a new unit cell to be devised (Table IV). It should be emphasized that usual crystal of DMDA cannot be completely polymerized, as has been done in this case.

Conclusions

1. All monomer epitaxial crystallization and polymerization is controlled by the substrate surface and results in new, highly oriented crystalline films with new structures and properties which are presently under investigation.

2. Defect-free single crystals of $(NPCl_2)_x$ and $(SN)_x$ have

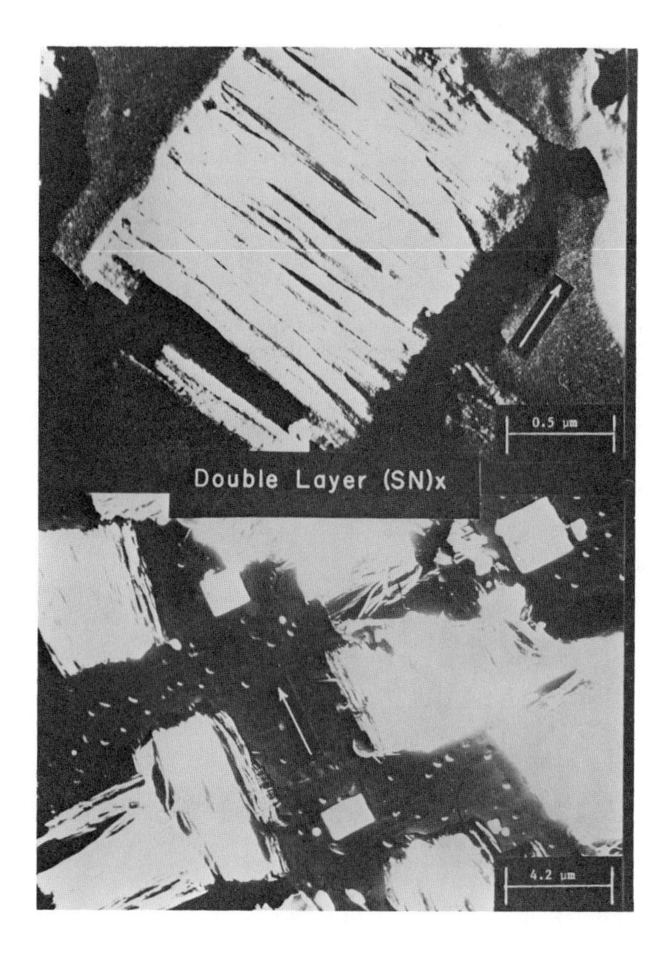

Figure 1. Photo at top: double-layer morphology of polythiazyl deposited on KI. The second layer preferentially grows on top of the first layer, and clearly shows a fibrous nature. Photo at bottom: electron micrograph as above, indicating the orientation of the chain is along both (110) directions of the substrate, but with only one direction per crystal. Some of the second deposit grows on previously unoccupied regions on the substrate as normal rectangular platelets (arrow points to one of three in micrograph).

TABLE I

Crystal Forms of $(SN)_x$ Produced by Epitaxial Polymerization.[a]

Form	Crystal Class	a	b	c	γ	Space Group	Substrate
α	Monoclinic	0.415	0.764	0.444	110	$P2_1/a$	KCl
β	Orthorhombic	0.920	1.072	0.493		$P2_12_12_1$	KI, NaF
δ	...	0.680	...	0.524		$P2_1$[b]	NaCl
ε	...	1.392	...	0.581		$P2_1$[b]	KBr

[a]Units are innanometers and degrees, and c is the chain axis.

[b]Full space group unknown.

TABLE II

STRUCTURES OF CHLORINATED PHOSPHAZENES

MATERIAL	a	b	c
normal trimer	1.415	1.299	.619
new trimer	.707	1.238	.458
normal polymer	1.299	1.111	.492
new polymer	.671	1.107	.494

Units are in nanometers.

b axes of new materials by x-ray diffractometer scan

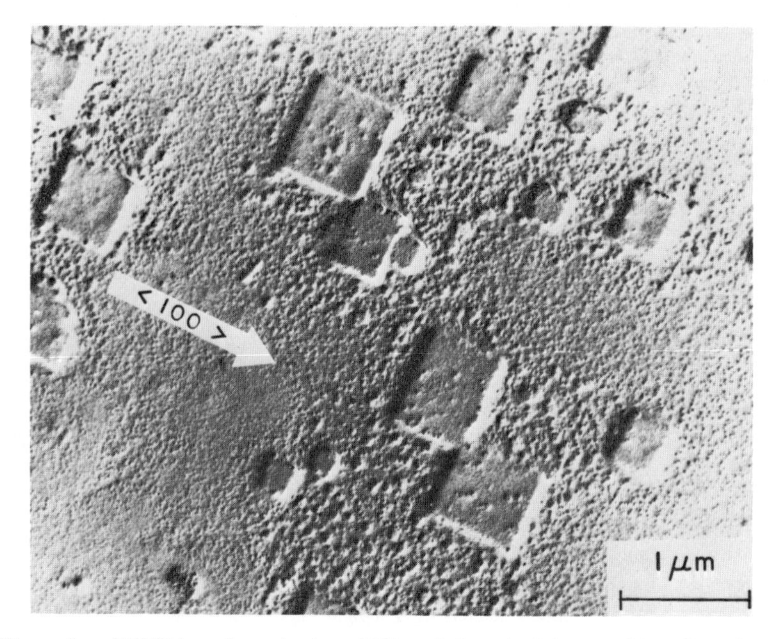

Figure 2. $(NPCl_2)_x$ polymerized on KCl and deposited from a 20 wt% solution in decane at 29 °C.

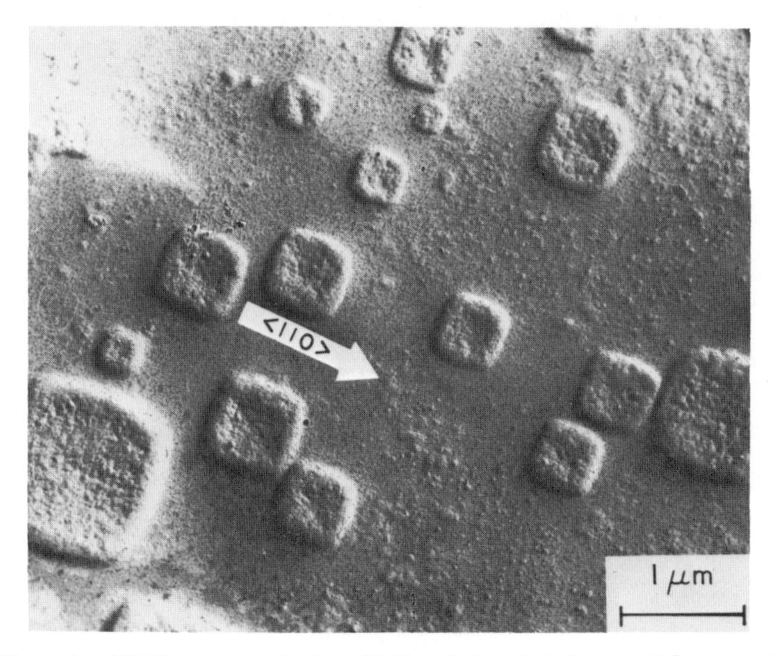

Figure 3. $(NPCl_2)_x$ polymerized on NaCl and deposited from a 10^{-5} torr sublimation at 70 °C.

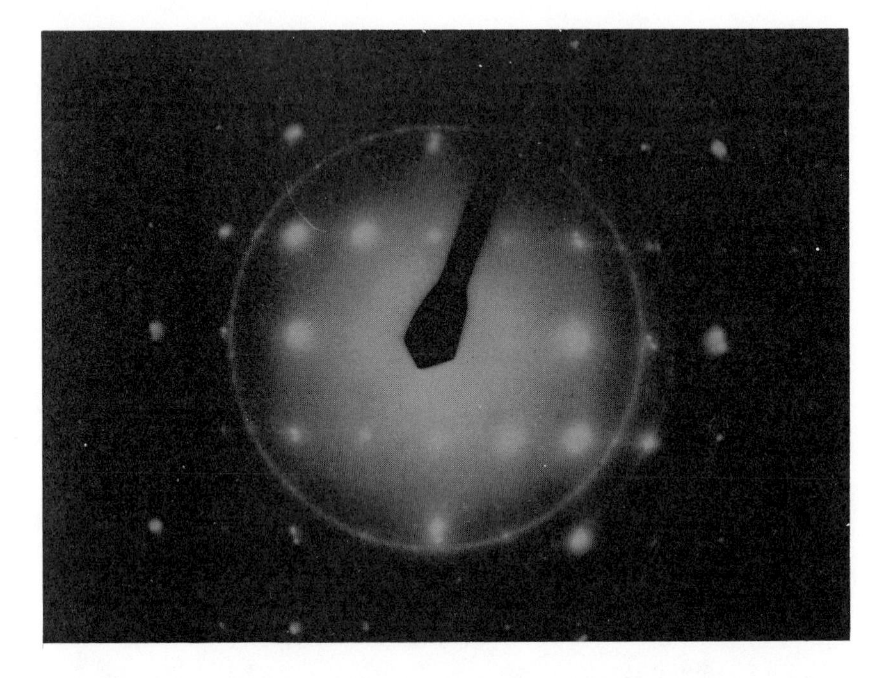

Figure 4a. Electron diffraction of $N_3P_3Cl_6$ from a 10^{-5} torr sublimation at 70 °C.

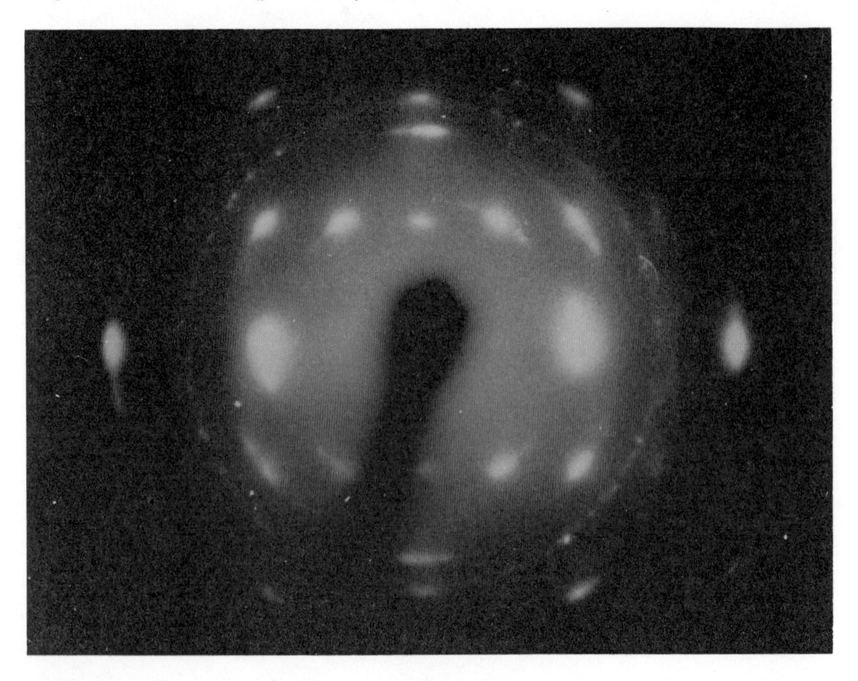

Figure 4b. Electron diffraction of $(NPCl_2)_x$ polymerized on NaCl and deposited from a 10^{-5} torr sublimation at 70 °C.

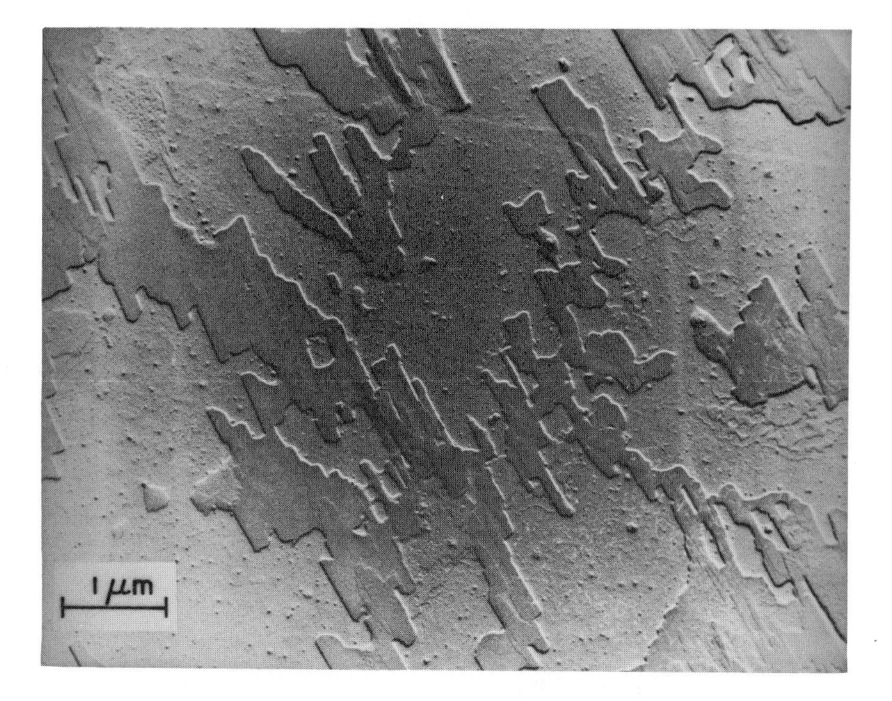

Figure 5. Poly(TCDU) polymerized on KCl and deposited from a 0.4 wt% ethyl acetate solution at 30 °C.

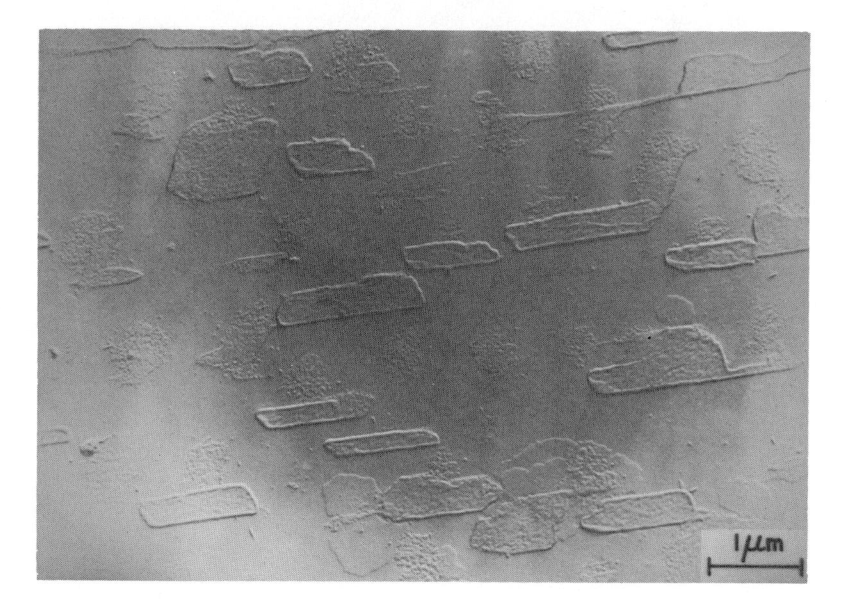

Figure 6. Poly(DMA) polymerized on KCl and deposited from a 1 wt% toluene solution at 78 °C.

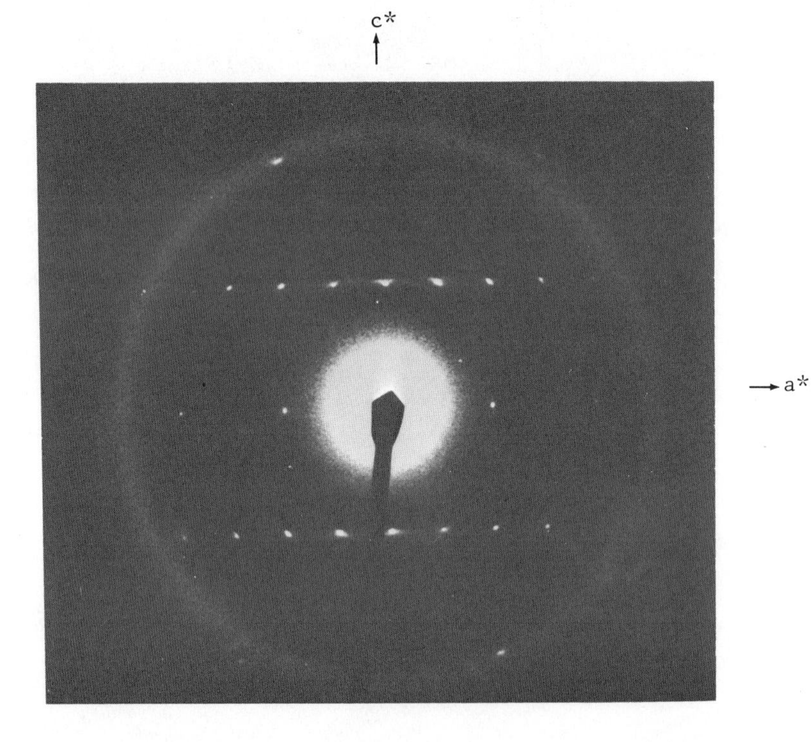

Figure 7. Electron diffraction of poly(TCDU) formed under conditions specified in Figure 5.

TABLE III

TCDU

	PHASE 1	PHASE 2
MONOMER	a = .708 nm	a = 1.160 nm
	b = 3.397 nm	b = 1.897 nm
	c = .523 nm	c = .519 nm
	β = 115.85°	γ = 91.0°
POLYMER	a = .623 nm	a = 1.184 nm
	b = 3.903 nm	b = 1.987 nm
	c = .491 nm	c = .495 nm
	β = 106.85°	γ = 94.94°

TABLE IV

DIOL UNIT CELL PARAMETERS

	a	b	c	β
monomer	.409	1.60	.477	106.3
polymer	.411	1.61	.482	106.1
polymer/KBr	.421	1.02		

units in nm

been produced through epitaxial polymerization for the first time. Such perfect crystals may exhibit significantly enhanced solid-state properties over the usual materials.

3. Epitaxial polymerization may allow complete topochemical control of a variety of solid state reactions to produce single crystals. This technique should continue to prove to be a powerful tool in the molecular engineering of thin polymer films.

Acknowledgment

The partial financial support of this work by the Army Research Office under contract number DAAG 29-80C-0099 is gratefully acknowledged.

Literature Cited

1. Lando, J. B. Contemporary Topics in Polymer Science 1977, 2, 189.
2. Mesrobrian, R. B.; Ander, P.; Ballantine, D. S.; Dienes, G. L. J. Chem. Phys. 1954, 22, 565.
3. Schulz, R; Henglein, A.; von Steinwehr, H. E.; Bambauer, H. Agnew. Chem. 1955, 67, 232.
4. Rickert, S. E.; Lando, J. B.; Hopfinger, A. J.; Baer, E. Macromolecules 1979, 12, 1052.
5. Rickert, S. E.; Ishida, H.; Lando, J. B.; Koenig, J. L.; Baer, E.; J. Appl. Physics 1980, 51, 5194.
6. Ishida, H.; Rickert, S. E.; Hopfinger, A. J.; Lando, J. B.; Baer, E.; Koenig, J. L.; J. Appl. Phys. 1980, 51 (10), 5188.
7. Boudelle, Ph.D. Thesis 1974, (University of Lyon, France).
8. Baughman, R. H.; Chance, R. R.; Cohen, M. J. J. Chem. Phys. 1976, 64, 1869.
9. Baughman, R. H.; Apgar, P. A.; Chance, R. R.; MacDiarmid, A. G.; Garito, A. F., J. Chem. Phys. 1977, 66, 401.
10. Young, R. J.; Baughman, R. H. J. Mater. Sci. 1978, 13, 55.
11. Allcock, H. R.; Arcus, R. A. Macromolecules 1979, 12, 1130.
12. Enkelmann, V.; Lando, J. B. Acta Cryst. 1978, B34, 2352.
13. Enkelmann, V., private communication.
14. Baughman, R. H. J. Appl. Phys. 1972, 43, 4362.

RECEIVED August 24, 1982

ORGANIC COATINGS AND PLASTICS CHEMISTRY

Friedel–Crafts Initiators in the Cationic Polymerization of *para*-Substituted α-Methylstyrenes

R. W. LENZ, J. M. JONTE, and J. G. FAULLIMMEL

University of Massachusetts, Chemical Engineering Department, Materials Research Laboratory, Amherst, MA 01003

The cationic polymerization of several para-substituted α-methylstyrenes initiated by various Friedel-Crafts catalyst-cocatalyst combinations has been studied for the effects of catalyst type, monomer substituent and reaction solvent polarity on polymer structure and properties. By using solvent mixtures, the tacticity of the resulting polymers could be varied over a wide range, the syndiotactic form being favored in the more polar mixtures. Some initiator systems produced bimodal molecular weight distributions (MWD), and some of these polymer samples were fractionated into their high (HMW) and low (LMW) molecular weight components for analysis, which revealed that the LMW fractions had higher isotactic triad contents. This structural difference is attributed to the presence of two different types of ion pair endgroups, which react simultaneously in the reaction mixture. By use of bicomponent cocatalyst combinations to create an excess of the counterion in these bimodal systems, a change in the relative area of the two peaks was observed and attributed to the common ion effect. By proper choice of the para-substituent, catalyst, cocatalyst and solvent, polymers of very narrow MWD could be obtained, suggesting the possibility of long-lived species in these carbonium ion polymerization reactions.

Although Friedel-Crafts or Lewis acid catalysts are often used to initiate carbocationic polymerizations and are very important from an industrial viewpoint, very little is known about the active intermediates involved. Such information is important because, in general for ionic polymerization reactions, small changes in the structure of the active center can result in large changes in molecular weight, molecular weight distribution (MWD),

0097-6156/83/0212-0103$06.00/0

polymer tacticity and copolymerization behavior. Much is known about these relationships in anionic polymerization reactions because of the long-lived character of carbanions, but very little of this type of information is available for reactions involving carbonium (carbenium) ions because of their very high reactivities and short lives (1, 2). That is, for polymerization reactions having carbanion or other anionic endgroups (such as alkoxide ions), a considerable amount of information is available on the effect of counterions, solvents, temperature and monomer substituents on ion pair structure and reactivity (3). Our objective is to obtain a similar understanding of polymerization reactions based on carbocation endgroups.

α-Methylstyrene, with both a methyl group and a phenyl ring attached to the charge-carrying carbon atom, is a good candidate for forming a more stable carbenium ion. Substituents at the *para*-position of the phenyl ring can also provide some control over the electron density at the active center, and, in a sense, can "fine tune" the reaction. Another advantage of the α-methylstyrene system is that the tacticity of the polymers produced can easily be analyzed by NMR (4), and through the use of tacticity information it should be possible to determine something about the nature of the active center which was involved in forming the polymer by the following reaction:

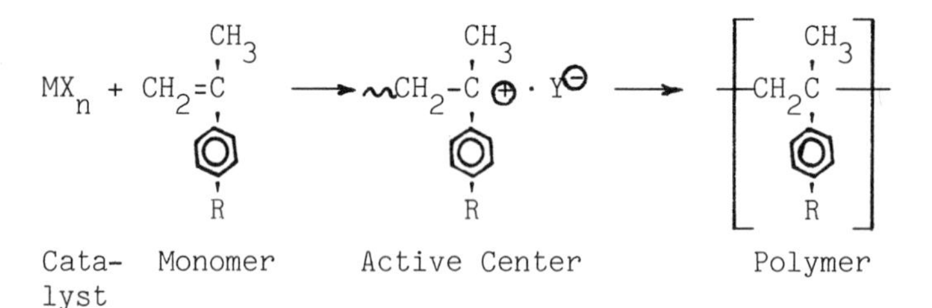

The active center in this reaction is presumably a carbonium ion ion pair, as shown above, which can vary in structure and reactivity from a free carbonium ion at one extreme to a contact ion pair (or even a readily dissociated covalent compound) at the other. The initiator, which consists of the catalyst shown in the equation and generally a cocatalyst, has a controlling effect on the structure of the ion pair because it provides the counterion, Y^-, for the active center. Hence, small changes in the composition of the initiator as well as in monomer structure, reaction solvent, and temperature can cause profound changes in both the rates of the propagation and termination reactions and in the structure of the polymer formed. For this reason, polymerization reactions have been referred to as "chemical amplifiers" in that the polymer molecule is formed by hundreds or thousands of propagation reactions followed by one termination reaction.

Polymer tacticity reveals a considerable amount of information about the stereochemistry of the hundreds or thousands of individual propagation reactions, and molecular weights and molecular weight distribution can provide information on effects of small changes in monomer structure and reaction conditions on reaction mechanisms. In the latter case, the data are very sensitive indications of the effect of active center changes on the one termination reaction among hundreds or thousands of propagation reactions. Hence, it was expected in the present investigation that the study of initiator and monomer substituent effects would reveal basic information on the nature of the carbonium ion involved in the cationic polymerization process. This information, in turn, could be of fundamental importance for controlling polymer structure and properties. The present investigation was undertaken, therefore, partly to help elucidate the nature of the active species in cationic polymerization reactions and partly to investigate the relationships between structure and properties in the resulting polymers.

Results and Discussion

Polymer Tacticity. Our initial results on the polymerization of several different p-substituted-α-methylstyrene monomers indicated that there was some relationship between polymer stereoregularity and both the type of initiator and substituent in these monomers (5). However, our recent investigations with a much wider variety of monomers, catalysts and cocatalysts revealed that the classical approach to analyzing substituent effects in organic reactions, the use of the Hammett $\rho\sigma$ relationship, gave no simple and self-consistent relationship between tacticity and the σ (or σ^+) constant for the *para*-substituent. These results are summarized in the data in Table I for the cationic polymerization of α-methylstyrene and a series of five p-substituted-α-methylstyrene monomers initiated with two different Friedel-Crafts catalysts, $TiCl_4$ and $SnCl_4$, either alone or with a cocatalyst: benzyl chloride (BC) or t-butyl chloride (TBC), in methylene chloride at -78°C. Where a cocatalyst was used, the initiator was presumably a carbonium ion formed by the following reaction:

$$TiCl_4 + C_6H_5CH_2Cl \longrightarrow C_6H_5CH_2^+ \; TiCl_5^-$$

$$SnCl_4 + (CH_3)_3CCl \longrightarrow (CH_3)_3C^+ \; SnCl_5^-$$

Where no specific cocatalyst was used, either adventitious amounts of water served that purpose or the catalyst underwent a self-ionization reaction as follows (6):

$$2TiCl_4 \rightleftharpoons TiCl_3^+ \cdot TiCl_5^- \quad \text{or} \quad TiCl_4 + H_2O \longrightarrow H^+ \cdot [TiCl_4OH]^-$$

Table I. Effect of Initiator and Monomer Substituent on Polymer Tacticity

p-Subst. R	σ^+	Syndiotactic Triad Content					
		$TiCl_4$		- (cat.) -	$SnCl_4$		
		none	BC	TBC -(cocat.)-	none	BC	TBC
OCH_3	-0.27	92	91	91	92	90	90
$CH(CH_3)_2$	-0.15	84	85	84	87	86	86
F	-0.07	73	73	73	83	83	83
H	0	86	86	90	92	92	91
Cl	+0.11	85	84	85	86	87	87
Bi	+0.15	82	81	82	88	89	88

It can be seen from the data in Table I that there was no direct relationship between the syndiotactic triad content and the inductive or resonance effects of the substituents as indicated by their σ^+ constants, which is the parameter normally chosen for application of the Hammett equation to a reaction involving carbonium ion intermediates. A more detailed consideration of the theory for the relationship between polymer tacticity and propagation rate constants confirms that, in fact, no simple relationship should be expected.

That is, the average or apparent rate constant for propagation is comprised of two competing rate constants, one for the average rate constant for isotactic dyad formation and the other for the average rate constant for syndiotactic dyad formation (the term average implies that there may be more than one type of active center involving different ions or ion pairs). Hence, the overall rate of monomer disappearance would be a function of the sum of these two rate constants, and the relationship between tacticity and σ^+ within a series of monomers would be quite complex. That is, this relationship would involve opposite substituent effects on the syndiotactic dyad placement rate constant, which should be directly proportional to σ^+, and on the syndiotactic dyad content itself, which should be inversely proportional to σ^+. Furthermore, in order to treat these relationships quantitatively, because it is likely that different types of active species (ion or ion pair end groups) give different tacticities, a complete analysis of the substituent effect will require knowledge of the rate constants and tacticities associated with each type of active center, so as yet there is not sufficient information from our work to quantitatively relate the effect of the nature of the counterion generated by the initiator to the ion pair structure and the latter, in turn, to tacticity.

As shown in the data above, the two different Friedel-Crafts catalysts, $TiCl_4$ and $SnCl_4$, with the same cocatalysts gave polymers with either the same or different tacticities depending on the monomer involved. For the p-chloro, p-bromo, and p-methoxy monomers, polymers with the same syndiotactic triad contents were obtained with both $TiCl_4$ and $SnCl_4$ catalysts. In contrast, with α-methylstyrene, to a small degree, and with the p-fluoro and p-bromo monomers, to a substantial degree, the $SnCl_4$ catalysts gave polymers of higher stereoregularities. In all cases, the type of the cocatalyst (TBC, BC or no added cocatalyst) had no effect. This last result, however, was not the case in the variation observed for molecular weights and molecular weight distributions with changes in type of initiator to be discussed below.

Molecular Weight and Distribution. In many cases, the polymers obtained in these polymerization reactions had multiple-peak molecular weight distributions when analyzed by GPC and the presence of two or more peaks suggests the existence of multiple active centers. In some cases, however, depending on the monomer and the catalyst-cocatalyst combination, surprisingly narrow single-peak distributions were observed. The reactant combinations for which such results were obtained are summarized in the data in Table II in which an indication of the molecular weight distribution is given by the ratio of the weight-to-number average molecular weights. This table also contains data on polymerization reactions initiated with triphenylmethyl chloride, TPMC, which is expected to form a relatively unreactive carbonium ion on reaction with the Friedel-Crafts catalyst, as follows:

$$(C_6H_5)_3CCl + SnCl_4 \longrightarrow (C_6H_5)_3C^+ \; SnCl_5^-$$

Table II. Effect of Variations in Catalyst-Cocatalyst Combinations on MWD for the Polymerization of Monomers with Different p-Substitutes

p-Subst. R	$\bar{M}_w/\bar{M}_n$ (a)					
	$SnCl_4$ + TBC	$SnCl_4$ + $\frac{2}{3}$TBC/$\frac{1}{3}$TPMC	$SnCl_4$	$TiCl_4$ + $\frac{2}{3}$TBC/$\frac{1}{3}$TPMC	$TiCl_4$ + TPMC	$SnCl_4$ + TPMC
F	2.4+	1.3	1.4	1.8	0	1.4
Cl	1.3	1.2	1.4	1.9+	0	1.7
Br	1.2	-	1.3	-	0	1.5
OCH_3	1.3	-	1.9	1.8	2.1+	1.4
$CH(CH_3)_2$	2.6+	3.7+	2.3+	1.2	0	2.8

(a) The symbol + indicates that the sample had either a bimodal or multiple peak distribution and 0 indicates no polymer obtained.

The narrow molecular weight distributions observed suggest the possibilities of the existence of "living polymers" in these polymerization reactions, as has been suggested previously for other cationic polymerization reactions (7). Indeed, in at least one of these reactions, that of the polymerization of the isopropyl monomer initiated with $TiCl_4$ and TPMC as the catalyst-cocatalyst pair, $\overline{M}_n$ was observed to increase nearly linearly with reaction conversion, as shown in Table III.

Table III. Effect of Reaction Time (or Conversion) on Molecular Weight in the Polymerization of p-isopropyl-α-methylstyrene with $TiCl_4$ + TPMC

Reaction time, min	Polymer Yield, %	$\overline{M}_n$	$\overline{M}_w$	$\overline{M}_w/\overline{M}_n$
10	6	5,200	7,100	1.4
20	16	9,700	13,000	1.4
30	27	12,500	17,600	1.4
40	28	12,800	18,000	1.4

It is doubtful, however, that a true living system without termination or transfer exists in these polymerizations; instead we believe that the narrow distributions may result from a combination of essentially instantaneous initiation, relatively long kinetic lifetime, and kinetic termination without transfer. H. Morawetz has derived an equation relating the molecular weight distribution to the relative rates of propagation and termination and the concentration of active species for such a case (8). The equation accounts for the possible occurrence of narrow distributions, and we are presently experimentally investigating these calculations and predictions for the polymerization of the p-isopropyl monomer.

Common Ion Effect. Still another indication of the possibility of living polymers, or at least of long-lived active centers as described above, was the effect observed on the partial addition to the initiator system of a second cocatalyst, TPMC, which is expected to generate an initiator species of relatively low reactivity. That is, the triphenylmethyl carbenium ion formed from this cocatalyst was found to be relatively unreactive for two of the monomers with electron-withdrawing groups, the p-fluoro and p-chloro monomers, so when it was substituted in part for an active cocatalyst, TBC, the number of active initiation and propagation centers was apparently reduced. If this was indeed the case, then for a given amount of monomer and conversion, the molecular weight would be expected to increase because of the decreasing initiator concentration, as is well known for living polymer systems. The molecular weight and distribution results obtained, as a function of the ratio of TPMC to TBC as cocatalysts, in the polymerization of these monomers with $SnCl_4$ catalyst are shown in Table IV.

All of the polymers in this table which were obtained from initiator systems containing TPMC had monomodal molecular weight distributions by GPC analysis, indicating that the use of this cocatalyst caused the elimination of the active centers which were

Table IV. Common Ion Effect on Molecular Weight and Molecular Weight Distribution

Fraction of TPMC in Cocatalyst	Monomer					
	p-Cl			p-F		
	Yield,%	$\overline{M}_n$	$\overline{M}_w/\overline{M}_n$	Yield,%	$\overline{M}_n$	$\overline{M}_w/\overline{M}_n$ [(a)]
0	98	45,000	1.3	97	34,000	2.4+
0.33	98	130,000	1.2	98	120,000	1.3
0.67	99	200,000	1.2	99	260,000	1.3
1.00	16	240,000	1.7	97	420,000	1.4

[(a)]See footnote of Table II.

responsible for forming the polymer fraction represented by the second peak of the distribution, the lower molecular weight fraction. A possible explanation for this effect is that if an ion pair, $P^+SnCl_5^-$, and a free ion, P^+, are both present as active centers, reacting simultaneously and in equilibrium, the excess $SnCl_4^+$ counterion formed from TPMC could displace the equilibrium increasingly toward the ion pair form. This effect has been frequently observed and is often used for kinetic studies, particularly in anionic polymerization reactions (9), and it is referred to by its general term, "the common ion effect" (10).

Indeed, in many of the polymerization reactions studied in these investigations when bimodal distributions were obtained, the lower molecular weight peak could be reduced in amount or completely eliminated by the replacement of either the TBC or BC cocatalyst with TPMC. These results also suggest that for reactions carried out in methylene chloride, free ion endgroups are present which have shorter kinetic lifetimes and form lower molecular weight polymers than the ion pair endgroups.

It would appear that this conclusion contradicts our earlier suggestion based on a study of the cationic polymerization of p-chloro-α-methylstyrene in a series of solvents of different polarities (methylene chloride, toluene and heptane)(5). In that case, the molecular weights of the polymers formed increased with increasing solvent polarity, but that effect may have been associated with the comparative reactivities of solvated ion pairs, not free ions.

Solvent Polarity Effects. To investigate the effect of solvent polarity on these systems, a series of polymerization reactions was performed using mixtures of 2 solvents with different dielectric constants (ε): methylene chloride (ε=14.8) and hexane (ε=2.0). As seen in Table V, the solvent polarity had a great effect on the tacticity of the resultant polymers. Predominately syndiotactic polymers were formed in the more polar solutions as previously found (5), and the level of isotacticity increased as the solutions became less polar. These results support the proposal that isotactic sequences come from backside attack on contact ion pairs and that syndiotactic sequences are produced by frontside attack on either solvated ion pairs or free ions (11), as illustrated in Figure 1.

A particularly interesting result in Table V is seen for the solutions containing about 66% methylene chloride. At this point a sharp break occurred in several properties: (1) the syndiotactic content decreased and the isotactic content increased, (2) the yield dropped drastically, (3) the molecular weight decreased, and (4) the molecular weight distribution broadened. These changes all point to a change in the reaction mechanism, probably from one of solvated ion pairs to one of predominantly contact ion pairs at that solvent concentration. Nevertheless, the tacticities of all the polymers formed in this series were found to fit on a Bovey plot, which indicates that they were formed by a Bernoullian process and no penultimate effect was present.

Table V. Effect of Solvent Composition on Polymer Yield, Tacticity and Molecular Weight in the Polymerization of p-Chloro-α-methylstyrene with $SnCl_4$

% CH_2Cl_2	% Hexane	% Yield	Triad Tacticity S	H	I	$\bar{M}_w$	$\bar{M}_w/\bar{M}_n$
100	0	97	92	8	1	1,100,000	1.2
97	3	98	96	3	1	800,000	1.4
67	33	85	92	7	1	910,000	1.2
65	35	10	77	15	8	280,000	1.5
63	37	2	73	22	6	180,000	1.6
50	50	5	68	23	9	160,000	1.6
30	70	2	61	31	8	97,000	2.2
3	97	3	55	35	10	91,000	2.8

Multiple Active Center Polymerizations. As mentioned above, it has been suggested that the two peaks observed in polymers with bimodal molecular weight distributions result from two different types of active centers present reacting simultaneously. As discussed above, these substituted α-methylstyrene monomers can propagate by either solvated or contact ion pairs to yield markedly different tacticities in the resulting polymers. To test the suggestion that the two modes were caused by the presence of two active centers, presumably solvated and contact ion pairs, several polymers with bimodal MWD were fractionated into their high (HMW) and low (LMW) molecular weight components by preparative GPC, and the two fractions were analyzed by NMR for tacticity. The results shown in Table VI indicate that the HMW fractions were highly syndiotactic whereas the LMW fractions had higher isotactic contents. These results are very similar to those obtained in the solvent polarity studies, leading to the suggestion that the HMW fractions were formed from solvated ion pairs and the LMW fractions from contact ion pairs.

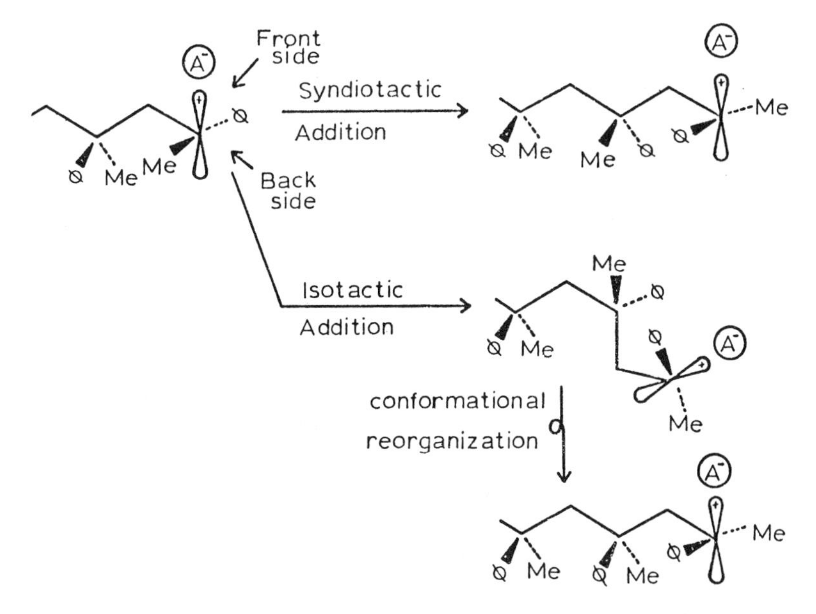

Figure 1. Schematic representation of syndiotactic and isotactic addition steps.

Table VI. Results of Fractionation of Polymers with Bimodal MWD's

Type	Tacticity S	H	I	σ	M_w	M_n	M_w/M_n
p-F,HMW	90	10	0	.051	640,000	315,000	2.0
p-F,LMW	71	20	9	.157	6,100	4,000	1.5
p-Br,HMW	94	6	0	.030	700,000	230,000	3.0
p-Br,LMW	60	28	12	.225	2,400	1,800	1.3

Narrow Molecular Weight Distributions. As was discussed above, for reactions which form polymers having a bimodal MWD, a change in reaction conditions can drastically alter the MWD of the resulting polymer. By proper choice of para-substituent, initiator, coinitiator and solvent, polymers of very narrow MWD could be obtained. An example is the reaction system p-chloro-α-methylstyrene/$SnCl_4$/"H_2O"/CH_2Cl_2 for which an MWD of 1.2 was obtained. Because a narrow MWD may be an indication of "living polymer" formation in an anionic polymerization, the question arose of whether these narrow MWD polymers were formed by a living cationic polymerization. A truly living system should show a linear increase of MW with time. Preliminary studies on these systems have shown such a linear increase over short periods. Another test of living character is the further polymerization of monomer added after the original monomer charge has been completely converted to polymer. Studies are now under way to assess that possibility as well as further studies of the increase of MW with conversion.

Experimental

Reagents. All solvents were purified by standard techniques (13), dried over calcium hydride and distilled under dry argon immediately prior to use. The para-substituted-α-methylstyrene monomers were dried over calcium hydride and distilled 3 times under vacuum. Purity was greater than 99.5% as determined by gas chromatography. The initiators, $TiCl_4$ and $SnCl_4$, were distilled three times under dry argon.

Polymerization Reactions. All polymerization reactions were carried out at -78°C under dry argon. The experimental apparatus, consisting of a 100 ml round bottom flask, fitted with a 25 ml addition funnel and a 3-way stopcock for maintaining an inert atmosphere, was flamed under vacuum 3 times and filled with argon. Monomer and solvent were added to make a 0.7 mole% solution. The flask was cooled in dry ice/acetone and a catalyst solution (0.1 mole% based on monomer) was added quickly from the addition funnel. The reaction was terminated after 5 minutes by the addition of 5 ml of methanol. The polymer was dissolved in methylene chloride, reprecipitated into 300 ml of methanol, filtered and dried under vacuum to constant weight.

Polymer Characterization. Molecular weight distributions (MWD's) were obtained from a Waters model 201 ALC/GPC using microstyragel columns with pore sizes of 500, 10^3, 10^4, 10^5 and 10^6 Å. Calibration was with polystyrene standards in THF solution. Fractionation of bimodal MWD polymers was achieved by use of a Knauer GPC using styragel columns of 10^3, 10^4, 10^5 and 10^6 Å. Tacticity information was obtained from NMR spectra using a 270 MHz Bruker instrument.

Acknowledgments

The authors are grateful to the National Science Foundation for the support of this work under Grant Number DMR-80-22273. The use of the facilities of the Institut für Makromolekulare Chemie, Freiburg, West Germany, for the polymer fractionations is also gratefully acknowledged. The high field NMR experiments were performed at the NMR Facility for Biomolecular Research located at the F. Bitter National Magnet Laboratory, M.I.T.

Literature Cited

1. Plesch, P. H. Makromol. Chem. 1974, 175, 1065.
2. Higashimura, T.; Kishiro, O.; Takeda, T. J. Polym. Sci., Polymer Chem. Ed. 1976, 14, 1089.
3. Szwarc, M. "Carbanions, Living Polymers and Electron-Transfer Processes"; John Wiley and Sons: New York, 1968.
4. Lenz, R. W.; Regel, W.; Westfelt, L. Makromol. Chem., 1975, 176, 781.
5. Lenz, R. W.; Westfelt, L. C. J. Polym. Sci., Polymer Chem. Ed., 1976, 14, 2147.
6. Longworth, W. R.; Plesch, P. H. J. Chem. Soc., 1959, 1887.
7. Higashimura, T.; Kishiro, O. Polymer J., 1977, 9, 87.
8. Morawetz, H. Macromolecules, 1979, 12, 532.
9. Smid, J. "Ions and Ion Pairs in Orgnaic Reactions"; Szwarc, M., Ed.; Wiley-Interscience: New York, 1972.
10. Benfey, O. T.; Hughes, E. D.; Ingold, C. K. J. Chem. Soc., 1952, 2488.
11. Matsuguma, Y.; Kunitake, T. Polym. J., 1971, 2, 353.
12. Higashimura, T.; Kishiro, O. J. Polym. Sci., Polymer Chem. Ed., 1975, 13, 1393.
13. Perrin, D. D.; Armarego, W. L. F.; Perrin, D. P. "Purification of Laboratory Chemicals"; Pergamon Press, 1966.

RECEIVED October 15, 1982

10

Ring Opening Polymerization: Make the Initiator Work for You

P. DREYFUSS

The University of Akron, Institute of Polymer Science, Akron, OH 44325

When you choose an initiator for ring opening polymerization, you may also unwittingly be choosing the properties of the polymer produced. Currently known initiation methods for ring opening polymerization are reviewed in a systematic way with special emphasis on their influence on the properties of the resulting polymer. The importance of the chemical elements that comprise each group of initiators is demonstrated and it is shown that the behavior of the initiators is related to the position of these chemical elements in the Periodic Chart of the Elements. The ring opening polymerization of tetrahydrofuran is used as a model for the review.

When you choose an initiator for ring opening polymerization, you may also unwittingly be choosing the properties of the polymer produced. Initiators have a large influence on the end-groups, processability, molecular weight, molecular weight distribution, long term stability, reactivity in different chemical/physical environments and even total cost of the resulting polymer. The goals of this short review are first, to illustrate the kinds of information that can and should be obtained from the literature when you want to make an initiator work for you in a ring opening polymerization and second, to provide some guidelines for organizing that information in a useful way.

There is such a vast literature in this field (See e.g. 1-5 and references therein) that it is not practical to try to examine simultaneously how different initiators work with all the kinds of rings that are known to undergo ring opening polymerization. Therefore, detailed discussion will be limited to only one group of rings; namely, cyclic ethers. Furthermore, since there is an effect of ring size, most of the discussion will be related to the single monomer, tetrahydrofuran (THF, a five-membered ring heterocycle with one oxygen atom in the ring) and its polymer, poly(tetrahydrofuran).

0097-6156/83/0212-0115$06.00/0

Initiation

In the present instance initiation is defined by the following equation:

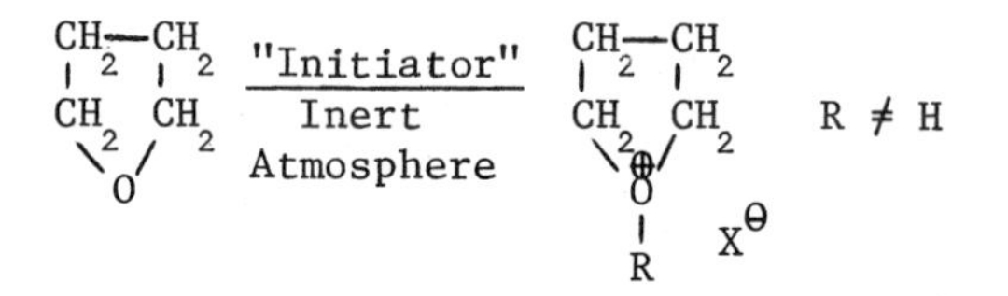

Pure, dry monomer interacts with a species to produce a tertiary oxonium ion. Often the so-called "initiator" is a precursor to the true initiating species. Formation of a secondary oxonium ion by proton addition has deliberately been excluded from this definition for the following reason. The propagating species in THF polymerizations is a tertiary oxonium ion and until such an ion forms a steady state is not present. (Formation of secondary oxonium ions by proton addition to THF is a fast reaction but addition of the next molecule of THF is slow.) Since the polymerization of THF proceeds via a cationic ring opening mechanism, prerequisites for polymerizations to occur are pure, dry monomer and an inert atmosphere such as nitrogen, inert gas or high vacuum. Also because the polymerization proceeds via cations, the total system also contains an equivalent number of anions, $X^{\ominus}$. Thus, making the initiator work for you requires knowledge not only about how to form the cation but also about the interaction of the cation formed with its corresponding anion and any other materials that may be present in the reaction mixture.

As shown by Figure 1, a bewildering number of different materials have been added to THF in an effort to induce its polymerization. Figure 1 shows just the elements in these materials and the position of those elements in the Periodic Chart. The nature of the compounds and especially the oxidation state of the elements in the compounds are very important if a given material is to be an initiator, but as we shall see, the kinds of active compounds that are useful and the manner in which active compounds can be used is related to the position in the Periodic Chart of the principal elements of which the compound is comprised.

Counterions and Endgroup Control by Termination

It is possible to carry out a THF polymerization so that a "living" polymerization results (1). Under these conditions the endgroups, but not the headgroups, can be controlled by termination. The headgroups in polymers with only one growing end are determined during initiation and will be discussed later. Typical examples of endgroup control by termination are shown below:

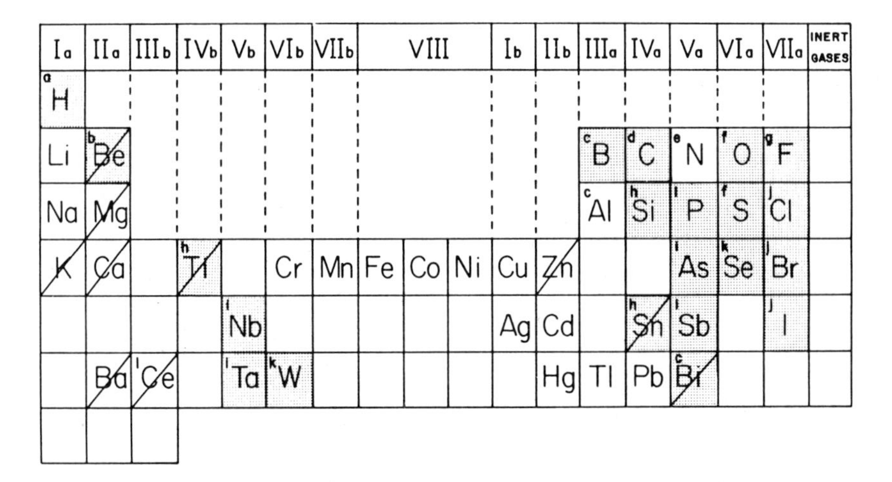

Figure 1. Periodic chart of elements used in initiation of THF polymerization. The elements in open squares were used as perchlorate salts (6). Those in shaded squares were used in the compounds given in the key below. A diagonal across the square indicates that, although the compounds containing the element have been examined, polymerization was not initiated (1).

Key: [a]*As* H^+ *with a variety of counter ions such as* ClO_4^-, PF_6^-, SO_3F^- . . . ; [b]*as* $BeCl_2$ (7); [c]*as* BCl_3, BF_3, $AlCl_3$, $AlBr_3$, AlR_3, $BiCl_3$, $Al(ClO_4)_3$ (6, 7, 8); [d]*as* R_3C^+, $RC(OR')_2^+$ (7, 8); [e]*as* NR_4^+, NO^+, NO_2^+; [f]*as* R_3O^+, R_3S^+, *or* Ar_3S^+ (7, 8, 10, 11), *also in numerous anions;* [g]*as* F^-, *very common as nonmetallic element in complex anions and as RF in conjugation with metal salts;* [h]*as* SiF_4, $SnCl_4$, $TiCl_4$ (7); [i]*as* PF_5, AsF_5, SbF_5, $SbCl_5$, $NbCl_5$, $TaCl_5$, NH_3, RNH_2, PR_3, . . . (8, 12–14); [j]*as* $(R)R'X^+$ (15–18); *as* Cl^-, Br^-, I^- *in complex anions;* ClO_4^-! *as RX in conjunction with metal salts;* [k]*as* SeF_6 *or* WCl_6 *or* Ar_2Se^+ (12, 19–21); [l]*Ce is element 58 and the first member of the lanthanide series. In all cases R,R' = alkyl or H; Ar = aryl; X = Cl, Br, I.*

$$\sim O^{\oplus}\text{(THF ring)}\; X^{\ominus} + \begin{matrix} HOH \\ ROH \\ NH_3 \\ RNH_2 \end{matrix} \longrightarrow \begin{matrix} \sim O(CH_2)_4OH \\ \sim O(CH_2)_4OR \\ \sim O(CH_2)_4NH_2 \\ \sim O(CH_2)_4NHR \end{matrix} + HX$$

Endgroup control by termination is only possible with those counterions that produce stable oxonium ion-counterion complexes. Simple anions like Cl^-, Br^-, I^- are not suitable counterions because the cation-anion complex is so unstable with respect to THF and alkyl halides that no polymerization at all occurs. Complex ions like PF_6^-, AsF_6^-, SbF_6^- are ideal. The cation-anion complex is stable under normal polymerization conditions and yet loose enough so that polymerization and reaction with a terminating species can occur. Notice that these anions all are derived from Lewis Acids of elements from Group Va in their higher valence state of +5 and the very electronegative halide, fluorine from Group VIIa. Lewis Acids and other compounds of elements from Group Va in their lower valence state of +3 terminate THF polymerizations. Ammonia, amines, and triphenylphosphine, for example, have all been used to introduce functional endgroups into PTHF (1). Complexes from the Lewis Acids derived from elements in groups left of Group Va and halides below fluorine in Group VIIa become increasingly less stable as the distance to the left or below increases. Thus when $SbCl_6^-$ counterions are present, both termination and transfer reactions with counterion occur; polymerizations containing BF_4^- suffer from termination reactions; and yields in polymerizations containing $AlCl_4^-$ counterions are low due to the rapid destruction of the growing centers.

Elements from Group VIa can form some very suitable counterions for use when endgroup control by termination is desired. Polymerizations containing the anions $SO_3CF_3^-$, SO_3F^- or Nafion (SO_3^-form) are "living". However, these anions are able to react reversibly with the THF tertiary oxonium ion to form esters. These esters have been called "sleeping" species (22) or in a state of "temporary termination" (23) because the rate of polymerization of the esters is negligible compared to that of free ions and ion pairs. When comparable numbers of growing centers are present, polymerization rates are lower if esters form (1). Perchlorate anion is similar.

Sulfonate anion is not suitable when control of endgroups by termination is desired. The THF tertiary oxonium ion reacts irreversibly with the sulfate anion to form sulfates and disulfates, which can be hydrolyzed but are otherwise stable.

The HX formed as a byproduct of the termination reaction can be a nuisance. If left in the polymer, these strong acids result in degradation and discoloration of the PTHF. Furthermore, the strong acids are also very corrosive to polymerization reactors. One anion that overcomes these problems is Nafion (SO_3^- form) (24), which is a polymeric resin produced by duPont. After termination Nafion can be filtered off. Choosing an "initiator" that will lead to a polymerization with suitable counterions is an important part of making the initiator work for you.

Endgroup Control by Transfer

Another way to control endgroups in THF polymerizations is by means of transfer reactions. These reactions usually lead to identical headgroups and endgroups. Some examples are shown in the following equations:

$$\text{THF} + HSbF_6 + (CH_3\overset{\overset{O}{\|}}{C})_2O \longrightarrow CH_3\overset{\overset{O}{\|}}{C}\!\left[O(CH_2)_4\right]_n O\overset{\overset{O}{\|}}{C}CH_3$$

$$\text{THF} + Et_3OPF_6 + (CH_3O)_3CH \longrightarrow CH_3\left[O(CH_2)_4\right]_n OCH_3$$

$$\text{THF} + NOPF_6 + CH_3\overset{\overset{O}{\|}}{C}Cl \longrightarrow CH_3\overset{\overset{O}{\|}}{C}\!\left[O(CH_2)_4\right]_n O\overset{\overset{O}{\|}}{C}CH_3$$

This is a very good way to obtain difunctional PTHF of any desired molecular weight and gives polymer with comparatively low molecular weight distributions in addition. $\overline{M}_w/\overline{M}_n$'s of 1.6 to 1.7 at moderate conversions to PTHF are quite commonly observed. Some examples of the degree of molecular weight control that is possible are shown in Tables I and II. Since molecular weight in this type of polymerization is controlled by the transfer agent rather than by the number of active centers generated by the initiator, smaller amounts of acid are generated on termination, more stable polymers form, and corrosion of equipment is reduced.

TABLE I

Molecular Weight Control by Acetic Anhydride[a] (24)

Polymerization Time (Hr)	Conversion (%)	$\overline{M}_n$ (g/mol)
1.5	47	2808
5.5	74	1270
21	82	1002

[a]The standard recipe for all experiments consisted of 5.58 g THF, Nafion Resin, 5.6 g acetic anhydride and 0.6 g acetic acid.

TABLE II

Control of Molecular Weight by Trimethylorthoformate (25)

mM Et_3OPF_6 / mole THF	mM$(CH_3O)_3CH$ / mole THF	$\overline{M}_n$ Calcd. for PTHF	$[\eta]$[a] dl/g
0.152	5.39	10,000	0.38
0.156	2.75	19,000	0.53
0.151	1.35	38,000	0.79

[a]Measured at 25°C in benzene

OH Endgroups During Polymerization

As stated above, initiation by Brønsted Acid is a two-step process:

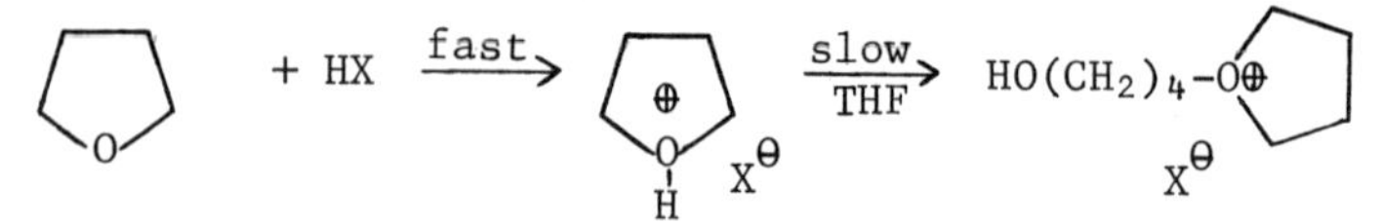

A polymer with an OH headgroup is produced, and chain coupling occurs repeatedly by reaction of the OH endgroup with the THF tertiary oxonium ion (11) until high molecular weight polymer is formed (26):

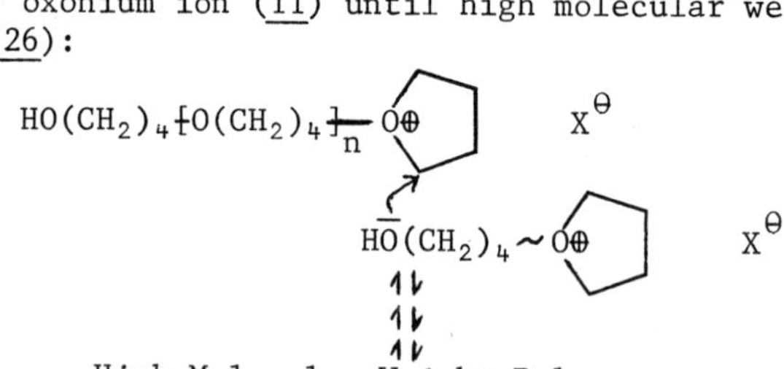

The combination of slow initiation and of chain coupling makes molecular weight control very difficult. At first glance this problem would appear to be one that is easily avoided by choosing a different "initiator". However, the problem is frequently encountered and the scientist needs to be aware of these reactions because many "initiators" lead to *in situ* generation of Brønsted Acid. Among these are ϕ_3CPF_6, *p*-Cl-ϕN_2PF_6, $NOPF_6$, and NO_2PF_6 alone, R_4NClO_4 plus electric current; $AgPF_6$ plus light or a free radical source and heat or light; $Ar_3S^{\oplus}$, $Ar_3Se^{\oplus}$, $R_2Br^{\oplus}$, $R_2Cl^{\oplus}$, and $R_2I^{\oplus}$ plus light (1). As a general rule carbocations do not add to THF while $NO^{\oplus}$ and $NO_2^{\oplus}$ may add; but in all these cases there is strong evidence for the formation of $H^{\oplus}$ and it is the $H^{\oplus}$ that starts the chain of reactions that leads to the polymerization of THF. Tetraalkylammonium cation does not add to THF but in the presence of an electric current its accompanying perchlorate anion undergoes a series of interactions with THF that again gives $H^{\oplus}$. $AgPF_6$ alone doesn't initiate THF polymerization but again there are reports that light or light/heat and a free radical source can lead to the formation of $H^{\oplus}$ and polymerization of THF. The cations derived from S and Se in Group VIa and the cations from Cl, Br, and I in Group VIIa do not add to THF either but in the presence of light, $H^{\oplus}$ is formed photolytically and polymerization occurs. The latter "initiators" were developed for the polymerization of epoxides in the form of thin films (11, 15, 16).

Controlling the Headgroup

Species that add to THF without any activation in the form of energy or other chemicals include $R_3O^{\oplus}$, $RC(\overline{\overline{}}OR')_2{}^{\oplus}$, $RC^{\oplus}{=}O$, super acid esters, PF_5, SbF_5, $SbCl_5$, $NbCl_5$, WCl_6 and superacid anhydrides (1). The cations and the superacid esters give alkyl or acyl headgroups depending on the nature of the R groups. In the few cases where the mechanism of initiation has been studied, the Lewis Acids give PTHF that is growing on both ends. The same is true for the superacid anhydrides. The endgroups in the latter polymerizations are determined by the species added for termination except in cases like $SbCl_5$, where some side reactions also occur. Notice that the cations that add are typically oxonium or carboxonium ions. An oxonium ion can add to heterocycle with sulfur as a heteroatom but the reverse does not occur. The Lewis Acids that add and lead to PTHF without further activation are primarily from Group Va. $NbCl_5$ is from Group Vb and WCl_6 and SeF_6 are from Group VI. Lewis Acids from Group IIIa will add to THF but polymerization does not occur unless a "promoter" such as ethylene oxide or epichlorohydrin is added also. Lewis Acids from elements in other groups do not initiation THF polymerization even in the presence of a promoter. Instead alcoholates form. $TiCl_4$ and epichlorohydrin, for example, give $(ClCH_2)_2CHOTiCl_3$ (1).

The Transition Metals and Halides

Soluble salts from the transition metals generally do not polymerize THF without some other activation. It has already been noted that $AgPF_6$ gives polymer in the presence of light or a free radical source plus heat or light. In this case the species that is responsible for polymerization is photolytically generated HPF_6. In the presence of an active alkyl halide the mechanism involves the $AgPF_6$ more directly:

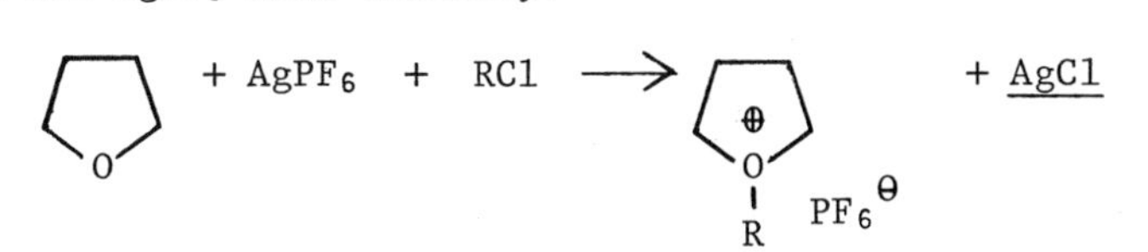

Other examples of combinations with halides that initiate the polymerization of THF include CH_3OCH_2Cl with $FeCl_3$, $CH_2{=}CHCH_2Cl$

with $AgPF_6$ or $Hg(ClO_4)_2$, CH_3COCl with $Pb(ClO_4)_2$ and CH_3COF with $NOPF_6$. Studies have shown that among the metal salts silver salts give the best results. $NO^{\oplus}$ is not a transition metal, of course, but it is included in this group in order to illustrate yet another way of making an "initiator" work for you. When the polymerization of THF is initiated with any combination of materials involving a transition metal salt, removal of catalyst residues is often very difficult. And if any metal residue remains the PTHF soon becomes discolored. The combination $NOPF_6$ and CH_3COF has the advantage that the NOF byproduct is a very low boiling gas that is readily removed or more accurately, spontaneously is evolved from the PTHF. Also CH_3COF is not a transfer agent and molecular weight depends on the amount of $NOPF_6$ added. Reactions involving active chloride/salt combinations are especially useful for the preparation of graft copolymers from hydrocarbons and PTHF (1). In fact a variety of graft copolymers that are not otherwise accessible can be prepared using this chemistry.

Concluding Comments

As we have moved across the periodic chart of the elements from right to left we have found first totally inert compounds in the form of the rare gases, next elements of Groups VIIa and VIa (except oxygen) that are useful mainly in counterions, then the very active Lewis Acids of Group Va pentavalent elements, and finally increasing need for activation of any soluble compounds combined with increasing instability of the counterions that might form. Nothing has been said so far about compounds of elements from Groups IIa and IIIb. These compounds appear to be inactive even in the presence of a very reactive halide like acetyl chloride. Compounds from Group Ia exhibit complex behavior in the presence of THF. The activity of H^+ is largely related to its accompanying anion and the same is probably true to a lesser extent of soluble lithium and sodium salts. $LiClO_4$, for example, can be a supporting electrolyte for THF polymerization in the presence of an electric current but it is again the $ClO_4^{\ominus}$ that gives the compound activity. Polymerization will occur in the presence of $LiPF_6$ but it has been suggested that the true initiator is PF_5. Salts of potassium are largely insoluble in THF.

By careful consideration of schemes like the foregoing, it is possible to predict how to make an initiator work for you in a ring opening polymerization.

Literature Cited

1. Dreyfuss, P. "Poly(tetrahydrofuran)", Gordon and Breach, New York, N.Y., 1982.
2. "Ring-Opening Polymerization", Saegusa, T.; Goethals, E. Eds., ACS Symposium Series 59, American Chemical Society, Washington, D.C., 1977.

3. "Ionic Polymerization, Unsolved Problems", Furukawa, J.; Vogl, O. Eds., Marcel Dekker, Inc., New York, N.Y., 1976.
4. "Polyethers", Vandenberg, E. J. Ed.,ACS Symposium Series 6, American Chemical Society, Washington, D.C., 1975.
5. "Ring-Opening Polymerization", Frisch, K. C.; Reegen, S. L. Eds., Marcel Dekker, Inc., New York, N.Y., 1969.
6. Eckstein, Y.; Dreyfuss, P. J. Inorg. Nucl. Chem. 1981 43, 23.
7. Furukawa, J.; Saegusa, T. "Polymerization of Aldehydes and Oxides", Interscience, New York, N.Y., 1965.
8. Dreyfuss, P. and Dreyfuss, M.P. Adv. Polym. Sci. 1967 4, 528.
9. Eckstein, Y.; Dreyfuss, P. J. Polym. Sci.: Polym. Chem. Ed. 1979 17, 4115.
10. Lambert, J. L.; Goethals, E. J. Makromol. Chem. 1970 133, 289.
11. Crivello, J. V.; and Lam, J. H. W. J. Polym. Sci.: Polym. Chem. Ed. 1979 17, 977.
12. Takegami, Y.; Ueno, T., and Hirai, R. J. Polym. Sci.: Al 1966 4, 973.
13. Muetterties, E. L. U.S. Patent 2,856,370 (Oct. 1958).
14. Muetterties, E. L.; Bither, T. A.; Farlow, M. W.; Coffman, D. D. J. Inorg. Nucl. Chem. 1960 16, 52.
15. Crivello, J. V., and Lam, J. H. W. Macromolecules 1977 19, 1307.
16. Crivello, J. V., and Lam, J. H. W. J. Polym. Sci.: Polym. Lett Ed. 1978 16, 563.
17. Smith, G. H. Belgian Patent 828,841 (Nov. 7, 1975).
18. Belgian Patent 837,782, to Imperial Chemicals Industries (June 22, 1976).
19. Binks, J. H., and Huglin, M. B. Makromol. Chem. 1966 93, 268.
20. Takegami, Y.; Ueno, T.; Hirai, R. Bull. Chem. Soc. Japan 1965 38, 1222.
21. Crivello, J. V.; Lam, J.H.W. J. Polym. Sci.: Polym. Chem. Ed. 1979 17, 1047.
22. Saegusa, T.; Kobayashi, S. in "Polyethers", Vandenberg, E. J. Ed., ACS Symposium Series 6, American Chemical Society, Washington, D. C., 1975, p. 150.
23. Matyjaszewski, K.; Diem, T.; Penczek, S. Makromol. Chem. 1979 180, 1817.
24. Vaughan, D. J. "Nafion, an Electrochemical Traffic Controller", E. I. duPont de Nemours and Co., Wilmington, Del. 19898.
25. Dreyfuss, P. Chemtech June, 1973 3, 356.
26. Pruckmayr, G.; Wu, T. K. Macromolecules 1978 11, 662.

RECEIVED August 24, 1982

11

Initiation Considerations and Kinetics of Formation of Poly(2-methyl-l-pentene sulfone)

M. J. BOWDEN and T. NOVEMBRE

Bell Laboratories, Murray Hill, NJ 07974

UV initiation provides a convenient reproducible technique for formation of poly(2-methyl-1-pentene sulfone). With acetone as solvent and $[olefin][SO_2] = 6.25\ mol^2 l^{-2}$, polymerization is homogeneous at temperatures down to -90°C. The polymerization rate and polymer molecular weight increase with temperature to about -70°C thereafter decreasing and becoming zero at the ceiling temperature (~-45°C). Polymerization in acetone is characterized by a variable induction period which is not apparent in methylene chloride. The induction period most likely results from the presence of aldol products formed by the acid catalyzed condensation of two acetone molecules. If the monomer is stored in the presence of small amounts of SO_2, impurities develop with time which have a retarding effect on the polymerization. These impurities are believed to result indirectly from redox reactions between SO_2 and oxidized olefin. Oxygen also acts as a retarder by quenching the excited state of the irradiated charge transfer complex.

Poly(2-methyl-1-pentene sulfone) (PMPS) is an alternating copolymer of 2-methyl-1-pentene (2MP) and sulfur dioxide. The formation of PMPS occurs only by a free radical polymerization mechanism and is complicated to a degree by ceiling temperature considerations. For all exothermic addition polymerization reactions there is a critical temperature called the ceiling temperature (T_c) above which no reaction occurs. The precise T_c depends upon the monomer concentration according to the expression (1)

$$T_c = \frac{\Delta H}{\Delta S^\circ + R\ln[2MP][SO_2]}$$

where ΔH and ΔS are the enthalpy and entropy changes for the reaction. For PMPS, in which the polymer has the repeat structure,

0097-6156/83/0212-0125$06.00/0

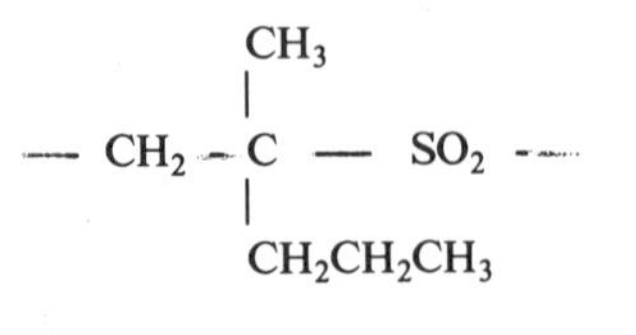

the ceiling temperature for $[2MP][SO_2] = 27mol^2l^{-2}$, is approximately -34°C (1).

The generation of free radicals at such low temperatures can be conveniently accomplished by redox reaction between organic hydroperoxides and sulfur dioxide. This is an effective method for initiation at temperatures as low as -80°C (2, 3). A plausible mechanism for radical formation is given below (4):

$$ROOH + SO_2 \rightarrow RO\text{-}O\overset{\overset{O}{\|}}{S}OH \rightarrow RO\cdot + HSO_3\cdot$$

There is some evidence to suggest that $HSO_3\cdot$ is the initiating species (5) in which case RO· must disappear, presumably by dimerization to give the dialkyl peroxide (3). Unfortunately our attempts at redox initiation were quite irreproducible from the kinetic standpoint, an observation also made by others (3).

Ultraviolet light provides an alternative initiation method. Although the exact mechanism of radical formation is not known, it presumably involves either direct photolysis of the charge transfer complex formed between the olefin and SO_2 or else may be related to the strong absorption by sulfur dioxide in the region of 3100Å. In any event, work by Dainton and co-workers (6) has shown that the 3650Å wavelength from a mercury lamp is efficient at effecting photopolymerization in poly(olefin sulfones).

Initiation with ultraviolet light proved to be a convenient reproducible technique for preparing PMPS. The following paper deals with some of the initiation and kinetic aspects of PMPS formation using UV initiation.

Experimental

Monomer Purification. 2-Methyl-1-pentene (Chemical Samples, 95-99% purity) was refluxed for ~1 hr over $LiAlH_4$ to destroy any peroxides present. The material was then distilled over argon and degassed by repeated freeze-thaw cycles (typically 3 or 4). The flask was then backfilled with argon to a pressure of 45 to 60 cm of Hg. In some instances, ~10ml of SO_2 were added to the flask prior to backfilling with argon.

Solvent preparation and storage. The solvents used in this study were acetone and methylene chloride. The latter was dried by stirring over molecular sieves (#3A, Aldrich) and then fractionally distilled onto molecular sieves in order to maintain anhydrous conditions. Acetone was fractionally distilled.

Reaction procedure. Solvent was first added to a 600ml 4-neck reactor flask equipped with a stirrer, well, bubbler and gas inlet, and degassed by bubbling argon through for ~30 minutes. The flask was transferred to a dry ice-acetone bath (-80°C) then connected to a vacuum line and evacuated. Next the requisite amount of SO_2 was condensed into the vessel which was then transferred to a low temperature methanol bath set at the desired polymerization temperature. The required volume of 2MP was added via a syringe and a stream of argon set bubbling through the reaction mixture.

Polymerization was initiated photochemically using a UV Penn Ray lamp equipped with a special filter to pass the 3650Å wavelength. The lamp was inserted into a N_2 flushed well which extended into the solution. (The lamp intensity was monitored using a Blak-Ray UV meter manufactured by UV Products. Values recorded after each polymerization ranged from 2800 to 3100 μ watts/cm^2).

Reaction kinetics were followed gravimetrically by withdrawing a small volume at varying time intervals into a weighed, evacuated, cooled, graduated test tube. An excess of methanol was added to the tube to precipitate the polymer. The methanol was then decanted and the samples dried under vacuum at 25°C for 18 hrs.

The intrinsic viscosity of each polymer was determined by dilute solution viscometry using distilled methyl ethyl ketone as the solvent at a constant temperature of 30°C. The MEK was distilled onto p-methoxyphenol to prevent peroxide formation.

Results and Discussion

General Considerations. Some difficulty was experienced initially in combining the reactants at the reaction temperature without a premature polymerization reaction occurring. There have been several reports in the literature of apparent spontaneous copolymerization of olefins with SO_2 (7, 8). Such "spontaneous" polymerization is generally attributed to hydroperoxide present as an impurity in either the olefin or solvent (9). Reaction between hydroperoxide and SO_2 gives rise to radical species which can initiate polymerization. An effective procedure for eliminating this difficulty involved adding a small amount of SO_2 (typically 10-15ml) to the olefin after it had been refluxed over $LiAlH_4$ and degassed, i.e., prior to sealing the flask under an atmosphere of argon. This procedure was typically carried out above T_c and was designed to scavenge any hydroperoxide which may not have been destroyed by $LiAlH_4$. Since the temperature was above T_c, any radicals so formed would decay in a non polymer-forming fashion. It was further felt that this procedure ensured continual "protection" of the olefin during storage.

An alternative procedure would be to mix the reactants at a temperature above the ceiling temperature for polymer formation. After a given period of time (during which peroxides would be destroyed harmlessly) the mixture would then be cooled to the desired polymerization temperature and polymerization initiated. This procedure is quite laborious and time consuming and was not given serious consideration. As will be shown later, the addition of a small amount of SO_2 to the olefin leads to kinetic complications. Subsequent modification of the purification apparatus and procedure proved to be simple and efficient and rendered the addition of SO_2 unnecessary.

Solubility Considerations. PMPS is insoluble in all monomer compositions at temperatures $< -35°C$. Thus it was desirable to use an inert solvent in order to maintain homogeneity. Both acetone and methylene chloride are suitable solvents although the latter must be rigorously dried to prevent phase separation of the small amount of water usually present in CH_2Cl_2 at the temperatures used in this study. Using acetone as solvent for example, the polymer remained in solution down to

-90°C for [2MP] = 0.76 mol l^{-1} and [SO_2] = 8.22 mol l^{-1}. In methylene chloride, phase separation of the polymer occurred at temperatures below -55°C after varying degrees of conversion for the same concentrations. The degree of conversion at which phase separation occurred decreased with decreasing polymerization temperature.

Effect of Oxygen. Oxygen acted as an inhibitor of the polymerization as seen in Figure 1. This is in contrast to the usual behavior of oxygen in the polymerization of other olefin-SO_2 mixtures whose ceiling temperatures lie above 0°C. Oxygen is actually an effective initiator for such systems, most probably via oxidation of the olefin to form hydroperoxide which is in turn reduced by SO_2 in a redox reaction producing radical intermediates. Since these oxidation reactions do not occur at low temperature, oxygen does not function as an initiator under such conditions (10). The observed inhibition most probably results from quenching of the excited state of the charge transfer complex (11).

Polymerization Kinetics. The initial experiments were carried out in methylene chloride with molar concentrations of SO_2 and 2MP equal to 11.5 mol l^{-1} and 1.07 mol l^{-1} respectively. The mole fraction of 2MP relative to SO_2 was 0.085. Plots of yield vs. time for several runs are shown in Figure 2. At -50°C the average rate of conversion was 3.6%/hr and the intrinsic viscosity was 312 cm^3/g. At −63°C, the polymer precipitated from the solution as a swollen gel after about 20 minutes. Thus, only the first few points are shown at −63°C but it appears the initial rate is comparable to that at −50°C. The rate of polymerization decreased to ~1.0%/hr at -45°C ([η] = 162 cm 3/g) and was zero at -40°C indicating T_c lies between -40 and -45°C. (An approximate calculation of T_c using Eq. 1 and adopting values of -14,400 cal mol^{-1} and -66.5 cal mol^{-1} for ΔH and $\Delta S°$ respectively (1) gave a value of -39°C under the prevailing conditions of monomer concentration and pressure).

In an attempt to overcome the phase separation problem resulting from solubility limitations of the polymer in methylene chloride/SO_2 mixtures, a series of runs were carried out in which the amount of solvent was doubled ([M][S] = 6.25 $mol^2 l^{-2}$) while maintaining the same mole ratio of reactants. The rate of conversion at −50°C was lowered to ~ 1%/hr for this condition (T_c ~ - 45°C). At −60°C however, phase separation of the polymer again occurred after about 4 hours and in less than 1 hr at −70°C.

In addition to the solubility limitations of methylene chloride, this solvent also tended to become turbid at temperatures below −40°C. This was due to the hygroscopic nature of CH_2Cl_2 leading to phase separation of absorbed water at low temperature with concomittant turbidity. This was not a problem with acetone in which the polymer was also soluble at temperatures down to −90°C. It was found however that polymerization in acetone was characterized by an induction period of anywhere from zero to several hours. A typical conversion curve is shown in Figure 3 for [M][S]=6.25 $mol^2 l^{-2}$ at -70°C. In this case the induction period was 1 1/2-2 hours after which polymerization occurred at a constant rate (4.3%/hr). No induction period had ever been observed using methylene chloride as the solvent suggesting that this phenomenon was attributable to the solvent. The polymer however remained soluble throughout the run and the reaction mixture thus remained clear and homogeneous.

An additional complicating factor was the observation that the rate of polymerization depended on the age of the monomer following purification as shown in Fig. 4 for a series of runs at −70°C. This primarily applied to monomer to which a small amount of SO_2 had been added to scavenge peroxide impurities. The highest

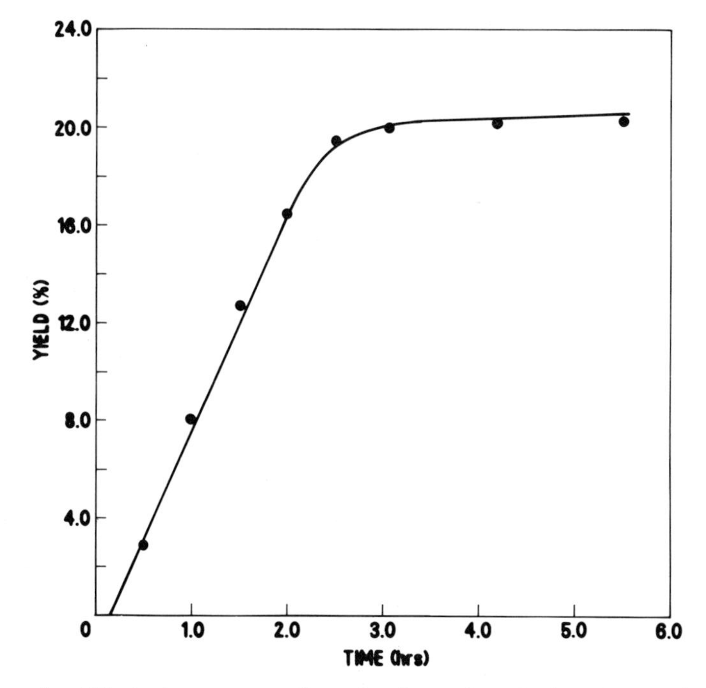

Figure 1. Effect of oxygen on the rate of copolymerization (oxygen bubbled through reactor after 2.5 h).

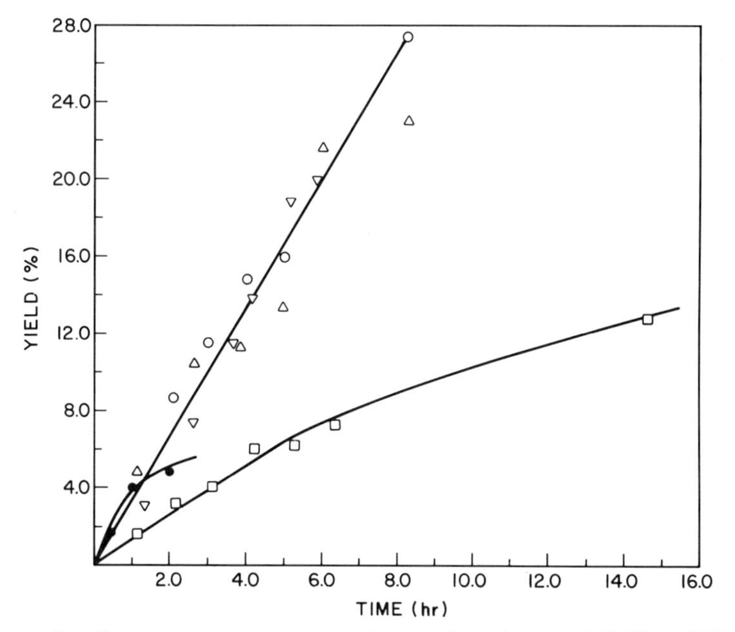

Figure 2. Conversion vs. time curves for copolymerization of 2MP and SO_2 in methylene chloride ($[2MP][SO_2] = 12.3\ mol^2L^{-2}$). Key: ○, △, ▽, −50 °C; □, −45 °C; and ●, −63 °C.

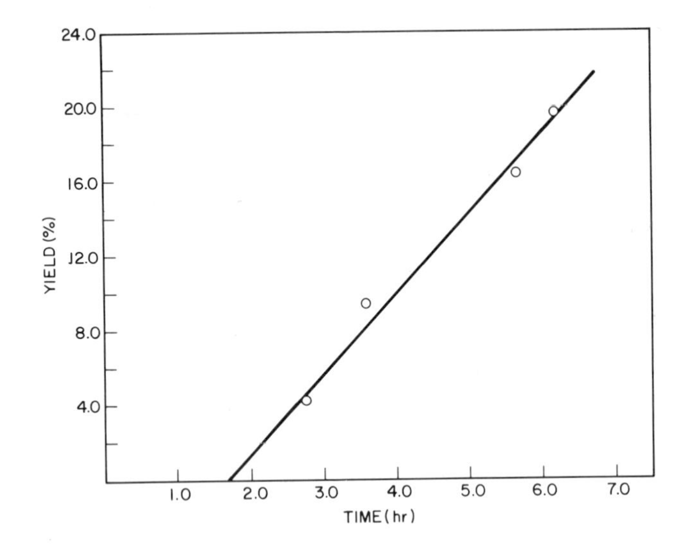

Figure 3. Plot of conversion vs. time for copolymerization of 2MP and SO_2 in acetone at −70 °C ([2MP][SO_2] = 6.25 mol^2L^{-2}).

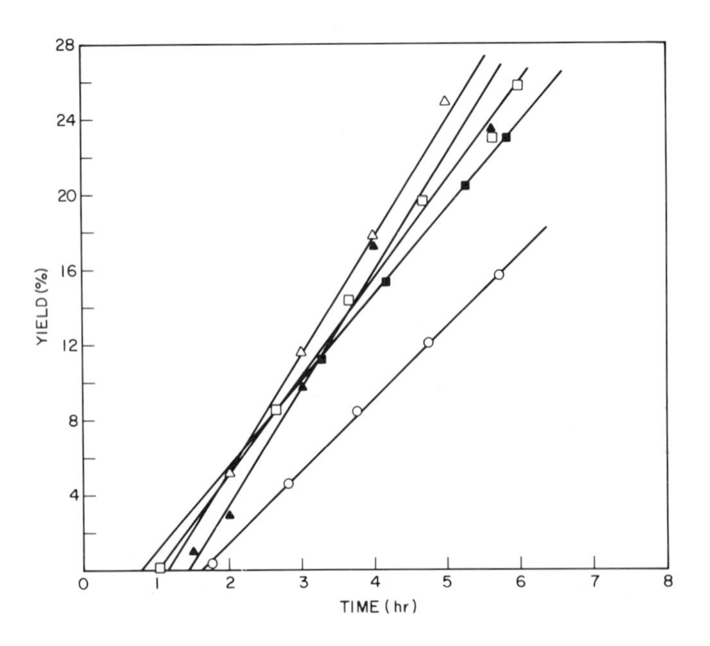

Figure 4. Effect of monomer age on the copolymerization rate of 2MP and SO_2 in acetone at −70 °C ([2MP][SO_2] = 6.25 mol^2L^{-2}). Key: △, ▲, freshly prepared; □, after 6 days; ■, after 12 days; ○, after 1 month.

rate was always observed for monomer used as soon as possible after purification. Thus at −70°C, the rate of polymerization varied from 6.2%/hr for freshly purified monomer to 3.8% for monomer stored for one month prior to use. We believe this may be due to by-products from the redox reaction of SO_2 with oxidation products of the olefin. Although the monomer is degassed and stored under a partial pressure of argon, some air apparently does leak into the flask either during syringe withdrawal or during storage, particularly after the septum has been used several times. This was confirmed somewhat indirectly by the fact that the olefin solution tended to develop a yellow color after a period of time. This color developed much more rapidly if the flask was opened to the air suggesting the same reaction occurred slowly over a period of time during storage. It will also be noted from Figure 4 that the length of the induction period appears to vary randomly and bears no apparent relationship to monomer age. Additional experiments confirmed that the induction period was due to impurities in the acetone and will be discussed later.

A series of polymerization runs were carried out at temperatures from -55 to -90°C for [M][S] equal to 6.25 mol^2l^{-2}. In all cases the monomer was used within a few days of preparation. The conversion curves are plotted in Figure 5 from which the rate of polymerization was calculated and plotted as an Arrhenius plot in Figure 6. The rate passed through a maximum at about -70°C and rapidly approached zero as the ceiling temperature (~-45°C) was approached. Again the conversion curves were characterized by varying induction times. The rates remained linear with conversion except for the run at -55°C where the rate began to decrease after about 15% conversion. This may be due to ceiling temperature considerations resulting from a lowering of the monomer concentration due to conversion. A value of 5 kcal/mole was obtained for the activation energy over the temperature range where propagation-depropagation equilibria are not significant. This value can only be considered to be approximate since only two points are involved. Nevertheless it does serve to show that the activation energy is low as would be expected for a photoinitiated polymerization. Dainton and Ivin (6) reported a value of zero for the activation energy for the UV initiated formation of poly(butene-1 sulfone).

The temperature dependence of polymerization rate was also mirrored by the temperature dependence of intrinsic viscosity [η] which increased with temperature to about -70°C thereafter decreasing and becoming zero at the ceiling temperature.

Effect of Monomer Storage Time. We noted earlier that the initial polymerization rate of monomer stored in the presence of a small amount of SO_2 decreased with time of storage. This may be due to products formed by reaction of SO_2 with trace hydroperoxide or alternatively, the products of this reaction such as H_2SO_4 might react over a period of time with the olefin resulting in the formation of retarding species. In order to test this hypothesis, a sample of purified olefin was stirred for 24 hr. at 35°C in the presence of O_2 while being irradiated with the full output of a Penray UV lamp. The presence of peroxides was confirmed by the KI test and when the peroxidized olefin was added to SO_2/acetone at -60°C, an immediate "spontaneous" polymerization occurred (curve A, Figure 7). Curve B in Figure 7 shows the normal yield-time conversion curve for the purified monomer (note that there was only a short induction period for this particular batch of acetone). When the peroxidized monomer was added to the SO_2/acetone mixture at -35°C (i.e., above T_c), held there for 1 hr and then cooled back to -60°C, no spontaneous initiation occurred (Curve C Figure 7). The peroxides must have been destroyed by reaction with SO_2 at -35°C and since this

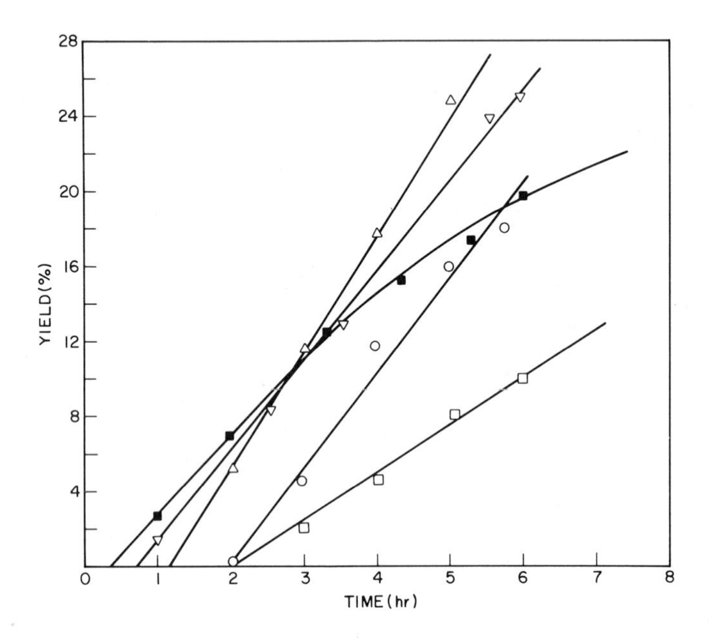

Figure 5. Conversion vs. time curves as a function of temperature for copolymerization of 2MP and SO_2 in acetone ($[2MP][SO_2] = 6.25\ mol^2L^{-2}$. Key: ■, −55 °C; ▽, −60 °C; △, −70 °C; ○, −80 °C; □, −90 °C.

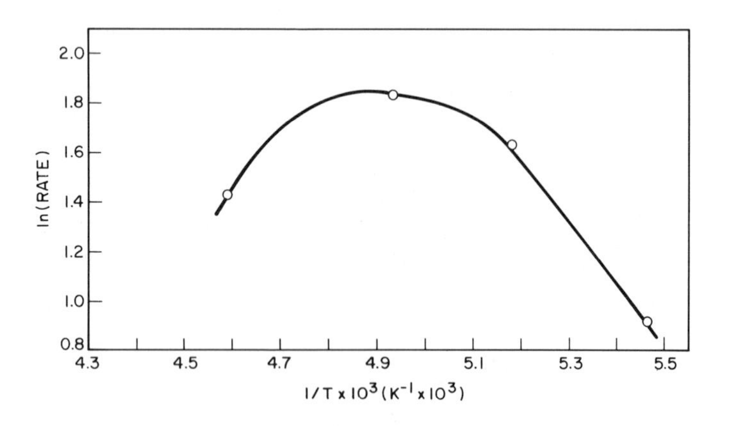

Figure 6. Arrhenius plot of data in Figure 5.

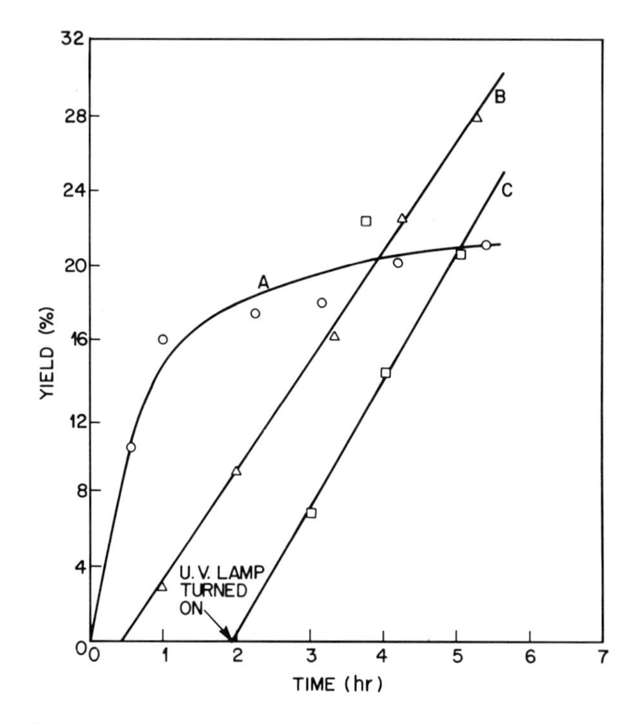

Figure 7. Yield vs. time curves for copolymerization. A, spontaneous initiation with peroxidized monomer at −60 °C; B, normal UV initiation with pure monomer at −60 °C; C, UV initiation of peroxidized monomer at −60 °C following storage for 1 h at −35 °C.

temperature is above T_c, no polymer can form. The reaction was essentially finished in less than an hour under these conditions. After cooling the sample to -60°C, polymerization was initiated only after the UV lamp was turned on. Again there was no induction period in spite of the fact that initiation was occurring in the presence of the primary products of the redox reaction. The latter can therefore play no part in causing the induction effect. Interestingly though, the polymerization rate was identical to the control. We therefore suggest that the effect of storage time on polymerization rate may be caused by species generated by subsequent slow reaction between the olefin and the primary redox reaction products.

Origin of the Induction Effect in Acetone. We have seen that although acetone is a preferable solvent for the copolymerization of 2-methyl-1-pentene and sulfur dioxide from the standpoint of solubility, there is a variable induction period associated with these polymerizations which is not evident using methylene chloride as the reaction solvent. Further, the length of the induction period varies randomly with no apparent relationship to monomer age. It seems unlikely that inhibition could be attributed to impurities in SO_2 since the same SO_2 cylinder was used throughout. Three other possibilities could be considered, viz., 1) trace oxygen in the polymerization vessel (as shown earlier, oxygen inhibits polymerization by quenching the excited state of the charge-transfer complex), 2) formation or presence of inhibitors in the solvent and 3) formation or presence of inhibitors in the olefin.

The apparent lack of an induction period in methylene chloride would appear to eliminate (1) and (3) as causes of this effect. Acetone, on the other hand, can contain a number of impurities which are not easily removed by distillation (12). For example, an aldol condensation can be induced under mildly acidic or basic conditions to 4-hydroxy-4-methyl-2-pentanone.

$$CH_3-\underset{\underset{O}{\|}}{C}-CH_3 + CH_3-\underset{\underset{O}{\|}}{C}-CH_3 \rightarrow CH_3-\overset{\overset{CH_3}{|}}{\underset{\underset{OH}{|}}{C}}-CH_2-\underset{\underset{O}{\|}}{C}-CH_3$$

The β-hydroxy ketones obtained from aldol condensations are very easily dehydrated, the major products having the carbon-carbon double bond between the α- and β-carbon atoms.

$$CH_3-\overset{\overset{CH_3}{|}}{\underset{\underset{OH}{|}}{C}}-CH_2-\underset{\underset{O}{\|}}{C}-CH_3 \rightarrow CH_3-\overset{\overset{CH_3}{|}}{C}=CH_2-\underset{\underset{O}{\|}}{C}-CH_3$$

In the case of acetone, the major product is mesityl oxide. We found that this compound has a drastic effect on the polymerization rate. Polymerization was totally inhibited for times in excess of 12 hr for a run in which acetone containing 1% added mesityl oxide was used as the solvent. Figure 8 shows a yield vs. time plot for polymerization using pure acetone. The initial rate was 4.6%/hr. After 3 hr, 0.3% mesityl oxide was added via syringe resulting in total inhibition of polymerization.

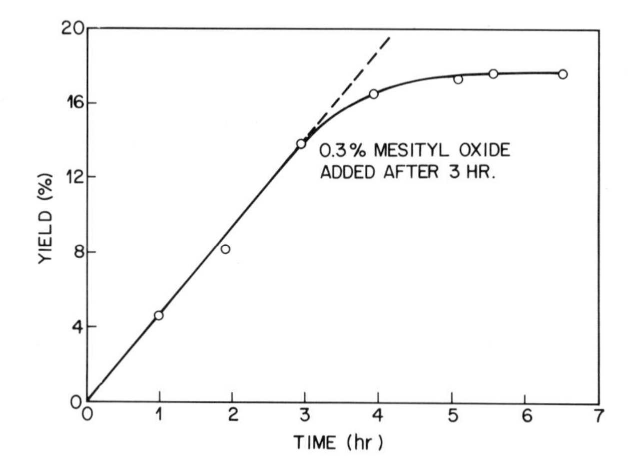

Figure 8. Effect of adding mesityl oxide on polymerization kinetics.

Acetone was also stirred over barium oxide for 12 hr and subsequent polymerization using this material gave a limiting yield of only 4% after 4 hr.

While not specifically detecting aldol condensation products in acetone, it is not entirely unreasonable that such products should form in acetone/SO_2 mixtures due to the acidic nature of the latter. In the presence of trace amounts of water, SO_2 forms a highly acidic species which could easily catalyze the aldol condensation reaction.

Literature Cited

1. Cook, R. E.; Dainton, F. S.; Ivin, K. J. J. Polymer Sci., 1957, 26, 351.
2. Gosh, P.; O'Driscoll, K. F. J. Macromol Sci., Chem.,, 1967, A1, 1393.
3. Oster, B.; Lenz, R. W. J. Polymer Sci., Polymer Chem. Ed., 1977, 15, 2479.
4. Schulz, R. C.; Banihaschemi, A. Makromol Chem., 1963, 64, 140.
5. Ivin, K. J.; Rose, J. B. Adv. Macromol. Chem., 1968, 1, 335.
6. Dainton, F. S.; Ivin, K. J. Proc. Roy. Soc., 1952, A212, 96, 207.
7. Iwatsuki, S.; Okada, T.; Yamashita, Y. J. Polymer Sci., 1968, 6, 2451.
8. Zutty, N. L.; Wilson, C. W.; Porter, G. H.; Priest, D. C.; Whiteworth, C. J. J. Polymer Sci., 1965, A3, 2781.
9. Sartori, G.; Lundberg, R. D. J. Polymer Sci.,PolymerLetters, 1972, 10, 583.
10. Fettes, E. M.; Davis, F. O. in "High Polymers" Vol. XIII, Eds., Raff, R. A. V.; Doak, K. Interscience, NY, 1962, p 225.
11. Turro, N. J. in "Modern Molecular Photochemistry" Ch. Benjamin/Cummings 1978, Ch. 14.
12. Riddick, J. A.; Bunger, W. B. in "Organic Solvents, Physcial Properties and Methods of Purification" Vol. 2 of Techniques of Chemistry, Weissberger, A., Ed., Wiley Interscience, 1970.

RECEIVED October 15, 1982

12

Rational Design of Catalysts with Interacting Supports

G. L. BAKER,[1] S. J. FRITSCHEL,[2] and J. K. STILLE

Colorado State University, Department of Chemistry, Fort Collins, C0 80523

Polymer bound catalyst systems have become highly sophisticated. For supports, commercial resins have in cases given way to custom tailored polymers designed to optimize a supported catalyst's performance. This progression from simple to complex systems is illustrated by advances in supported catalysts used for the asymmetric hydrogenation of enamides. Early supports were designed so that the support had no influence on the catalysts' activity. An optically active support, designed to interact with the catalyst and improve its performance, has now been synthesized and used in asymmetric hydrogenations. The influence of the support on the catalyst was surprisingly low. This result can be interpreted in terms of the mechanism of asymmetric hydrogenation.

Soluble homogeneous catalysts can perform useful chemical transformations under mild conditions. One of their chief disadvantages is that their solubility in the reaction medium makes the separation, recovery, and recycling of the catalysts difficult. This would be a minor concern, but for the fact that the most useful catalysts often use costly metals such as platimum, palladium and rhodium.

The solution to this problem has been to attach these catalysts to polymer supports. The ideal polymer-bound catalyst must satisfy a formidable list of requirements. It should be easily prepared from low cost materials. The support must be compatible with the solvent system employed, and be chemically and thermally stable under the reaction conditions. The catalyst should show minimal losses in reaction rate or selectivity when bound to the support, and should be able to be recycled many times without loss of activity. Finally, the interactions between the catalytic site and the support must be either negligible or beneficial. The development of polymer supported rhodium-phosphine catalysts for the asymmetric hydrogenation of amino acid precursors illustrates the incremental process which has led to supports which approach the ideal support.

Early catalysts were bound to crosslinked polystyrenes (*1,2,3*). and shared the same swelling characteristics as polystyrene. These catalysts proved inferior to their

[1] Current address: Bell Laboratories, Murray Hill, NJ 07974.

[2] Current address: E. I. duPont de Nemours and Company, Polymer Products Department, Wilmington, DE 19898.

0097-6156/83/0212-0137$06.00/0

homogeneous analogs, chiefly due to the incompatibility of the support with the solvent of choice, ethanol. Reactions demanding a non-polar medium, such as asymmetric hydrosilylation, however, could be carried out (*3*).

To surmount this problem, monomers were prepared which contain the desired ligand, and upon polymerization with suitable comonomers, the ligand is incorporated into the polymer. This has been accomplished with monomers such as **1** and **2** (Figure 1) to give polymers containing either DIOP (*4,5,6*) or BPPM-type (*7,8*) ligands respectively. The comonomers can be chosen to give the optimum ligand density, crosslink density, and swelling characteristics for the supported catalyst. In addition, this route assures a high degree of ligand purity, since few reactions are carried out on a crosslinked support. Using catalysts derived from these polymers, products can be obtained in optical purities comparable to those achieved with their homogeneous analogs.

Interacting Supports

Although most supports have been designed to minimize catalyst-support interactions, an appropriate designed support might be expected to interact with the catalytic site in a favorable way, enhancing enantioselectivity. This can be tested by providing an additional optical center at the catalyst and then observing any change in the enatiomeric excess of the product. A previous approach to this problem demonstrated that such an effect was indeed possible (*7*). A polymer containing an optically active ligand and methyl ketones was reduced by asymmetric hydrosilylation to give the optically active support. Unfortunately the enantioselectivity of this reduction could not be evaluated, and catalysts derived from this polymer gave low optical yields.

A superior method for preparing polymer bound phosphines of high optical purity is to polymerize an optically pure phosphine monomer with the desired comonomers. This approach was extended to the comonomer. An optically active comonomer suitable for use in preparing a polymer containing optically active pendant alcohols should be available in both the R and S enantiomers so that the chirality of the alcohol and that of the catalyst may be matched to provide a synergistic effect. A suitable starting material for the synthesis of optically active comonomers is 2,3-butanediol, since the R,R isomer is commercially available, and the S,S isomer can be synthesized in a straightforward manner from tartaric acid (9). Optically active acrylates *3a-c* were prepared from the diols to give optically pure monomers (Fig. 2). An inactive monomer was synthesized from racemic 2,3-butanediol for use as a standard.

Two types of supported phosphine polymers were prepared. Phosphinopyrrolidine-containing polymers were prepared by copolymerizing **2** with acrylates *3a-c* and ethylene dimethacrylate to give white free-flowing powders. In a similar fashion **1** was copolymerized with *3a-c* and ethylene dimethacrylate. Treatment of these polymers with a large excess of sodium diphenylphosphide gave polymers containing DIOP-type ligands.

All of the polymers swell in both tetrahydrofuran and ethanol. Since ethanol would be expected to compete with the polymer bound alcohols for sites at the catalyst, tetrahydrofuran was chosen as the reaction solvent. The polymer bound catalysts were prepared by stirring the polymer and μ-dichlorobis(1,5-cyclooctadiene)dirhodium(I) in tetrahydrofuran for several hours, and after filtration the yellow catalyst was then transferred under argon to the reaction vessel containing the substrate. Solvent was added, and the reaction vessel was pressurized with hydrogen. At the end of the reaction, the pressure was released and the product was iso-

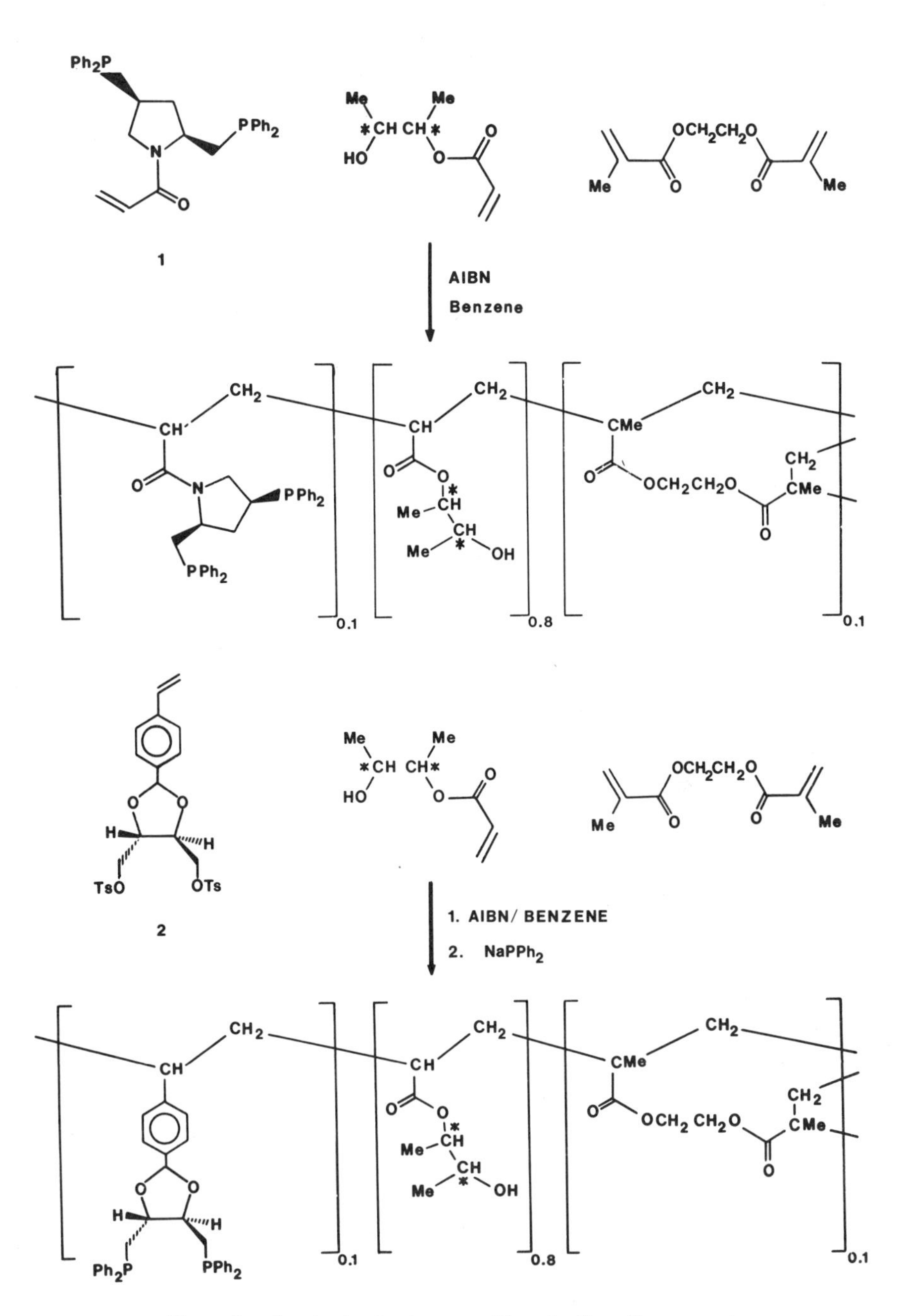

Figure 1. Synthesis of polymers with optically active supports.

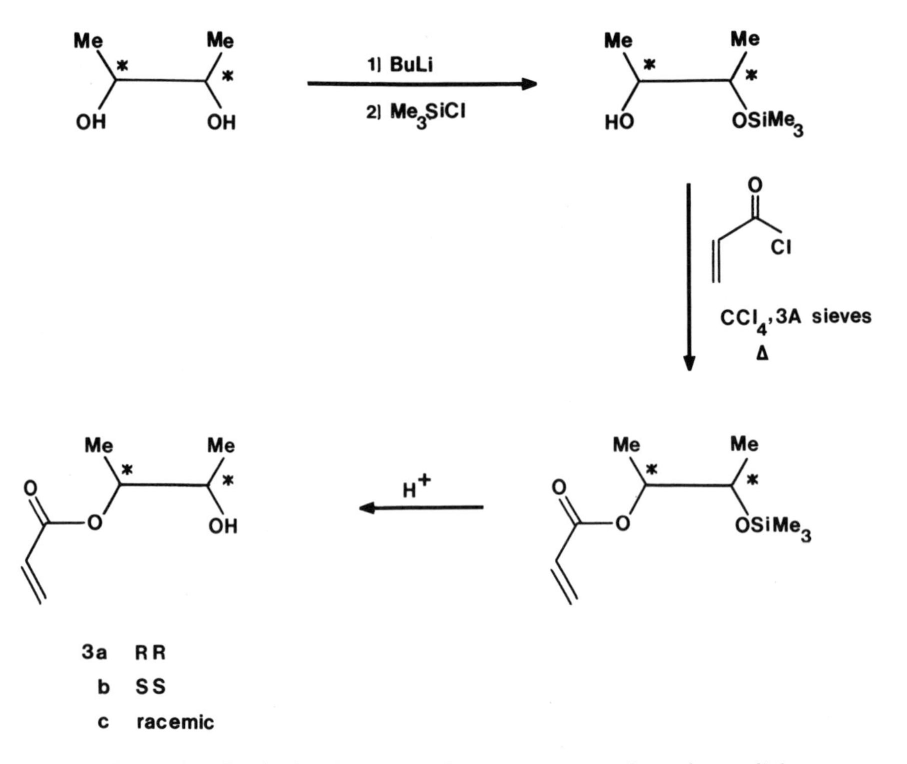

Figure 2. Synthesis of monoacrylate comonomers from butanediols.

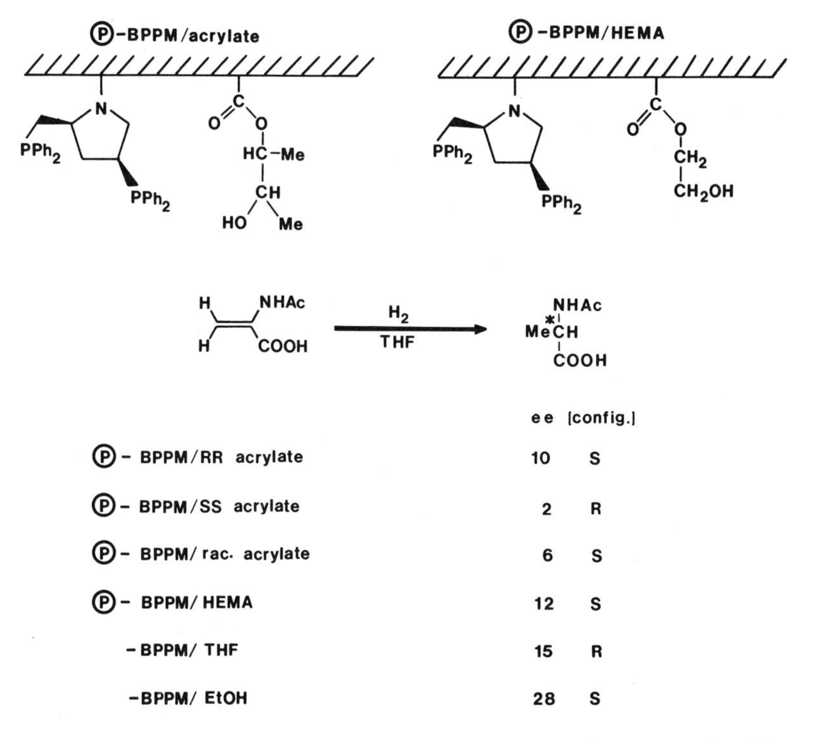

Figure 3. Rhodium-catalyzed hydrogenation of 2-acetamidoacrylic acid with a BPPM-type phosphine on an optically active support.

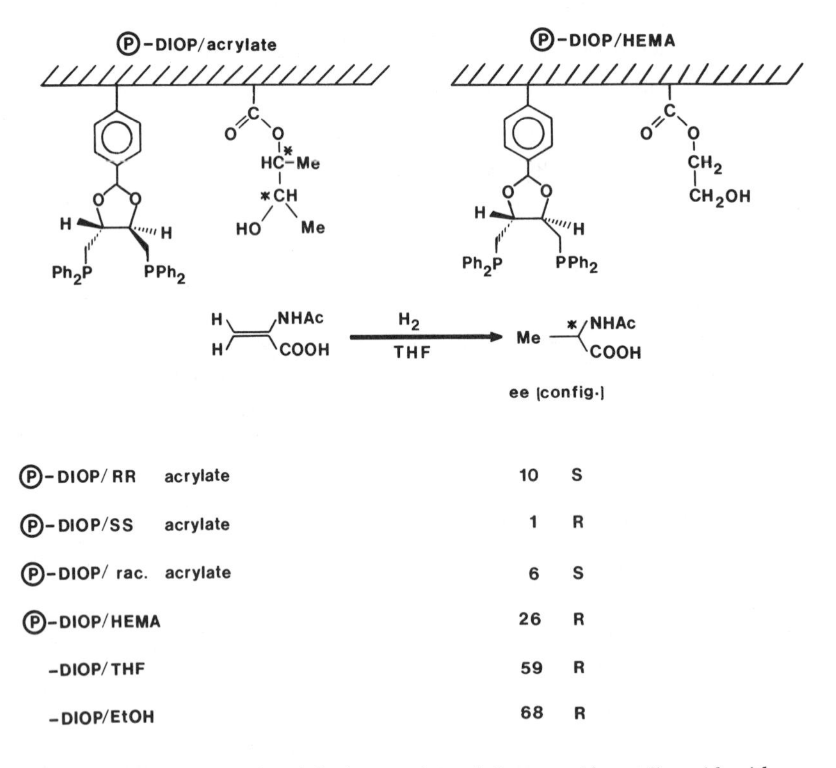

Figure 4. Rhodium-catalyzed hydrogenation of 2-acetamidoacrylic acid with a DIOP-type phosphine on an optically active support.

lated. The workup consisted of filtration to remove the supported catalyst and evaporation of the solvent to dryness. The substrate chosen to probe the effect of the optically active alcohols was 2-acetamidoacrylic acid. The hydrogenation product, N-acetylalanine, was converted to the methyl ester by diazomethane and was then analyzed by GLC using a column containing a chiral stationary phase.

The effect of the additional optical center was smaller than anticipated (Figure 3). Different enantiomeric excesses were observed for BPPM type catalysts containing R,R, S,S or racemic alcohols, with the racemic support falling between the optically active supports. Far greater was the solvent effect due to the alcohols, shifting the enantiomeric excesses to a position midway between those obtained homogeneously in ethanol and those obtained with tetrahydrofuran as the solvent.

The hydrogenation reaction is also sensitive to the structure of the alcohol comonomer, with the primary alcohols of the hydroxyethyl methacrylate polymer interacting with the catalyst to give results more closely resembling those found when ethanol is used as the solvent. Similar results were found with DIOP-type ligands (Figure 4).

The weak effect of the secondary optical center can be explained by a consideration of the mechanism of asymmetric hydrogenation. It was demonstrated that the rate determining step is the oxidative addition of hydrogen to the rhodium-olefin complex (*10*). During this step, solvent would not be expected to be coordinated to the complex. Thus any influence of the secondary optical center would not be through direct coordination to the metal, but rather by a lesser effect in the surrounding medium. This suggests that it will be difficult to prepare a system in which a secondary optical center will have much influence in rhodium catalyzed hydrogenations of enamides.

One important observation is the ability of the polymer to greatly affect the solvent environment at the catalytic site. This can occur even though the polymer is highly swollen in solvent. Although this is a complicating factor at times, it is likely that a carefully designed system would be able to exploit this characteristic of supported catalysts.

Acknowledgment

This work was supported by Grant DMR-77-14447 from the National Science Foundation, and by the Phillips Petroleum Company.

Literature Cited

1. Strukal, G.; Bonivento, M.; Graziani, M.; Cernia, E.; Palladino, N. Inorg. Chim. Acta 1975, 12, 15.
2. Krause, H. W. React. Kinet. Catal. Lett. 1979, 10, 243.
3. Dumont, W.; Poulin, J.-C.; Dang, T.-P.; Kagan, H. B. J. Am. Chem. Soc. 1973, 95, 8295.
4. Fritschel, S. J.; Ackerman, J. J. H.; Keyser, T.; Stille, J. K. J. Org. Chem. 1979, 44, 3152.
5. Takaishi, N.; Imai, H.; Bertelo, C. A.; Stille, J. K. J. Am. Chem. Soc. 1978, 100, 264.
6. Masuda, T.; Stille, J. K. J. Am. Chem. Soc. 1978, 100, 268.
7. Achiwa, K. Chem. Lett. 1978, 905.

8. Baker, G. L.; Fritschel, S. J.; Stille, J. K. J. Org. Chem. 1981, 46, 2954.
9. Schurig, V.; Koppenhoefer, B.; Buerkle, W. J. Org. Chem. 1980, 45, 538.
10. Chan, A. S. C.; Pluth, J. J.; Halpern, J. J. Am. Chem. Soc. 1980, 102, 5952.

RECEIVED October 15, 1982

13

An Overview of the Polymerization of Cyclosiloxanes

J. E. McGRATH, J. S. RIFFLE, A. K. BANTHIA, I. YILGOR, and G. L. WILKES

Virginia Polytechnic Institute and State University, Department of Chemistry and Department of Chemical Engineering, Polymer Materials and Interfaces Laboratory, Blacksburg, VA 24061

A review of the polymerization of cycloorganosiloxanes is provided. Cycloorganosiloxanes such as octamethyl cyclotetrasiloxane are the principal intermediate for the formation of high molecular weight polyorganosiloxane polymers. The ring opening reaction can be initiated through the use of suitable basic or acidic catalysts. The transformation of the cyclosiloxanes into linear chains is an equilibrium process characterized by a very low heat of polymerization. At equilibrium conversions one produces 12-15% of cyclic oligomers, which are predominantly but not exclusively the cyclic tetramer. Equations from the literature which define these equilibrium are reviewed. The principal mechanisms involved in the anionic ring opening polymerization via initiators such as potassium hydroxide are discussed. A host of other related initiator types have also been used, including the so-called transient catalysts which are based upon quaternary silanolates. The transient catalysts are so described since they rapidly decompose above 130°C to yield inactive byproducts. The anionic polymerization of siloxanes is relatively well understood and currently accepted mechanisms are presented. Cationic polymerization of organosiloxanes is also well known but is much less understood. In general, molecular weight vs. conversion curves for the two processes are considerably different and these aspects are reviewed. Molecular weight is often controlled by disiloxane or low molecular weight siloxane molecules terminated with triorganosiloxy groups. These materials (which are known as end blockers) control the molecular weight due to the

0097-6156/83/0212-0145$08.00/0

fact that their silicon-oxygen bonds can exchange with and incorporate the growing chain. By contrast, the silicon-carbon bonds in such materials are incapable of reacting. Thus at equilibrium one incorporates most of the linear chains between the two triorganosiloxy terminals. Such a process is possible with either anionic or cationic intermediates. This approach has recently permitted the introduction of functional groups such as aminopropyl, carboxy propyl, epoxy propyl, etc. Such oligomers are interesting intermediates for novel segmented copolymers.

Perhaps the first syntheses of the major types of the polyalkylsiloxanes were performed by Friedel, Ladenburg, and Crafts during the period 1865-71 (1-6). However, it was not until the early 1900's that F. S. Kipping and his group demonstrated the siloxane polymeric structure (3). This group prepared a large number of linear and cyclic polymers of type $(R_2SiO)_N$ and $HO(R_2SiO)_NH$, as well as crosslinked architectures $(RSiO_{3-n}H_{3-2N})_m$. Kipping's studies disclosed for the first time that the Si-O-Si group differs dramatically from the C-O-C group in its reactivity and that the Si-O-Si group could be easily cleaved by acids, bases, or Lewis acids. Moreover, he was the first to observe the catalytic effect of acids and bases for the now commercially important siloxane redistribution and ring opening polymerization reactions. Following this period, commercial production of siloxane polymers was restrained by the absence of convenient methods for monomer synthesis. In the 1930's, Rochow at General Electric (4,7,8) and R. Muller (nine months later) in Germany apparently independently discovered what is known as the "direct process" (9). This process was essential for the industry since it allowed for the economical manufacture of the family of methylchlorosilanes necessary for siloxane production directly from silicon and alkyl (mostly methyl) chlorides. Controlled hydrolysis of the various organohalosilanes then provided the cyclic trimers and tetramers which could be used for polymerization. Some of the important terminology and nomenclature for siloxanes is reviewed in Tables 1 and 2. Some important physical constants (10) for chlorosilanes are also given in Table 3.

Although there are a variety of routes presently available for siloxane production, only two are of commercial importance. These include hydrolytic reactions of organohalosilanes or organoalkoxysilanes and redistribution type polymerizations of

TABLE 1. COMMON TERMINOLOGY OF SILOXANES

Formula	Equivalent formula	Symbol
$(CH_3)_3SiO_{0.5}$	$Me_3SiO_{0.5}$	M
$(CH_3)_2SiO$	Me_2SiO	D
$CH_3SiO_{1.5}$	$MeSiO_{1.5}$	T
$CH_3(C_6H_5)SiO$	MePhSiO	D'
$(C_6H_5)_2SiO$	Ph_2SiO	D'
$(CH_3)HSiO$	MeHSiO	D'
SiO_2	SiO_2	Q

TABLE 2. EXAMPLES OF NOMENCLATURE

Structural formula[a]	"MDT" formula	Systematic name[b]
$Me_3SiOSiMe_3$	NM	hexamethyldisiloxane
$Me_3SiOSiMe_2OSiMe_3$	MDM	octamethyltrisiloxane
$Me_3SiO(SiMe_2O)_2SiMe_3$	MD_2M	decamethyltetrasiloxane
$Me_2Si—O—Si—Me_2$ | | O O | | $Me_2Si—O—SiMe_2$	D_4	octamethylcyclotetra-siloxane (methyl-tetramer)
$Ph_2Si—O—Si—Ph_2$ | | O O | | $Ph_2Si—O—Si—Ph_2$	D'_4	octaphenylcyclotetra-siloxane (phenyl-tetramer)
MePhSi—O—Si—MePh | | O O | | MePhSi—O—Si—MePh	D'_4	1,3,5,7 tetramethyl-1,3,5,7-tetraphenyl-cyclotetrasiloxane (methylphenyltetramer)
$(Me_3SiO)_3SiMe$	M_3T	1,1,1,3,5,5,5-heptamethyl-3-trimethylsiloxy-trisiloxane
$Me_3SiOSiHMeOSiMe_3$	MD'M	1,1,1,3,5,5,5-hepta-methyltrisiloxane
$Me_2PhSiOSiMePhOSiPhMe_2$	M'D'M'	1,1,3,5,5-pentamethyl-1,3,5-triphenyl-trisiloxane

[a]Abbreviated [b]IUPAC

Example: $Me_3Si—O—SiMe_2—O—SiPh—(OSiMe_3)_2$

sometimes is termed $MDT'M_2$ where T', in this case, is $C_6H_5SiO_{1.5}$.

TABLE 3. PHYSICAL PROPERTIES OF METHYLCHLOROSILANES AS A FUNCTION OF STRUCTURE

Compound	Boiling point, °C	Density, d^{20}	Refractive index, n_D^{20}	Assay, %
$(CH_3)SiCl_3$	66.4	1.273	1.4088	95-98
$(CH_3)_2SiCl_2$	70.0	1.067	1.4023	99-99.4
$(CH_3)_3SiCl$	57.9	0.854	1.3893	90-98
CH_3SiHCl_2	41.0	1.110	1.3982	95-97
$(CH_3)_2SiHCl$	35.0	0.854	1.3820	

cyclic monomers. Both of these types of reactions involve a ring-chain equilibrium distribution of products. The principles applicable to the distribution are probably identical in both cases.

The discussion following will deal mainly with the important equilibrations and polymerizations of cyclic siloxanes. For further information concerning these methods and for excellent discussions of other types of siloxane polymerizations, the reader is referred to Voronkov (4), Noll (7), and Meals (10). In addition, Noll (7), Eaborn (11), and Arkles and Peterson (12) offer reviews of the general chemistry of silicon compounds.

A Review of the Polymerization of Cyclic Siloxanes

Linear polysiloxane can be synthesized by both the anionic and cationic polymerization of cyclic siloxanes. Molecular weight is regulated through the incorporation of controlled amounts of monofunctional endblockers into the system (7,13). As a result of the nature of the configurations of siloxane chains coupled with the similar reactivity of siloxane bonds in the linear as compared to cyclic species, the anionic or cationic catalysts attack both the rings and chains during the polymerization. These so called "redistribution" or "equilibration" polymerizations involved reactions such as those listed in equations 1-4 occurring throughout the process. At

$$-D_x- + D_4 \longrightarrow -D_{(x+4)}- \quad (1)$$

$$-D_x- + MM \longrightarrow MD_xM \quad (2)$$

$$MD_xM + MM \longrightarrow MD_{(x-5)}M + MD_5M \quad (3)$$

$$MD_xM + MD_yM \longrightarrow MD_{(x+w)}M + MD_{(y-w)}M \quad (4)$$

(In siloxane nomenclature (cf. Tables 1 and 2), "M" denotes a monofunctional siloxane unit whereas "D" refers to a difunctional siloxane unit. "D_4" therefore represents the cyclic siloxane tetramer while "MM" is the linear dimer. "D" nomenclature is normally associated with dimethylsiloxy units).

thermodynamic equilibrium these reactions result in a Gaussian distribution of molecular weights among the chain molecules together with an approximately monotonically decreasing distribution of ring species as ring size increases (13-20). Rates of the various processes depend upon factors such as catalyst type and concentration, temperature, pressure, and the use of various types and amounts of promoters and can undoubtedly be controlled to some extent. However, with the exception of using organolithium catalysts in conjunction with the D_3 monomer (21-22), significant amounts of redistribution cannot be avoided.

The technique involved in these types of polymerizations is normally quite facile. The general method potentially lends itself well to the laboratory synthesis of diorganofunctional siloxanes through variance of the structures of "R" and "R'" shown in equation 5. The R group is principally methyl, but other groups such as phenyl, vinyl, hydrogen, cyanoethyl and trifluoropropyl may also be used. The degree of polymerization at equilibrium of 3 is basically a function of the ratio of the monomer (2) to the end-blocker (1).

In order to precisely synthesize difunctional siloxanes of varied but controlled molecular weights by this method, an understanding of the mechanisms and ring-chain equilibria must be considered. Jacobson and Stockmayer published their now classic paper (14) in 1950 describing the theory of molecular weight

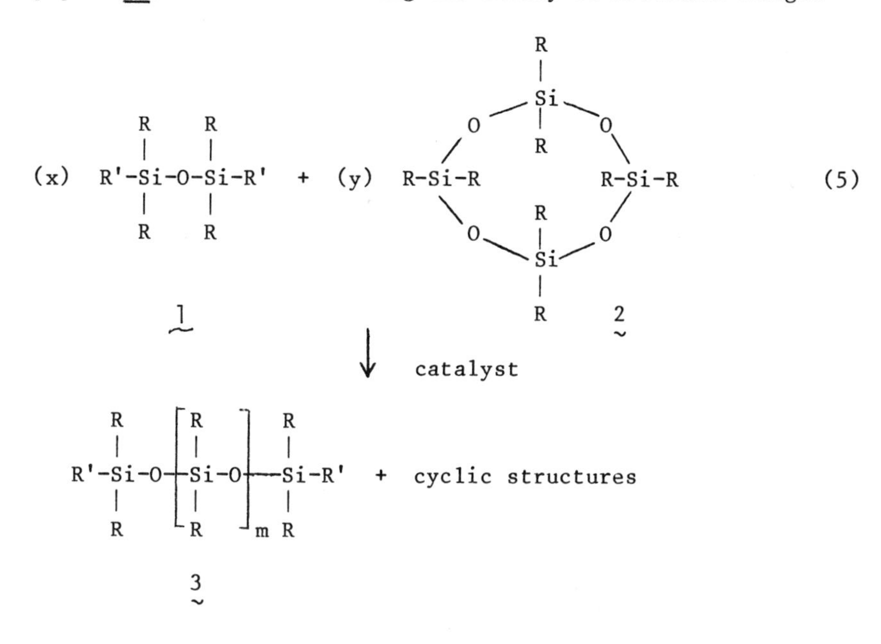

distributions in linear/cyclic equilibrated polymer systems. Up to that time, theories describing molecular weight distributions had not considered the presence of cyclic structures. Jacobson and Stockmayer's theory is based on the premise that the proportion of a macrocyclic species of units in equilibrium with linear components is related to the probability of coincidence of the terminal atoms of the sequence of x units. In development of this theory, the following processes were considered:

Process 1: $-M_y- \rightleftharpoons -M_{y-x}- + -M_x-$

Process 2: $-M_x- \rightleftharpoons C-M_x$

M = monomer unit
C-M = cyclic form of the monomer unit
x and y = degrees of polymerization

Process 3 denotes the sum of processes 1 and 2.

Process 3: $-M_y- \rightleftharpoons -M_{x-y}^- + C-M_x$

An equilibrium constant, K_x, for ring formation (for process 3) was derived from entropy relationships and is shown in equations 6, 7, and 8 for processes 1, 2, and 3, respectively. The parameter "P" in these equations was described by Jacobson and Stockmayer as the probability that a chain would be in an appropriate configuration for ring formation (equation 9). Jacobson and Stockmayer actually set $\langle \vec{r}_x^2 \rangle$ equal to $(\nu x b^2)$ which meant implicitly that $\langle \vec{r}_x^2 \rangle$ was to be set equal to $\langle \vec{r}_x^2 \rangle_o$, the mean-square end-to-end distance of the unperturbed chain. The term "$W_x(\vec{r}=o)(\vec{r})$" in equation 9 denotes the Gaussian distribution function expressing the density of end-to-end vectors, r, in the vicinity of $\vec{r} = 0$. The existence of this function in the entropy term

$$\Delta s_{(1)} = k \ln \left[\frac{V}{\sigma_A v_s} \right] \tag{6}$$

$$\Delta S_{(2)} = k \ln \left[P \frac{\sigma_A}{\sigma_{R_x}} \right] \tag{7}$$

$$\Delta S_{(3)} = k \ln \left[\frac{VP}{\sigma_{R_x} v_s} \right] \tag{8}$$

V = Volume of the system.

σ_A = Symmetry number for the chain species (σ_A = 2 for organosiloxanes).

σ_{R_x} = Symmetry number for the ring species of "x" repeating units (σ_{R_x} = 2x for organosiloxanes).

$$P = \int_0^{v_S} W_x(\vec{r})\, dr \cong \left[\frac{\beta^3}{\pi^{3/2}}\right] v_s = \left[\frac{3}{2\pi\nu x}\right]^{3/2} \frac{v_s}{b^3} \qquad (9)$$

$$\nu x b^2 = \langle \vec{r}_x^{\,2} \rangle_o = \langle \vec{r}_x^{\,2} \rangle$$

ν = Number of links per repeat unit.

b = Average "effective link length"

$\langle \vec{r}_x^{\,2} \rangle$ = The mean-square end-to-end length averaged over all configurations of chain size x.

obviously implies a Gaussian distribution of end-to-end vectors. This point may, in fact, be a poor assumption for the cases of very short chains (which, in turn, form small rings). The term "v_s" was defined as the volume element within which two termini must meet in order to form a bond. It appears in the denominator of Equation 6 due to the fact that the two atoms which formed the bond broken in process 1 were constrained to the volume element v_s prior to the occurrence of process 1. Since $_0\int^{v_s} W_x(\vec{r})$ is the probability of the termini of a molecule meeting in the volume range, v_s, this factor also appears in the entropy term describing process 2. The enthalpy terms for the first two processes presumably cancel each other with the assumption that the ring formed is not small enough to be sterically strained. This was rationalized by noting that the intramolecular bond formed in process 2 was similar in nature to the one severed in process 1. The equilibrium constant for process 3 (expressed in moles/liter) was derived from equation 8 (equations 10 and 11). The definition of "p" in this context (equation 11) differs

$$K_x = \frac{\left(\frac{3}{2\pi\nu}\right)^{3/2}}{2b^3} N_A^{-1} x^{-5/2} = \frac{\left(\frac{3}{2\pi\nu}\right)^{3/2}}{2\langle \vec{r}_x^{\,2} \rangle_o^{3/2} N_A} x^{-1} \qquad (10)$$

$$\langle \vec{r}_x^{\,2} \rangle_o = \nu x b^2$$

$$K_x = \frac{[C\text{-}M_x]\,[-M_{y-x}-]}{[-M_y-]} = \frac{[C\text{-}M_x]}{p^x} \qquad (11)$$

slightly from its "normal" definition (i.e. fractional extent of reaction). Here, "p" is defined as the extent of reaction of the functional endgroups in the chain portion of the system. Moreover, "p" could also be identified as the ratio of the concentrations of acyclic species of sizes x to x-1 (12).

The combination of equations 10 and 11 predicted several interesting points:

1. The concentration of x-mer rings was shown not to be a function of dilution. The number of x-mer rings was predicted to increase linearly with dilution, thereby making the proportion of rings in the system greater as it is diluted. Upon incrementally adding solvent, this process eventually results in a "critical dilution point" above which only rings are present.
2. The rings formed are small species. As $p \rightarrow 1$, $K_x \rightarrow [C\text{-}M_x]$ and $[C\text{-}M_x]$ approaches proportionality to $x^{-5/2}$.

Jacobson and Stockmayer also reasoned that each subset of species (e.g. rings vs. chains) within the overall distributions must be in equilibrium with itself at thermodynamic equilibrium. Therefore, they predicted the normal distribution for the chain species whether or not ring species were formed. Figure 1 (reproduced from their 1950 paper) illustrates the predicted distribution for ring and chain species.

One problem of Jacobson and Stockmayer's interpretation is observed in the cases of very small but unstrained rings, where much higher concentrations of rings were formed than were predicted. Flory and Semlyen (15) explained this deviation by suggesting that not only did two termini have to meet within a volume "v_s" in order to establish a bond, they also had to approach each other from a specified direction. This direction was specified by a solid angle fraction $\delta\omega/4\pi$. They explained that this term should appear in the entropy expression for process 1 in the inverse form, i.e. $4\pi/\delta\omega$. In process 2, if the chains were sufficiently long, there would be no correlation between the probability for two termini of a molecule to meet within v_s and to approach within the solid angle $\delta\omega$. In this case, the term $\delta\omega/4\pi$ would be valid for inclusion into the entropy term for process 2 and, hence, when the entropies for processes 1 and 2 were summed, these terms would cancel (and the equation for $\Delta S_{(3)}$ from Jacobson and Stockmayer's theory should be valid). However, for short chains, the probability of approach

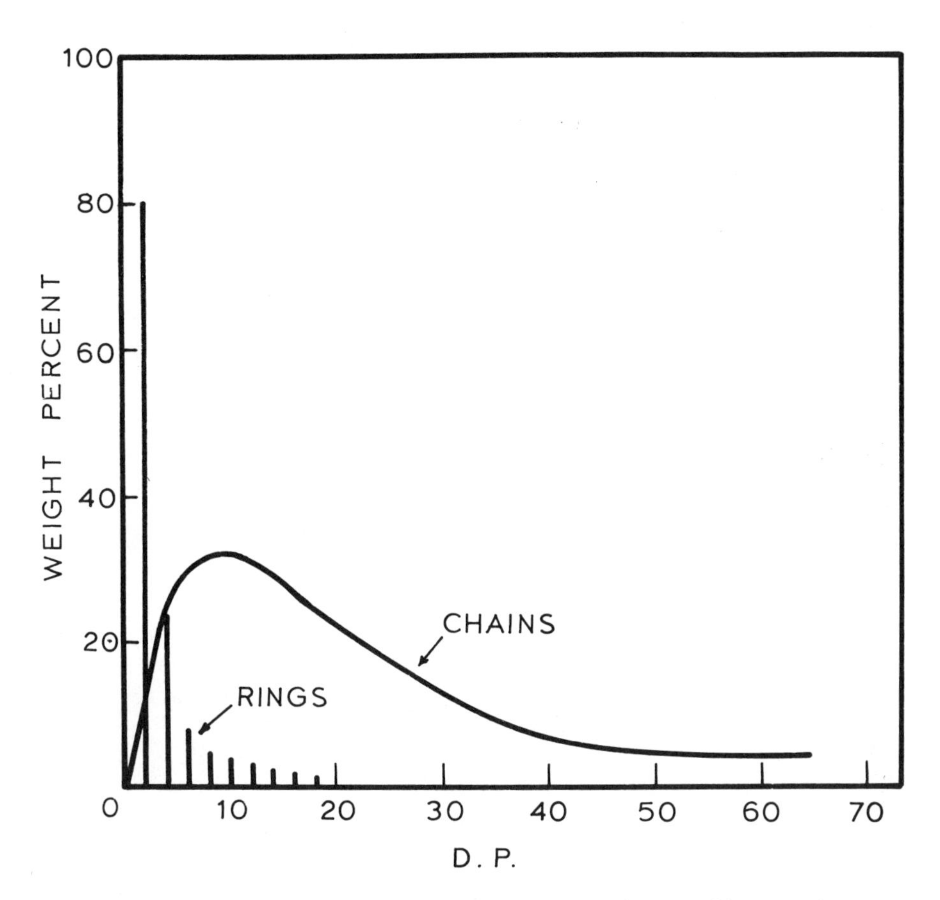

Figure 1. Typical molecular distribution for a ring-chain equilibrium polymer (14).

of two termini of one chain within v_s would depend on bond angles, steric factors, etc., particular to the specific system, and would deviate from $\delta\omega/4\pi$. In that case the probability "P" in Jacobson and Stockmayer's theory in the entropy term would be incorrect.

There have been several studies conducted investigating equilibrium distributions of polyorganosiloxanes experimentally. Theoretical and empirical weight fractions of cyclics and their distributions with variances in x are of particular interest to the synthetic polymer chemist. Equation 10 can be rearranged to give the weight fraction, w_r, of cyclics (equation 12) for high extents of reaction (the empirical comparisons were all made

$$w_r = \left(\frac{3}{\pi}\right)^{3/2} \frac{M_o}{2^4 \ell^3 N_A c} \sum_{x=4}^{\infty} (x\, C_x)^{-3/2} \qquad (12)$$

c = Total siloxane concentration in g/ℓ

M_o = Molecular weight of a repeat unit

ℓ = The length of a siloxane bond.

$\langle \vec{r}_x^{\,2} \rangle_o = C_x 2x\ell^2$

at $p \rightarrow 1$). Contributions from the weight fraction of the strained cyclic trimer which is not applicable to the theory are neglected. Wright and Semlyen compared their own (17) values together with Brown and Slusarczuk's (18) calculated and experimental values of K_x for polydimethylsiloxane (PDMS). Calculated values were based on calculations of $\langle \vec{r}_x \rangle_o$ derived from Flory's rotational isomeric state model for PDMS (23). Empirical values came from measurements of the concentrations of x-meric rings according to Jacobson's and Stockmayer's relationship, $K_x = [C\text{-}M_x]/p^x$. Brown and Slusarczuk (18) equilibrated PDMS in toluene at 110 degrees centigrade and a concentration of 0.22 g/mℓ of siloxane. Values from that equilibration and Wright's and Semlyen's bulk equilibration (17) as a function of x are compared with theoretical values in Figure 2. As predicted, the K_x values in the range x = 11-40 were experimentally independent of dilution. In direct contrast, the cyclization constants for x = 4-10 increase with dilution (the increase becoming more pronounced with decreases in x). Siloxane chains for x greater than approximately 15 agreed well with theoretical values.

These same authors (16) also compared experimental K_x values as a function of x for a series of bulk equilibrates of the structure $-\!\!\left[R(CH_3)Si\text{-}O\right]\!\!-_x$ wherein R equalled H, CH_3, CH_3CH_2, $CH_3CH_2CH_2$, and $CF_3CH_2CH_2$ in order to assess the effect of the substituent size on the equilibrium distribution. The K_x values

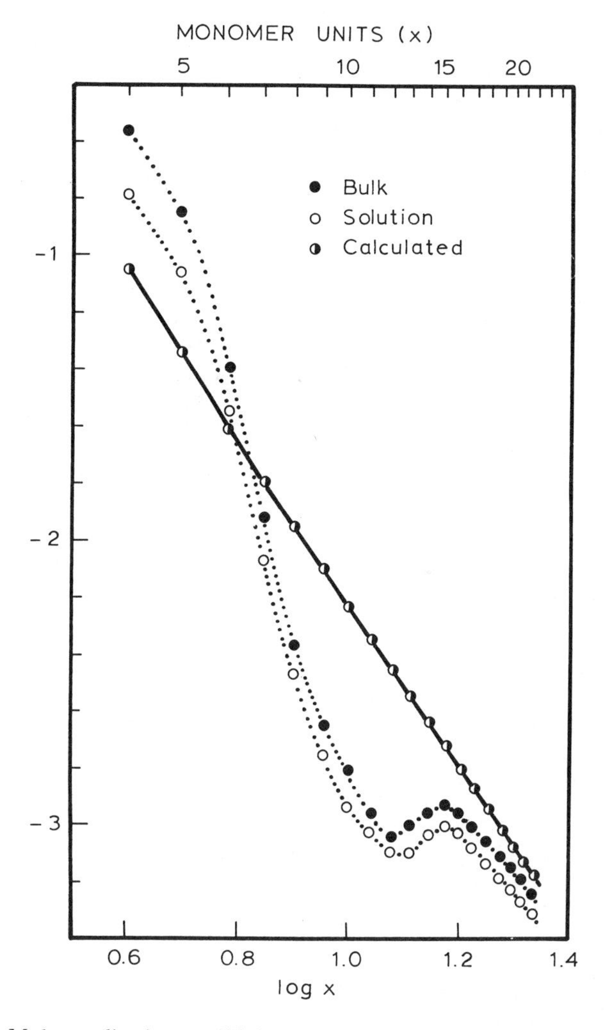

Figure 2. Molar cyclization equilibrium constants of dimethylsiloxane at 110 °C (17).

for the smallest unstrained rings (x=4 or 5) were found to increase along the series $R = H<CH_3<CH_3CH_2<CH_3CH_2CH_2<CF_3CH_2CH_2$. By contrast, the corresponding values for the large cyclics decrease along the same series. The total weight fraction (experimental) of cyclics in these bulk equilibrates are listed in Table 4. The effect of dilution with cyclohexanone in the syntheses of these same polymers is illustrated in Figure 3. In all cases, a critical dilution point was reached as predicted but it was reached at a lower diluent volume than the theoretical value. Molar cyclization equilibrium constants for undiluted poly(methylphenyl)siloxane were later measured (19) and found to be similar to those previously obtained in the case of poly(ethylmethyl)siloxanes.

As predicted by theory, the position of the ring/chain equilibrium was found to be independent of the nature of the redistribution catalyst employed (acid or base) (4,13,24-25) and of the specific inert solvent used (26). Russian authors (4,27) equilibrated mixtures of cyclosiloxanes comprised of dimethylsiloxane (75 mole %) and either trifluoropropylmethyl, cyanoethylmethyl, or cyanopropylmethyl siloxane (25 mole %) in acetone at a siloxane repeating unit concentration of 0.833 moles/ℓ. They measured the dipole moments of the respective cyclosiloxanes, $[(CH_3)_2SiO]_3[Si(CH_3)RO]$, to be 2.76 for R = trifluoropropyl, 3.45 for R = cyanoethyl, and 3.58 in the case of R = cyanopropyl. The equilibrium weight fraction of rings, 54.7%, 96.2%, and ~100% respectively, was found to increase with the dipole moment, and, hence, with the polarity of the substituent.

Carmichael et al. performed a series of bulk equilibrations of hexamethyldisiloxane and D_4 using sulfuric acid activated fuller's earth as the catalyst (28-29). These workers varied the molar ratio of endblocker to D_4 so as to obtain calculated molecular weights (on the basis of monomer ratio) of 459, 904, and 1348 g/mole. They found the total weight fractions of cyclics in these polymers to be 4.86%, 7.69%, and 8.92% respectively whereas they had obtained a corresponding value of 12.8% for analogous polymer with $<M_n> = 10^6$ (24). They explained this molecular weight dependence by variances in "p" according to Jacobson's and Stockmayer's equation for cyclization equilibrium constants, $[C\text{-}M_x] = K_x p^x$, by assuming the equilibrium distribution of linear x-mers to be random (i.e. $[C_x] = Ap^x$ wherein $[C_x]$ = concentration of linear x-mer (moles/ℓ), A = a normalization constant, and p = fraction of unreacted endgroups in the linear portion of the polymer). Carmichael at al. derived "p" for their polymers from the slopes of plots of ln $[C_x]$ vs. x. They used these values together with measured concentrations of cyclics as x was varied to calculate K_x values. In all cases their values of K_x derived as explained above agreed well with values of K_x calculated for a high molecular weight polymer.

TABLE 4.

Cyclic Contents of Undiluted Polysiloxane Equilibrates

Substituent Group R of Monomeric Unit $(R(CH_3)SiO)_4$	Temperature T(K)	Weight Percent Cyclics in High Molecular Weight Equilibrates (p = 1)			
		x = 3-5	x = 6-18	$x = 19-\infty^{+}$	Total
H	273	4.5	3.4	4.6	12.5
CH_3	383	10.0	3.6	4.7	18.3
CH_3CH_2	383	17.0	4.9	3.9	25.8
$CH_3CH_2CH_2$	383	27.0	*	-	
$CF_3CH_2CH_2$	383	71.1	8.9	2.7	82.7

+ These values have been computed by extrapolation of experimental K_{18} values assuming that K_x is proportional to $x^{-2.5}$.

* 31.0% for $[CH_3CH_2CH_2(CH_3)SiO]_x$ with x = 3-8.

Source: Reproduced with permission from Ref. 16.

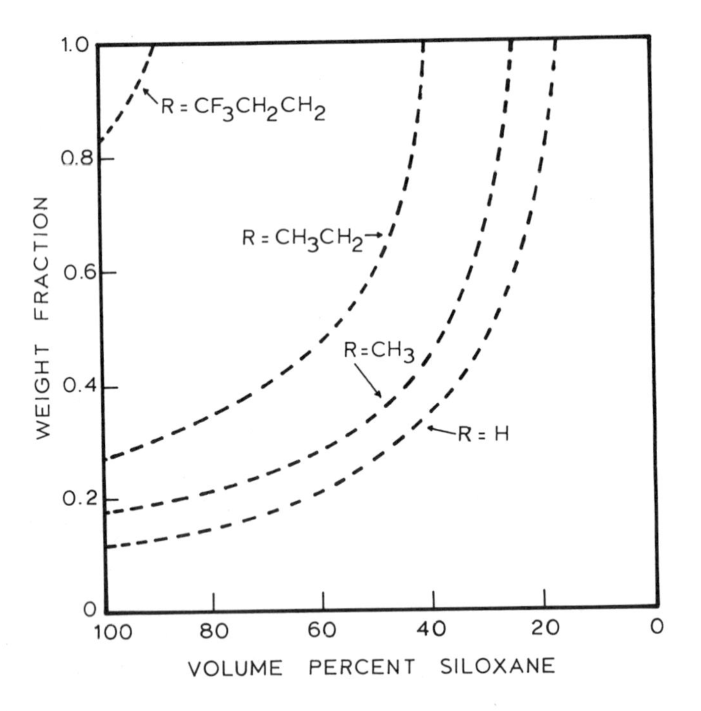

Figure 3. Weight fractions of cyclics $[R(CH_3)SiO]_x$ in high molecular weight ($p \cong 1$) polysiloxane equilibrated at 383 K ($R = CH_3$, CH_3CH_2, $CF_3CH_2CH_2$) and 273 K ($R = H$) as a function of the volume % siloxane in cyclohexane ($R = CF_3CH_2CH_2$) and toluene ($R = H$, CH_3, CH_3CH_2) (16).

The mechanism illustrated in equations 13-15 was proposed by Grubb and Osthoff (30) for the anionic ring opening polymerization. There is, however, considerable question as to the nature of the ion pair formed in equations 13 and 14.

$$-\overset{|}{\underset{|}{Si}}-O-\overset{|}{\underset{|}{Si}}- \; + \; KOH \longrightarrow -\overset{|}{\underset{|}{Si}}-O^{-}\;{}^{+}K \; + \; HO-\overset{|}{\underset{|}{Si}}- \qquad (13)$$

$$-\overset{|}{\underset{|}{Si}}-O^{-}\;{}^{+}K \rightleftharpoons -\overset{|}{\underset{|}{Si}}-O^{-} \; + \; K^{+} \qquad (14)$$

$$-\overset{|}{\underset{|}{Si}}-O^{-} \; + \; \text{(cyclic Si–O–Si)} \; \longrightarrow \; -\overset{|}{\underset{|}{Si}}-O-\overset{|}{\underset{|}{Si}}\sim\sim\sim\overset{|}{\underset{|}{Si}}-O^{-} \qquad (15)$$

These anionic ring opening polymerizations are usually carried out either in bulk or in solution. A host of catalyst types are active. For synthetic references using specific catalysts, the reader is referred to several excellent sources (4,7,31,32). Representative catalysts include hydroxides, alcoholates, phenolates, silanolates, siloxanolates, mercaptides of the alkali metals, organolithium and potassium compounds, and quaternary ammonium and phosphonium bases and their silanolates and siloxanolates. Some physical characteristics of linear oligomers are given in Table 5 (10).

The activities of the often used hydroxides and siloxanolates decreases in the order Cs > Rb > K > Li (33). At equal molar concentrations, the rates using the hydroxide or siloxanolate of the same metal have been found to be similar (30). Tetramethylammonium siloxanolates exhibit activities close to the cesium siloxanolates (33). Tetramethylammonium hydroxide, silanolate, and siloxanolate rapidly decompose above 130°C yielding methanol, methoxy trimethylsilane, or methoxysiloxane respectively and trimethylamine (34-35). In the cases of the first two compounds listed above, the catalyst breakdown products are fugitive at the decomposition temperature. Thus, the usual need for catalyst neutralization and removal following polymerization is eliminated. They are often termed "transient" catalysts (34).

At low catalyst concentrations and in the absence of "end-blockers", the degree of polymerization is approximately

TABLE 5. SOME PHYSICAL CHARACTERISTICS OF LINEAR OLIGOMERS

Symbol	Melting point, °C	Boiling point, °C	Density, d^{20}	Refractive index, n_D^{20}
MM	-67	99.5	0.7636	1.3774
MDM	-80	153	0.8200	1.3840
MD_2M	-76	194	0.8536	1.3895
MD_3M	-80	229	0.8755	1.3925
MD_4M	-59	245	0.8910	1.3948
MD_5M	-78	270	0.9012	1.3965
MD_6M	-63	290	0.9099	1.3970
MD_7M		307.5	0.9180	1.3980

Source: Reproduced with permission from Ref. 10.

inversely related to the catalyst concentration (36). Russian authors (4) have noted a deviation in the linearity of the $1/X_n$ vs. catalyst concentration function at high catalyst loadings. It was suggested that this deviation could be attributed to the presence of stable associated structures similar such as 4.

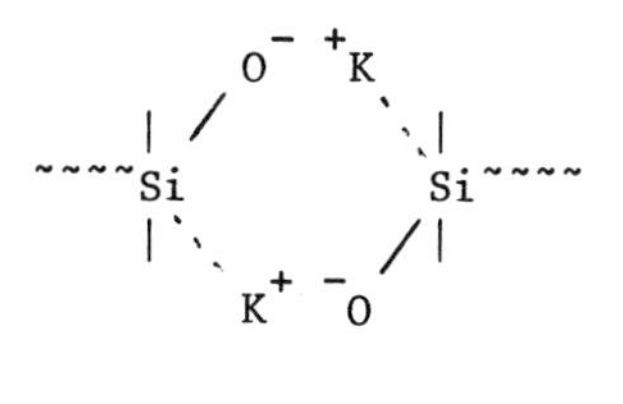

4

Practically, polymerization temperatures are selected on the basis of the activities of the catalyst, cyclosiloxane, and endblocker used with the aim of arriving at thermodynamic equilibrium within an acceptable time period. In the bulk polymerization of D_4, modest temperature changes reportedly do not affect the final equilibrium number average molecular weight of the polymers (37).

Rates of anionic polymerization are influenced by the number of siloxane units present in the monomer rings. Some characteristics are given in Tables 6 and 7 (10). Due to ring strain in the three unit rings, all of the cyclotri- siloxanes polymerize faster than the cyclotetrasiloxanes. In the dimethylsiloxanes, D_3 reportedly polymerizes approximately 50 times faster than D_4 (4).

Several investigations have been performed concerning the relative reactivity of cyclic tetramers substituted with varying amounts of phenyl and methyl groups (38-42). Andrianov et al. (42) studied anionic copolymerizations of D_4 with varying amounts of octaphenylcyclotetrasiloxane (10-70 mole percent). They found that the rate of copolymer formation, the viscosity of the resulting copolymers, and the equilibrium yield of linear species all decreased regularly as the mole percent of phenyl tetramer was increased in the reaction mixture. They also noted that in the early stages of conversion, although both monomers had become incorporated in the linear portion to some extent, the polymers formed were enriched with diphenyl units. On the basis of electronic factors, it was reasoned that in the diphenyl substituted tetramer, the silicon atoms would be more susceptible to nucleophilic attack. Conversely, the phenyl substituted siloxanolate anion, once formed, would be less reactive than the methyl substituted analog. These same authors (42) also studied the structure of the cyclics as equilibration proceeded. Redistribution type steps had evidently occurred even in the

TABLE 6. INFLUENCE OF RING SIZE ON THE CHARACTERISTICS OF CYCLIC SILOXANES

Symbol	Melting point, °C	Boiling point, °C	Density, d^{20}	Refractive index, n_D^{20}
D_3	64.5	134		
D_4	17.5	175.8	0.9561	1.3968
D_5	-44	210	0.9593	1.3982
D_6	-3	245	0.9672	1.4015
D_7	-32	154[a]	0.9730	1.4040
D_8	31.5	290	1.1770[b]	1.4060

[a] At 20mm Hg. [b] Crystals

Source: Reproduced with permission from Ref. 10.

TABLE 7. SOME PHYSICAL PROPERTIES OF CYCLIC SILOXANES

Formula	Boiling point, °C, (mm Hg)	Density d^{20}, gm/cc	Refractive index, n_D^{20}	Melting point, °C
$(Ph_2SiO)_4$	335_1			200
$(CH_3PhSiO)_4$	$237_{1.5}$	1.183	1.5461	99
$[(CF_3CH_2CH_2)CH_3SiO]_4$	134_3	1.255	1.3724	
$[(CH_2{=}CH)(CH_3)SiO]_4$	111_{10}	0.9875	1.4342	-43.5
$(CH_3HSiO)_4$	134	0.9912	1.3870	-69
$[(CH_3)_2SiO]_4$	175.8	0.9561	1.3968	17.5

Source: Reproduced with permission from Ref. 10.

initial stages since mixed cyclics (diphenyl and dimethyl units in the same cyclic molecules) were observed soon after the beginning of the reaction.

In equilibration reactions, hexaorganyl disiloxane or low molecular weight siloxanes terminated with triorganylsiloxy groups in the reaction mixture can regulate molecular weight by acting as chain transfer agents or "end-blockers". The efficiency of these agents depends on the relative amount of positive charge on their silicon atoms. Results of viscosity measurements as a function of time using hexamethyldisiloxane endblocker, D_4, and tetramethylammonium hydroxide catalyst (13) are shown in Figure 4. The initial large viscosity increase was found to be a result of the faster reaction of D_4 as compared to hexamethyldisiloxane under these conditions. Contrasted with the above results was the analogous reaction with the exception that a sulfuric acid catalyst instead of tetramethylammonium hydroxide (see Figure 5) was used (13). The shape of the curve in Figure 5 represents approximately equal reactivity of D_4 and hexamethyldisiloxane towards an acid catalyst.

Many compounds with an electron donor character are reported to have an accelerating effect on the anionic polymerizations. No doubt, these alter the nature of the ion pair. Representative "promoters" reported include tetrahydrofuran (43-45), dimethylformamide (46), sulfoxides (47,48), and cryptates (60).

Cationic polymerization of cyclosiloxanes is well known but used much less frequently than anionic reactions. The most frequently used catalysts include sulfuric acid and its derivatives (4,49-52). Trifluoroacetic acid has also been used to polymerize D_4 in bulk (53,61).

The mechanism of acid catalyzed polymerization is postulated to be that schematically illustrated in equations 16 and 17. However, according to the recent text by Odian (54), there is no evidence for the rearrangement of the siliconium ion shown in these equations. Alternatively, one could propose that the propagating species is the tertiary oxonium ion.

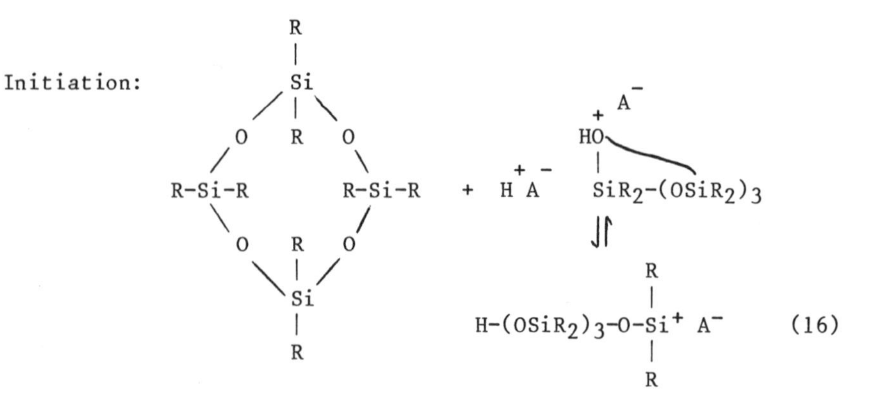

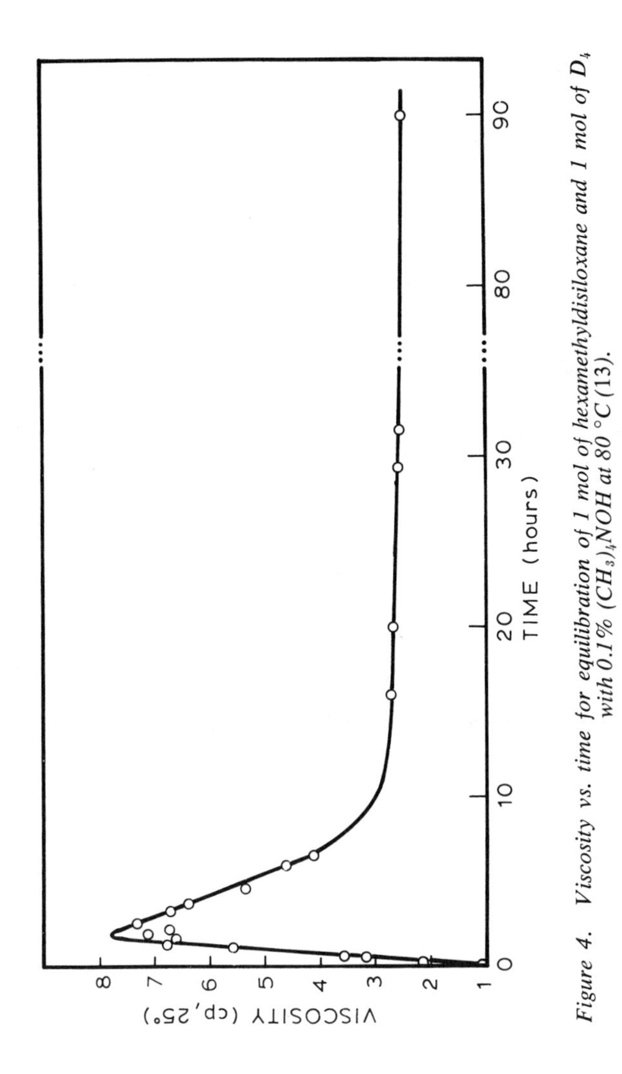

Figure 4. Viscosity vs. time for equilibration of 1 mol of hexamethyldisiloxane and 1 mol of D_4 with 0.1% $(CH_3)_4NOH$ at 80 °C (13).

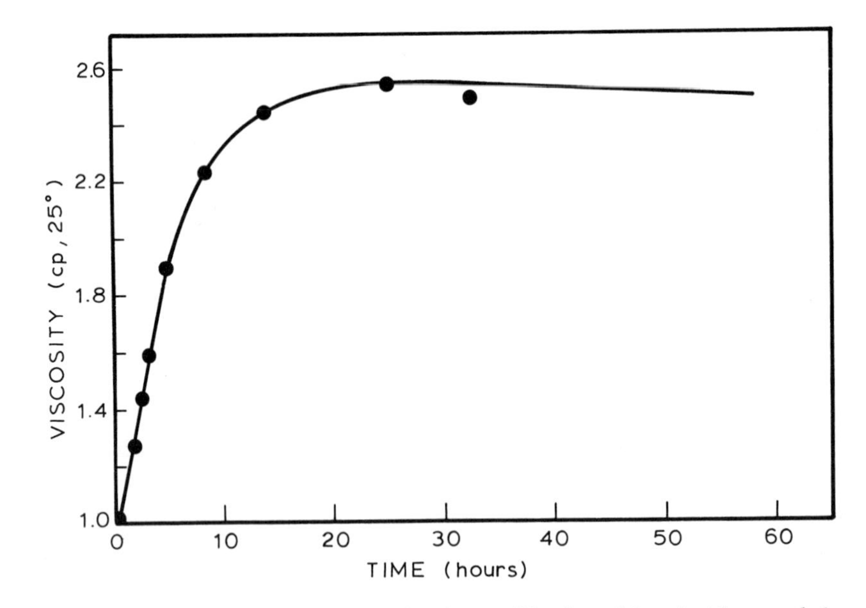

Figure 5. Viscosity vs. time of reaction for equilibration of 1 mol of hexamethyldisiloxane and 1 mol of D_4 with 4% H_2SO_4 at room temperature (13).

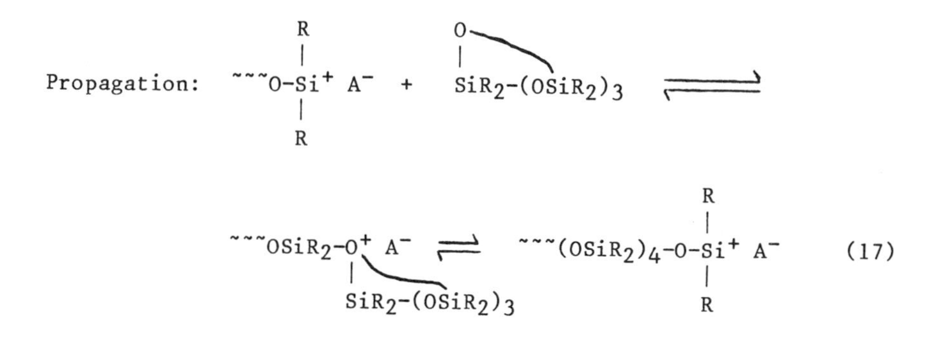

Our own research (55-58, 62-67) has recently been concerned with the polymerization of D_4 and of combinations of D_4 and D''_4 (octaphenylcyclotetra- siloxane) in the presence of organofunctional endblockers used as molecular weight regulators. We have successfully obtained dimethyl amino (62), aminopropyl, glycidoxypropyl, carboxypropyl, hydroxybutyl, and hydroxyphenyl-propyl terminated siloxane oligomers of controlled molecular weights by these methods.

Transient siloxanolate anionic catalysts prepared by reacting four moles of D-4 with one of tetramethyl ammonium hydroxide at 80°C are effective for equilibrating "neutral" systems such as the epoxy (59), "basic" dimethyl-amino (64) or aminopropyl (59,67) end-blockers and D-4. With "acidic" functionality on the end-blocker, we have successfully utilized trifluoroacetic acid for the equilibrations. Further details of the oligomer synthesis and their utilization in segmented copolymers will be described in future publications.

Literature Cited

1. R. West and T. J. Barton, J. Chem. Ed., 1980, 57 (3), 165.
2. C. Friedel and J. M. Crafts, Ann., 1965, 136, 203.
3. F. S. Kipping, Proc. Roy. Soc., A, 1937, 159, 139).
4. M. G. Voronkov, V. P. Mileshkevich, and Y. A. Yuzhelevskii, The Siloxane Bond, Plenum Press, N. Y., 1978, 2.
5. A. Ladenburg, Ann. Chem., 1971, 159, 259.
6. C. Friedel and J. M. Crafts, Ann. Chim. Phys., 1970, 19 (5), 334.
7. W. Noll, Chemistry and Technology of Silicones, Academic Press, N.Y., (1968).
8. B. B. Hardman and R. W. Shade, Mat. Technol, 26, Spring, 1980.
9. R. J. H. Voorhoeve, Organohalosilanes: Precursors to Silicones, Elsevier, N.Y., (1967).
10. R. Meals, Encyclopedia of Chemical Technology, 18, 2nd edition, 1969, 221-260.

11. C. Eaborn, Organosilicon Compounds, Butterworths Scientific Publications, London, (1960).
12. B. C. Arkles and W. R. Peterson, Jr., ed., Silicon Compounds, Register and Review, Petrarch Systems, Levittown, Pa., (1979).
13. S. W. Kantor, W. T. Grubb, and R. C. Osthoff, J. Amer. Chem. Soc., 1954, 76, 5190.
14. H. Jacobson and W. H. Stockmayer, J. Phys. Chem., 1950, 18, 1600.
15. P. J. Flory and J. A. Semlyen, J. Amer. Chem. Soc., 2966, 88, 3209.
16. P. V. Wright and J. A. Semlyen, Polymer, 11 (9), 1970, 462.
17. P. V. Wright and J. A. Semlyen, Polymer, 10, 1969, 543.
18. J. F. Brown and G. M. Slusarczuk, J. Amer. Chem. Soc., 1965, 87, 931.
19. M. S. Beevers and J. A. Semlyen, Polymer, 1971, 12 (6), 373.
20. T. C. Kendrick, J. Polym. Sci., 1969, A-2, 7, 297.
21. W. C. Davies and D. P. Jones, Polym. Prepr., 1970, 11, 447
22. J. C. Saam, D. J. Gordon, and S. Lindsey, Macromolecules, 1970, 3, 4.
23. P. J. Flory, V. Crescenzi, and J. E. Mark, J. Amer. Chem. Soc., 1964, 86, 146.
24. J. B. Carmichael and R. Winger, J. Polym. Sci., A, 1965, 971.
25. J. B. Carmichael, Rubber Chem. Technol., 1964, 40, 1084.
26. J. B. Carmichael and D. J. Gordon, J. Phys. Chem., 1967, 71, 2071.
27. Y. A. Yuzhelevskii, E. G. Kagan, and E. B. Dmokhovskaya, Khim. Geterotsikl. Soedin., 1967, 951.
28. J. B. Carmichael and J. Heffel, J. Phys. Chem., 1965, 2218.
29. J. B. Carmichael, J. Macromol. Chem., 1 (2), 1966, 207.
30. W. T. Grubb and R. C. Osthoff, J. Amer. Chem. Soc., 1955, 77, 1405.
31. V. Bazant, V. Chvalovsky', J. Rathousky, Organosilicon Compounds, 1, Academic Press, N.Y., 15 (1965).
32. K. A. Andrianov, Metalorganic Polymers, Interscience, N.Y., (1965).
33. D. T. Hurd and R. C. Osthoff, J. Amer. Chem. Soc., 1954, 76, 249.
34. A. K. Gilbert and S. W. Kantor, J. Polym. Sci., 1959, 40, 35.
35. A. Noshay, M. Matzner, and T. C. Williams, Industrial and Engineering Product Research and Development (I. & E. C. Product Research and Development), 1973, 12, 268.
36. C. L. Lee and O. K. Johanson, J. Polym. Sci., 1966, A-1 (4), 3013.
37. M. Kucera, J. Polym. Sci., 1962, 58, 1263.
38. K. A. Andrianov and S. E. Iakushkina, Polym. Sci, U.S.S.R., 1960, 1, 221-228.

39. Z. Laita and M. Jelinek, Polym. Sci., U.S.S.R., 1964, 5, 342-353.
40. K. A. Andrianov, S. Ye. Yakushkina, and L. N. Guniava, Vysokomol. soyed., 1966, 8 (12), 2166-2170.
41. K. A. Andrianov et al., Vysokomol. soyed., 1970, A14(6), 1268-1276.
42. K. A. Andrianov et al., Vysokomol. soyed., 1972, A14(5), 1294-1302.
43. B. Suryanarayanan, B. W. Peace, and K. G. Mayhan, J. Polym. Sci., Polym. Chem. Ed., 1974, 12, 1089.
44. B. Suryanarayanan, B. W. Peace, and K. G. Mayhan, J. Polym. Sci., Polym. Chem. Ed., 1974, 12, 1109.
45. W. A. Fessler and P. C. Juliano, Polym. Prepr., 1971, 12, 151.
46. J. B. Gangi and J. A. Bettelheim, J. Polym. Sci., 1964, A-2, 4011.
47. J. G. Murray, Polym. Prepr., 1965, 6, 163.
48. G. D. Cooper and J. R. Elliott, J. Polym. Sci., A-1 (4).
49. B. Kanner, B. Prokai (Union Carbide Corp.) Ger. Offen. 2,629,138 (Cl.C08G 77/78), 13 Jan., 1977, U.S. Appl. 592,129, 30 Jun. 1975.
50. G. Rossmy, R. D. Langenhagen (Goldschmidt), Ger. Offen. 2,714,807, 20, Oct. 1977, Brit. Appl. 76/14,391, 08 Apr. 1976, C. A. 87: 202569z.
51. P. Rosciszeqski, E. Jagielska, K. Bartosiak, Pol. 92,926, 15 Dec. 1977, C. A. 89:75864f.
52. R. E. Moeller (General Electric Co.), Braz. Pedido PI 7,703,261, 20 Feb. 1979, C. A. 90: 188640u.
53. D. T. Hurd, J. Amer. Chem. Soc., 1955, 77, 2998.
54. G. Odian, Principles of Polymerization, 2nd Edition, Wiley, N.Y., 549 (1981).
55. J. S. Riffle, R. G. Freelin, A. K. Banthia, and J. E. McGrath, J. Macromol. Sci. - Chem., 1981, A15(5) 967-998.
56. J. S. Riffle, Ph.D. Thesis, VPI & SU, Blacksburg, Va., Dec. 1980.
57. J. E. McGrath, et al., Proceedings of the International Rubber Conference, Kharagpur, India, 1980.
58. J. E. McGrath, J. S. Riffle, I. Yilgör, A. K. Banthia and P. Sormani, Org. Coatings and Plastics Preprints, 1982, 46, 693.
59. J. S. Riffle, I. Yilgör, A. K. Banthia, G. L. Wilkes, and J. E. McGrath, "Epoxy Resins", R. S. Bauer, Editor, ACS Symp. Vol., in press, 1982.
60. S. Boileau, in "Anionic Polymerization: Kinetics, Mechanism and Synthesis," J. E. McGrath, Editor, ACS Symposium Volume 1982, Series No. 166.
61. L. Wilczek and J. Chojnowski, Macromolecules, 1981, 14 (1), 9.
62. T. C. Ward, D. P. Sheehy, J. S. Riffle, and J. E. McGrath, Macromolecules, 1981, 14 (6), 1791.

63. Anionic Polymerization: Kinetics, Mechanism and Synthesis, J. E. McGrath, Editor, ACS Symposium Volume Series No. 166, (1981).
64. J. E. McGrath, J. S. Riffle, D. W. Dwight, D. C. Webster and T. F. Davidson, Macromolecules, in preparation, 1982.
65. T. F. Davidson, M. S. Thesis, VPI & SU, Blacksburg, Va., 1980.
66. S. Tang, E. Meinecke, J. S. Riffle, and J. E. McGrath, Rubber Chem. and Tech., 1980, 54 (5), 1160.
67. J. E. McGrath, J. S. Riffle, A. K. Banthia, I. Yilgor, and G. L. Wilkes, To be published.

RECEIVED October 15, 1982

14

Cationic Photoinitiation Efficiency

L. R. GATECHAIR
CIBA-GEIGY Corporation, Plastics and Additives Division, Ardsley, NY 10502

S. P. PAPPAS
North Dakota State University, Polymers and Coatings Department,
Fargo, ND 58105

The results of these photochemical studies form guidelines for the choice of sensitizers, onium salts and other additives potentially useful in the cationic curing of coatings. The sensitized photochemistry of diphenyliodonium hexafluoroarsenate and triphenylsulfonium hexaflurorarsenate was investigated at 366 nm. Product quantum yields are compared to relative rates of photoinitiated cationic polymerization of an epoxy resin.

Sensitized photolysis of both salts resulted in the formation of the same major products as was observed in direct irradiation, iodobenzene, or phenylsulfide and acid. Quantum yields of organic and acidic products were measured using nine sensitizers. Product quantum yields were relatively high (< 1) under conditions where electron transfer sensitization was expected to occur. Quantum yields > 2 were obtained in solvents likely to be substrates for hydrogen abstraction, indicating that a chain reaction was involved. Triplet energy sensitization resulted in low product quantum yields.

Many recent publications and patents give evidence of the growing interest in photoinitiated cationic polymerization using onium salts.(1-7) The objective of this study was to better understand the photochemistry of onium salts in order to improve their efficiency as photocuring agents. The results of these studies form guidelines for the choice of sensitizers, onium salts and other additives potentially useful in the cationic curing process. Sensitizers which produce easily oxidizable free radicals resulted in the highest quantum efficiencies for Diphenyliodonium hexafluoroarsenate salts ($\phi > 3$). Electron transfer sensitization

0097-6156/83/0212-0173$06.00/0

was the most efficient process with triphenylsulfonium hexafluoroarsenate ($\Phi \leq 1$).

Since epoxy functional resins contribute many desirable characteristics to coatings such as toughness and abrasion resistance, it would be desirable to have UV curable coatings based on epoxy chemistry. In conventional UV curable coatings, the photoinitiator absorbs light and produces free radical species generally resulting in addition polymerization of acrylate or unsaturated polyester components. Epoxy resins, however, are not curable by free radical polymerization (except epoxyacrylates). In order to cure an epoxy resin based coating, the photoinitiator must produce an acidic species capable of catalyzing the cationic polymerization of the epoxy resin (see Figure 1).

$$\text{Photoinitiator} \xrightarrow{\text{Light } (h\nu)} H^+$$

$$H^+ + \text{Epoxy Resin} \longrightarrow \text{Crosslinked Polymer}$$

Figure 1.

Many photoinitiator systems have been developed which are capable of producing acidic products. Most of these cationic photoinitiators are based on some form of onium salt which (in the absence of light) is stable in the presence of epoxy functional resins. Following irradiation, an acidic catalyst is produced which can initiate the formation of polymer.

Photochemistry of Aryldiazonium Salts. Aryldiazonium salts have long been known to be a class of photoactive chemicals (see Figure 2). When aryldiazonium salts having nucleophilic anions (such as chlorine or bromine) are irradiated, the resulting product is nonacidic. Acidic products may, however, be formed during photolysis in the presence of water.(17) Such chemistry is often useful in the area of diazo copying, more commonly known as blueprints. If the diazonium salt has a nonnucleophilic anion such as PF_6^-, BF_4^-, AsF_6^-, PCl_6^-, the resulting product (i.e., PF_5) is a Lewis acid capable of catalyzing the polymerization of epoxy functional resins. This chemistry was patented by the American Can Company for use in various kinds of coatings.

The class of diazonium salts, in general, have the advantage of very strong absorption curves which extend into the visible region of the spectrum. Thus, some diazonium salts are sensitive to a wide range of wavelengths. They also have disadvantages, in that many of the compounds are highly colored, and are, to some degree, thermally unstable, limiting the package stability of a formulated coating.

Photochemistry of Diaryliodonium Salts. The photochemistry of diaryliodonium salts having nucleophilic anions is illustrated in Figure 3.(19) This photochemistry also results in the formation of nonacidic products. Photolysis in alcohol, however, was reported to form acid.(13)

Crivello and Lam, at the General Electric Company, found that the photolysis of diaryliodium salts having nonnucleophilic anions results in the formation of acidic species capable of polymerizing epoxy resins.(18) Their evidence, based on product analysis, supported the formation of protons as the initiating species, rather than the formation of a Lewis acid as in the photochemistry of diazonium salts (see Figure 4). The photolysis quantum efficiency (Φ) ranges from 0.2-0.4 for the formation of organic products. (15,18)

Photochemistry of Triarylsulfonium Salts. A similar mechanism was proposed for the photolysis of triarylsulfonium salts having nonnucleophilic anions. The arylsulfonium and to a lesser extent, the aryliodonium salts have improved thermal stability as compared to the diazonium salts, however, their absorption maximum occurs at a much shorter wavelength since the aromatic rings are isolated by the heteroatom, and therefore not conjugated.

Recognizing this deficiency, the workers at General Electric, and separately, Smith at 3M (7), discovered that iodonium and sulfonium salts could be sensitized to longer wavelengths by various classes of aromatic hydrocarbons, aromatic carbonyl compounds, and some classes of dyes.

Sensitization of Onium Salts

In the curing of coatings, the use of photosensitizers offer several advantages over direct excitation of the onium salts. Sensitizers added to onium salts may dramatically increase the polymerization rate of reactive monomers as compared to unsensitized experiments.(5) The choice of sensitizer will largely determine what portion of the available radiation will be utilized in the curing process. Some sensitizers may result in a dramatic increase in cure rate simply because they have greater absorptivity than the onium salt at excitation wavelengths. Alternatively, the absorbed light may be used more efficiently in the sensitized system.

There are many commercially available photosensitizers. The overall cost of coating formulation may be reduced by the use of small amounts of a highly absorbing photosensitizer in combination with a reduced concentration of the onium salt.

In order to choose appropriate sensitizers for onium salts, it is necessary to understand the possible mechanisms by which sensitization can occur.

Energy Transfer Sensitization. Energy transfer is a process where an excited state donor molecule is returned to the ground state, with simultaneous promotion of an acceptor molecule to its

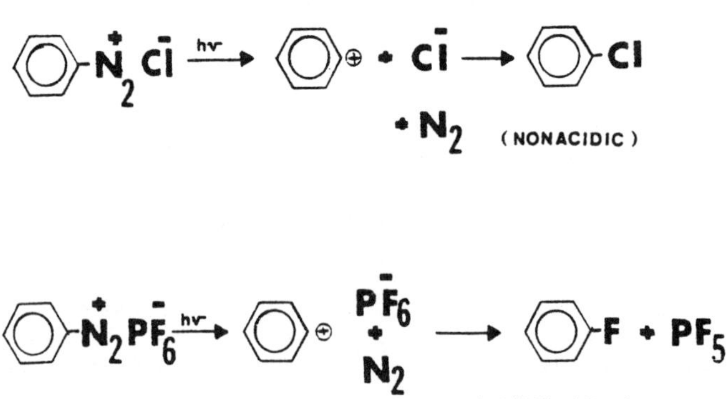

Figure 2. Photolysis of aryldiazonium salts. Photochemistry of these salts is dependent on the nucleophilicity of the anion.

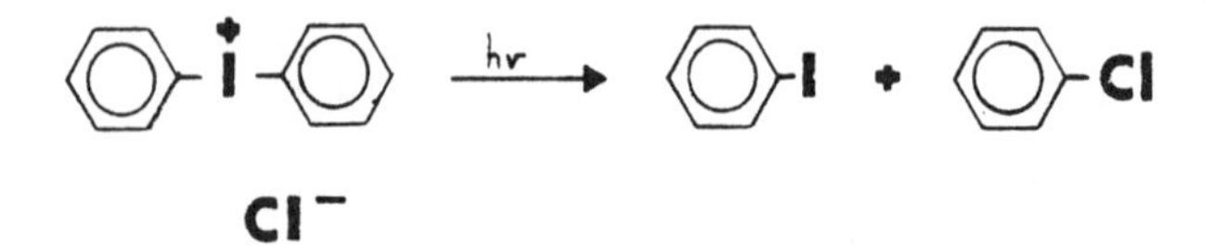

Figure 3. Photolysis of diaryliodonium salts having nucleophilic anions results in nonacidic products.

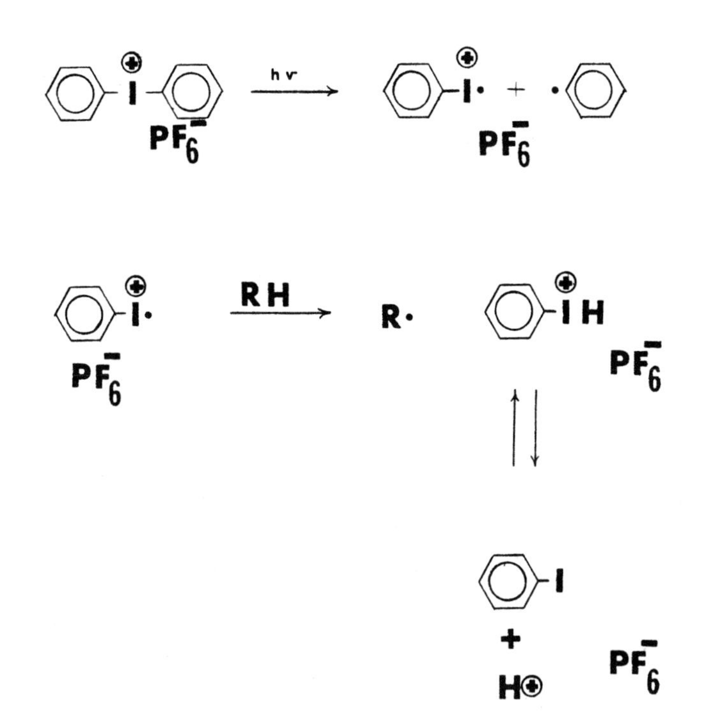

Figure 4. Photolysis of diaryliodonium salts having nonnucleophilic anions results in the formation of strong acids (HPF_6). Nonnucleophilic acids such as HPF_6 can initiate cationic polymerization of vinyl ethers and cyclic ethers.

excited state. This process normally involves triplet state donors due to their longer lifetimes than excited singlet states. Conversion of the acceptor to an excited triplet state results in net spin conservation. Triplet energy transfer would be expected to approach diffusion controlled rates under conditions where the triplet energy of the donor is approximately five kcals/mole greater than the triplet energy of the acceptor.

Acetophenone and meta- and para- cyano substituted acetophenone were chosen as sensitizers for this study (see Tables 1 and 2). Their high triplet energies (74, 73 and 70 kcal/mole, respectively) favor energy transfer, while their high oxidation energies prevent electron transfer. The triplet energies of both salts were estimated to be below 72 kcal/mole.(15) Triplet sensitization experiments consistently result in low quantum yields of both acidic and organic products from both salts. The results of triplet quenching experiments were also consistent with the conclusion that the triplet states of $\phi_3S\ AsF_6$ and $\phi_2I\ AsF_6$ are much less reactive than their respective singlet excited states.(15) Thus, sensitizers which operate only via triplet energy transfer, are not expected to be useful with aryliodonium or arylsulfonium salts.

Sensitization Via Electron Transfer. Pappas and Jilek analyzed the energetics of energy and electron transfer sensitization.(5,8) Their evidence indicated that in most cases, sensitization could be explained by an electron transfer mechanism (see Figure 5). Sensitization is believed to involve the transfer of an electron from an excited photosensitizer (S*) to an onium molecule, which may involve the formation of an excited state complex. Product analysis and irreversible polarography previously determined from electrochemical studies, indicate that the reduced iodonium salt undergoes homolytic bond cleavage to form a phenyl radical and iodobenzene.(9) The reduced sulfonium salt produces phenylsulfide and a phenyl radical.(10)

The photosensitizer radical cation ($S^{\cdot}+$) may abstract a hydrogen atom and regenerate the sensitizer by releasing a proton. In the presence of monomer, $S^{\cdot}+$ may act directly as the initiating species.

An understanding of the theory of electron transfer sensitization is necessary in order to choose sensitizers which can photoreduce the onium salt. For electron transfer to be energetically favorable, the excitation energy of the photosensitizer (E*) must be greater than the net energy required to oxidize the photosensitizer, and reduce the onium salt.(8)

The free energy of this photosensitized electron transfer (ΔG) can be calculated from the oxidation energy (E^{ox}) and excitation energy (E*) of the sensitizer, and the reduction energy of the onium salt (E^{Red}).(12,16) These equations are shown in Figure 6. Figure 7 shows model calculations for sensitization by thioxanthone.

Table 1

Sensitized Photolyses $\phi_2I\ AsF_6 \longrightarrow$ Iodobenzene + H^+

Quantum Yields (Φ), Relative Polymerization Rates (RPR)

and Free Energy of Electron Transfer (ΔG)

Sensitizer	Φ_{IB}	Φ_{H^+}	RPR[e]	ΔG[a]
Anthracene	.56	.6	1.1[c]	-45
Perylene	.28	.3	1.0	-41
Phenothiazine	.42	1.0	0.7[c]	-39
Xanthone	.41	.5	1.0	-27
Thioxanthone	.32	.7	2.3[c]	-22
Acetophenone	.05	.1	0.5	-7
Benzophenone	.01	.2	0.8[c]	-3
Benzophenone + THF (4M)	3.3	8.	3.7[c]	d
Benzophenone + Isopropanol (5M)	2.4	5.	d	d
m-Cyanoacetophenone	.04	.1	.09	+5
p-Cyanoacetophenone	.04	.1	0.5	+9

Notes:

Photolyses in air except as noted. $[\phi_2I\ AsF_6] = 3 \times 10^{-2}$ M.

Error: $\Phi_{IB} = \pm 10\%$, $\Phi_{H^+} = \pm 25\%$

a) Kcal/mole, see text
b) measured in vacuum
c) data from reference 5
d) not determined
e) relative to perylene + $\phi_2I\ AsF_6$

Table 2

Sensitized Photolyses $Ø_3S\ AsF_6 \longrightarrow$ Phenylsulfide + H^+

Quantum Yields (Φ), Relative Polymerization Rates (RPR)

and Free Energy of Electron Transfer (ΔG)

Sensitizer	Φ_{PS}	Φ_{H^+}	RPR[b]	ΔG[a]
Anthracene	.59	1.0	1.0[c]	-22
Perylene	.20	.3	1.0	-18
Phenothiazine	.69	1.0	1.1[c]	-16
Xanthone	<.02	.1	< 0.1	-4
Thioxanthone	<.02	.3	< 0.1[c]	+1
Acetophenone	<.02	.1	< 0.1	+16
Benzophenone	<.02	.3	< 0.1[c]	+20
Benzophenone + THF (4M)	d	d	0.14[c]	d
m-Cyanoacetophenone	<.02	.1	< .1	+28
p-Cyanoacetophenone	<.02	.2	< .1	+32

Notes:

Photolyses in air except as noted. $[Ø_3S\ AsF_6] = 3 \times 10^{-2}$ M.
Excitation wavelength = 366 nm.

Error: $\Phi_{PS} = \pm 10\%$, $\Phi_{H^+} = \pm 25\%$

a) Kcal/mole, see text
b) relative to perylene + $Ø_3S\ AsF_6$
c) data from reference 5
d) not determined

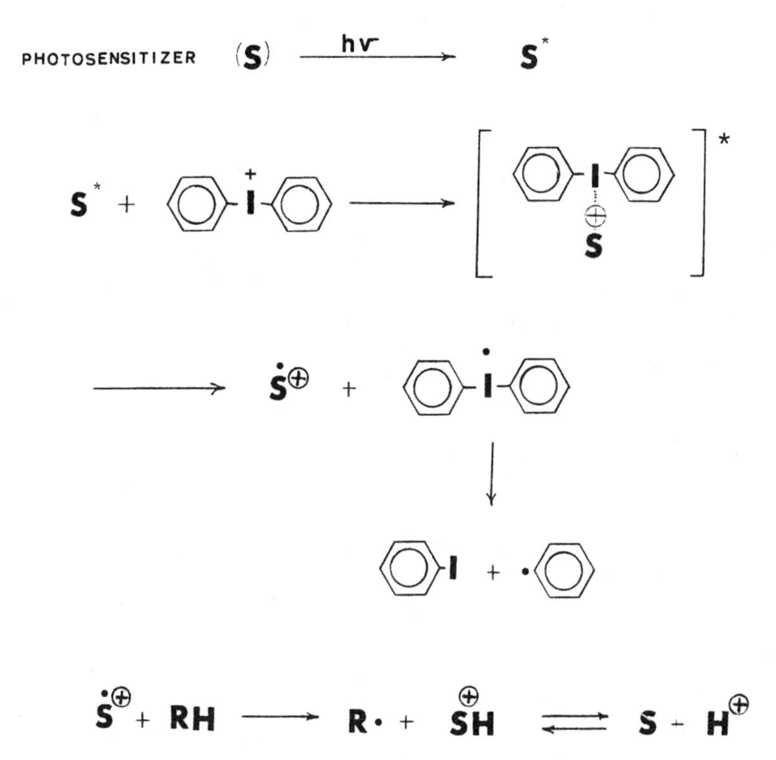

Figure 5. Sensitization of diaryliodonium salts via electron transfer results in formation of the sensitizer radical cation ($S^{\bullet +}$). Hydrogen abstraction produces a proton. Both species may initiate cationic polymerization. Note: The nonnucleophilic anion (AsF_6^-) has been deleted for clarity.

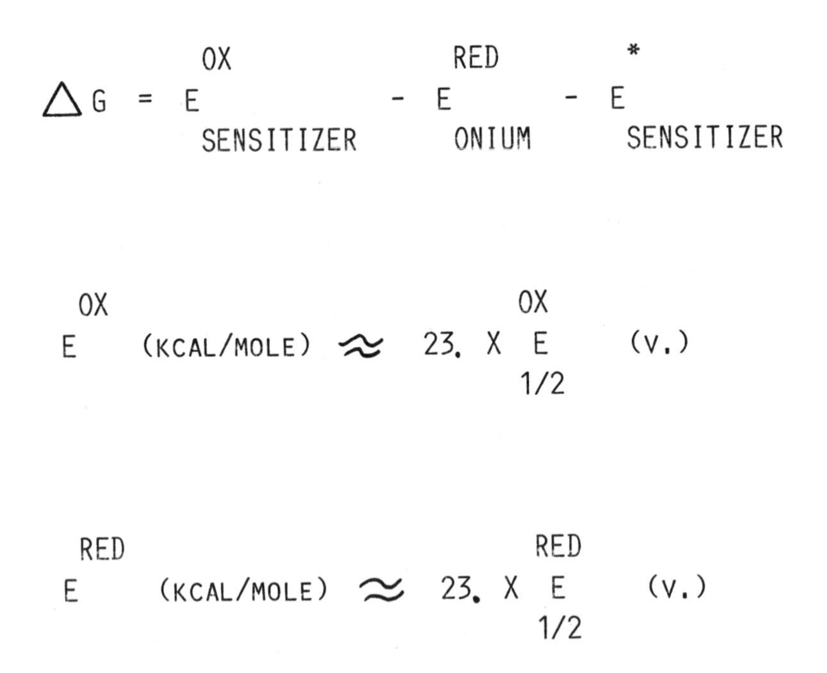

Figure 6. Free energy of electron transfer. In order for electron transfer to occur from the excited sensitizer to the onium salt, ΔG must be negative (i.e., the electron transfer must be exothermic). ΔG can be calculated using half-wave redox potentials and excitation energies (see text).

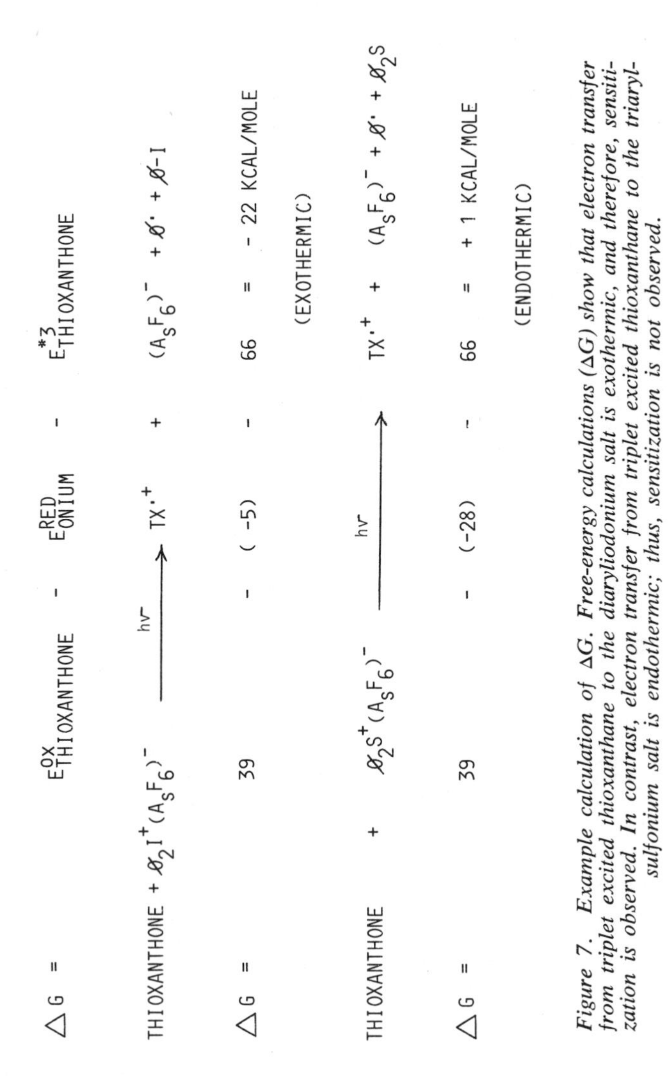

Figure 7. Example calculation of ΔG. Free-energy calculations (ΔG) show that electron transfer from triplet excited thioxanthane to the diaryliodonium salt is exothermic, and therefore, sensitization is observed. In contrast, electron transfer from triplet excited thioxanthane to the triarylsulfonium salt is endothermic; thus, sensitization is not observed.

The reduction energy of diphenyliodonium hexafluoroantimonate was calculated to be -5 kcal/mole (corresponding to -0.2 v vs a Standard Calomel Electrode, SCE). The reduction energy for triphenylsulfonium hexafluoroarsenate was calculated to be = -28 kcal/mole (corresponding to -1.2 v vs SCE). Details of these calculations were published previously.(8-11,15) Oxidation and reduction potentials of organic compounds are available (14).

The rate of electron transfer approaches diffusion control when ΔG is more negative than -10 Kcal/mole. ΔG was calculated for all sensitizers used in this study with both $\phi_2I\ AsF_6$ and $\phi_3S\ AsF_6$. An examination of Tables 1 and 2 shows that sensitizer systems with $\Delta G < -10$ exhibit relatively large quantum yields of photolysis and high relative polymerization rates (RPR).

The relatively high quantum yields of acid relative to organic products indicate the presence of a side reaction which is not yet clearly understood. Although hydrolysis had been suggested to explain high yields of acid relative to organic products in a previous study,(13) control onium samples (not irradiated) failed to produce significant yields of acid. In addition, undried acetonitrile solutions of $KAsF_6$ (3×10^{-2} M) refluxed for 1/2 hour showed negligible hydrolysis.

Chain Reactions in the Photoinitiation Process

The addition of THF to benzophenone sensitized polymerization of epoxy resin CY-179 results in a dramatically increased relative polymerization rate in the presence of $\phi_2\ IAsF_6$, and a small increase in the presence of $\phi_3S\ AsF_6$ (compare Tables 1 and 2). Similarly, the addition of THF or isopropanol results in large product quantum yields ($\Phi > 2$) for benzophenone sensitized decomposition of $\phi_2I\ AsF_6$, but only small increases in the decomposition of $\phi_3S\ AsF_6$ ($\Phi \approx 0.2$). Quantum yields > 1 suggest that a chain reaction is involved in these experiments.

A mechanism consistent with these observations is presented in Figure 8. Excited state aromatic carbonyl molecules having N,π^* triplet states are known to abstract hydrogen from alcohols, ethers and other substrates. In contrast, π,π^* excited states (i.e., perylene and anthracene) are not observed to participate in this reaction. The addition of THF to perylene and anthracene resulted in only slight increases in polymerization rates and product yields.(5)

The triplet excited benzophenone abstracts a hydrogen atom from THF producing benzhydryl and THF free radicals. Both radicals are easily oxidized by $\phi_2I\ AsF_6$ resulting in the formation of possible initiating species for cationic polymerization. Hydrogen abstraction by phenyl radicals propagates the chain.

The proposed chain reactions require initiation via some hydrogen abstracting process (i.e., most aromatic carbonyl sensitizers) and a substrate which will form electron rich (oxidizable) free radicals. The halfwave oxidation potentials of common free

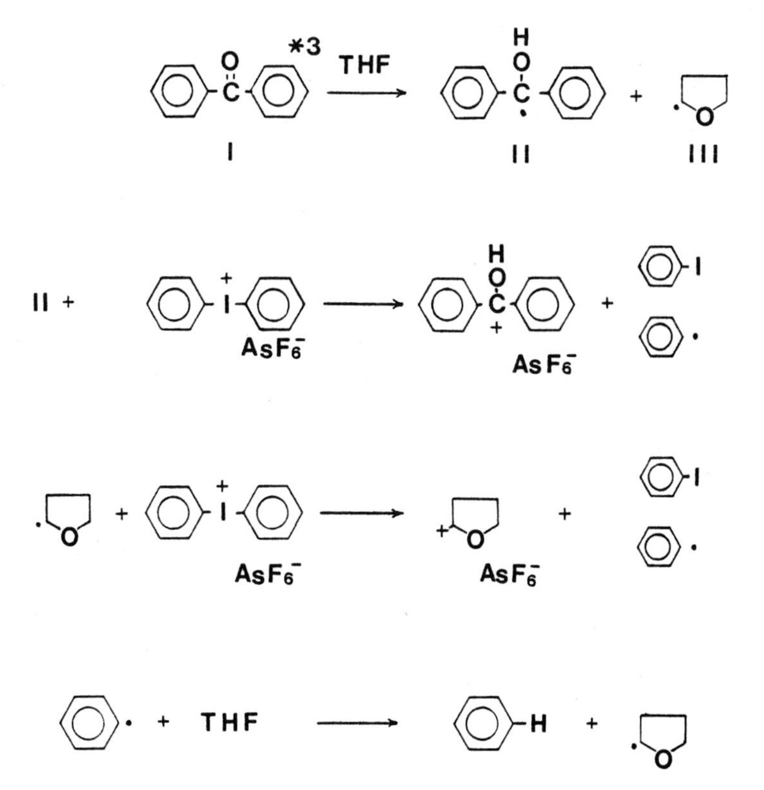

Figure 8. Chain reaction photoinitiation. The efficiency of photoinitiation can be increased by chain reactions. Hydrogen abstraction by triplet excited benzophenone forms a THF free radical. Subsequent oxidation by the aryliodonium salt produces the THF cation capable of initiating polymerization, and a phenyl radical. Hydrogen abstraction by the phenyl radical produces the THF free radical completing the chain.

radicals (i.e., acetone ketyl) are about -1 v vs. SCE. Thus, only salts having reduction potentials above this level (i.e., more positive) would be expected to be useful for chain enhanced photoinitiated cationic polymerization.

Onium salts are also reduced by photoinitiators such as hydroxy cyclohexyl phenyl ketone (IRGACURE 184), which efficiently form easily oxidizable free radicals as is shown in Figure 9.

Ledwith and coworkers have studied the cationic polymerization of THF initiated by onium salts in combination with free radical proginators.(20-23) When 2-2-dimethoxy-2-phenylacetophenone (IRGACURE 651) was irradiated in THF, the experiment containing an aryliodonium salt had a seven fold greater polymerization rate than a comparable experiment containing an arylsulfonium salt (21).

The sulfonium salt was shown to spontaneously oxidize highly stabilized free radicals such as the triphenylmethyl radical to form the corresponding triphenylmethyl carbonium ion.(20) It would also appear that the dimethoxybenzylic free radical (a Norrish Type I photocleavage product of 2,2-dimethoxy-2-phenylacetophenone) is similarily oxidized by the arylsulfonium salt (21).

Because the free radical formed by hydrogen abstraction from a cyclic ether is stabilized by only one oxygen, a stronger oxidizing agent is required. Thus, THF or possibly epoxy free radicals can be oxidized by strong oxidizing agents such as silver (22) or aryl iodonium salts in a chain reaction as in Figure 8.

The sulfonium salt will not participate in these chain reactions due to its larger reduction potential (-1.2 v. vs SCE compared to -0.2V). The slight increase in relative polymerization rates (RPR, Table 2) when THF was added to the benzophenone/sulfonium system may indicate that oxidation of the benzhydryl free radical is possible by $\phi_3S\ AsF_6$ resulting in the formation of acid by non-chain processes.

Experimental

Quantum yields of acid, and iodobenzene or phenylsulfide were measured by irradiating .03 molar acetonitrile solutions of diphenyliodonium hexafluoroarsenate and triphenyl sulfonium hexafluoroarsenate. Photosensitizers were included at identical optical density (1.0) at the wavelength of irradiation (366 nm). Relative polymerization rates were calculated as a function of the gelation time of a solution of 50% acetone, and 50% cycloaliphatic diepoxide resin ARALDITE CY-179 (v/v) containing the photosensitizer and onium salt as above. Experimental details were published previously.(8,15) IRGACURE and ARALDITE are registered trademarks of the Ciba-Geigy Corporation.

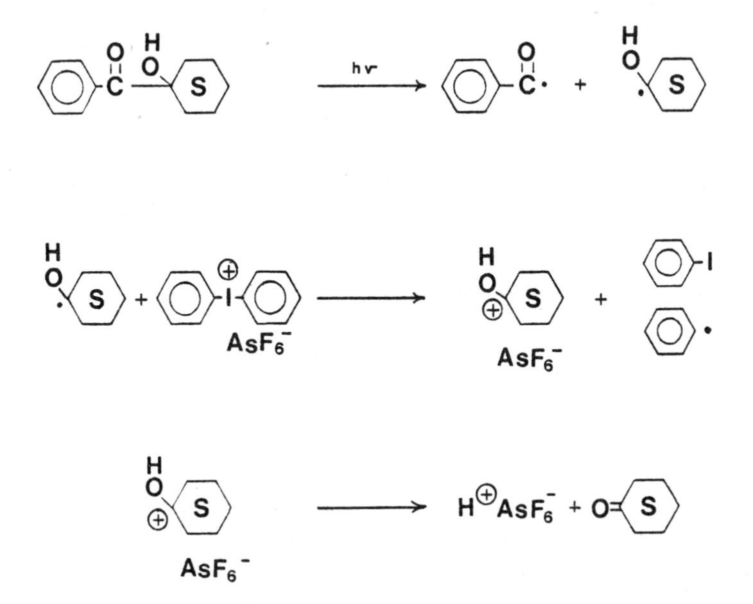

Figure 9. Photoreduction of onium salts by free radicals. Photoinitiators which directly form easily oxidized free radicals can efficiently sensitize aryliodonium salts, resulting in the formation of acid ($HAsF_6$). Having higher reduction potentials, the arylsulfonium salts do not oxidize most free radicals.

Conclusion

Sensitization of aryliodonium and arylsulfonium salts having nonnucleophilic anions can greatly increase the quantum efficiency of photolysis. This results in the more rapid curing of epoxy based coatings.

Electron transfer sensitization occurs with both classes of salts, with product quantum yields below unity. Sensitizers which produce easily oxidizable free radicals in the presence of the iodonium salt, can result in higher product quantum yields and higher observed polymerization rates.

Having a much higher reduction potential, the sulfonium salt is not sensitized efficiently by this mechanism. Neither class of onium salts were effectively sensitized under conditions where only triplet energy transfer was possible.

Literature Cited

1. For a discussion of the technology of the use of Radiation in the Curing of Coatings, see "U.V. Curing: Science and Technology", (S. P. Pappas, Ed.) Technology Marketing Corp., Norwalk, Conn. (1978)
2. Crivello, J. V. and Lam, J. H. W., J. Polym. Sci., Polym. Chem. Ed., 1979, 17, 977.
3. Crivello, J. V., Lam, J.H.W., and Volante, C. N., J. Radiation Curing, 1977, 4 (3) 2.
4. Crivello, J. V. and Lam, J. H. W., J. Polym. Sci., Polym. Lett. Ed., 1979, 17, 759-764.
5. Jilek, J. H., Ph. D. Thesis, North Dakota State University (1979).
6. Crivello, J. V. and Lam, J. H. W., J. Polym. Sci., Polym. Chem. Ed., 1978, 16, 2441.
7. Smith, G. S., U. S. Patent 4,069,054.
8. Pappas, S. P. and Jilek, J. H., J. Photogr. Sci. and Eng., 1979, 23 (3) 140.
9. Bachofer, H. E., Beringer, F. M., Meites, L., J. Amer. Chem. Soc., 1958, 80, 4269.
10. McKinney, P. S. and Rosenthal, S., J. Electroanal. Chem., 1968, 16, 261.
11. E^{ox} and E^{Red} values have been converted to the units of kcals/mole by multiplying half wave potentials (Volts-vs-SCE) x 23.06.
12. Arnold, D. R. and Maroulis, A. J., J. Amer. Chem. Soc., 1976, 98, 5931.
13. Knapczyk, J. W., Lubinkowsky, J. J., and McEwen, W. E., Tetrahedron Lett., 1972, 35, 3739.

14. a. Weinberg, N. L. (Ed.), "Techniques of Chemistry" Vol. V., part II, pp 667-1055. Values vs. Ag are corrected to SCE Reference (Neutral pH conditions) by adding 0.3 v.
    b. Loutfy, R. O. and Loutfy, R. O., J. Phys. Chem., 1973, 77, (3) 336-339.
15. Gatechair, L. R., Ph. D. Thesis, North Dakota State University (1980).
16. Rehm, D. and Weller, A., Ber. Bunsenges. Phys. Chem., 1969, 73, 834; see also Ref. 12.
17. Jacobson, K. I. and Jacobson, R. E., "Imaging Systems", J. Wiley and Sons, N.Y. (1976), p. 159.
18. Crivello, J. V., pp. 24-77 in Reference 1.
19. Irving, H. and Reid, R. W., J. Chem. Soc., 1960, 2078.
20. Knapczyk, J. W., et. al., Tetrahedron Lett., 1972, 35, 3739.
21. Abdul-Rasoul, F. A. M, Ledwith, A., and Yagaci, Y., Polymer, 1978, 19, 1219.
22. Abdul-Rasoul, F. A. M, Ledwith, A., and Yagaci, Y., Polymer Bulletin, 1978, 1, 1-6.
23. Ledwith, A., Polymer, 1978, 19, 1217.

RECEIVED October 15, 1982

15

The Influence of Hydroxyls on the Cr/Silica Polymerization Catalyst

M. P. McDANIEL and M. B. WELCH

Phillips Research Center, Bartlesville, OK 74004

The activity and termination rate of the Phillips Cr/silica polymerization catalyst have been examined in relation to the surface hydroxyl population. Results indicate that, far from being a necessary part of the active center as has sometimes been proposed, these hydroxyls may actually interfere with the active site. Hydroxyls could be removed by chemical as well as by thermal means to improve activity and melt index potential, such as by treating the catalyst in carbon monoxide, sulfur, or halides.

This paper examines some factors which affect not only the overall activity, but also the rate of termination of polyethylene chains growing on the Phillips Cr/silica polymerization catalyst. Although the theme of this symposium is not the termination but the initiation of polymer chains, the two aims are not inconsistent because on the Phillips catalyst the initiation and termination reactions probably occur together. They are both part of a continuous mechanism of polymerization. One possibility, proposed by Hogan[1], is shown below. The shift of a beta hydride simultaneously terminates one live chain while initiating another:

(1)

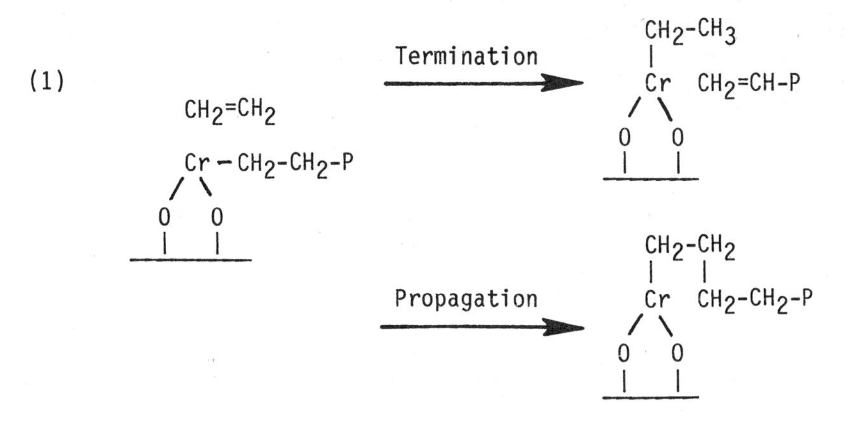

0097-6156/83/0212-0191$06.00/0

Or a chromium hydride might be formed as an intermediate, followed by initiation as a separate step. Either way each chain is expected to have a terminal vinyl group and this agrees with actual infrared measurements on the polymer. Or propagation can also occur if the coordinated monomer inserts into the active bond. Thus the molecular weight of the polymer gives an indication of the rate of termination of chains relative to propagation.

At the other end of the chain a methyl group has been found. This is as would be expected from the above mechanism (1) since each chain is initiated with a hydrogen atom from the previously terminated chain. But it raises the question: From where did the original hydrogen atom come, which initiated the very first chain? It does not help to say that it comes from the solvent or from a cocatalyst because these catalysts are quite active without cocatalyst or, in the gas phase processes, without solvent. One possible answer, which is shown below, has been proposed recently by Schuit and his coworkers from isotopic labeling experiments.[2] They suggest that the first initiating proton comes from a hydroxyl group on the surface of the silica. And that later termination just gives it back again to form the double bond. They found that when the silica surface was deuterated, the deuterium eventually found its way into the polymer. Thus to be active each catalyst site would have to consist of a chromium center and a neighboring hydroxyl to act as the initiator. A chromium center by itself would not be expected to be active.

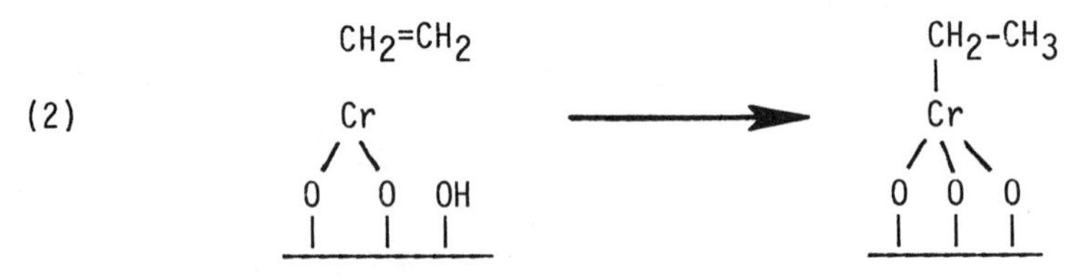

The Phillips catalyst is usually prepared by depositing chromium (VI) oxide onto a fully hydrated silica and then calcining this mixture. The catalyst is not active until it has been calcined. The calcining step, or activation, binds the chromium to the silica by esterification to hydroxylated sites:[1,3,4]

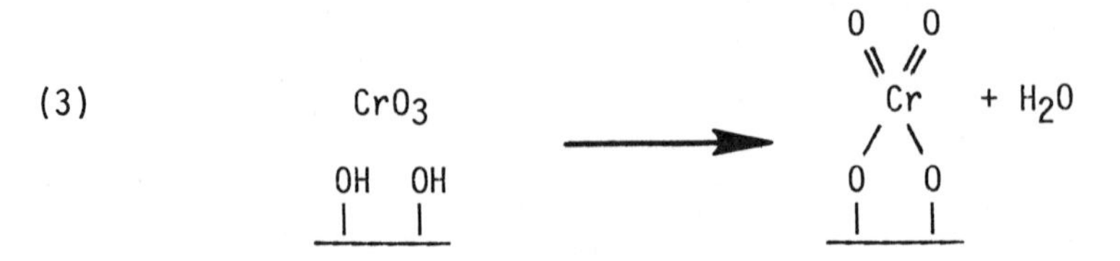

The chromium loading is always low enough that only a few of these hydroxyls are occupied by chromium. The others remain as a background hydroxyl population, some of which condense during activation to release water. Conceivably those remaining could affect the polymerization reaction, either directly as in previous mechanism (2), or indirectly by affecting the electronic environment around the chromium active center. The final step before polymerization is to reduce the hexavalent chromium to the lower valent active form (mostly likely Cr+2).[5,6,9] This is normally accomplished in the reactor by the ethylene monomer and Hogan[1] has even proposed that the initiating hydrogen could come from the "refuse" of this redox reaction. But any number of reducing agents will work, some of which like CO contain no hydrogen, so that explanation cannot be correct.

Therefore in this paper we have tried to determine if these hydroxyls could be directly involved in the polymerization. The hydroxyl population has been studied under various activation conditions to see how it correlates with the overall activity of the catalyst and also with the termination-initiation rate.

Experimental

Catalysts. Two types of silica support were used in these experiments. Davison grade 952 silica had a pore volume of 1.6 cc/g and a surface area of about 280 m^2/g. The other support was a coprecipitated silica-titania (3.3 wt% TiO_2) having a pore volume of 2.5 cc/g and a surface area of about 450 m^2/g. Ordinarily both supports were first treated with chromium (III) acetate to yield 1 wt% Cr. Activation was accomplished in a shallow bed fluidized by air or another gas predried through alumina columns. Gases other than air were also deoxygenated through columns of specially reduced Cr/silica-alumina catalyst.

Polymerization. Ethylene polymerization runs were made in a two liter stirred autoclave, having a pressurized jacket containing boiling alcohol for accurate control of the reactor temperature. After the catalyst had been charged to the reactor under dry N_2, one liter of isobutane was added as diluent and the ethylene (both Phillips polymerization grade) was supplied on demand of 550 psig. At 100-110 C polyethylene was obtained as a slurry, and after one or two hours the isobutane was flashed off leaving several hundred grams of polymer powder.

Relative Melt Index Potential. Melt indices were obtained from polymer samples by the standard test (ASTM D 1238-73) at 190 C using a weight of 2160 grams. However, melt index values are not just affected by activation parameters, but also by reactor conditions, such as the temperature, monomer concentration, and residence time. Therefore for clarity in this report we have normalized melt index values against those of a reference catalyst run under the same reactor conditions. We call the normalized value the relative melt index potential (RMIP) because it is independent of reactor conditions or catalyst type. Its value reflects only the activation parameters. The RMIP was defined as unity for each catalyst after activation in air at 870C.

Results and Discussion

Activity. Although pure CrO_3 begins to decompose above 200 C into O_2 and eventually Cr_2O_3,[1] a certain amount (0.4 Cr/nm^2)[2] is stabilized on silica up to 900 C in O_2.[4] And lower valent salts of chromium are easily oxidized on silica to the hexavalent form.[4] This is probably due to the formation of a stable chromate ester on the silica surface as in (3).

While heat is necessary to effect the esterification, this is probably not the only purpose of the high temperature activation step. Figure 1 shows how the ethylene polymerization rate of CrO_3/silica developed as the activation temperature was increased. A respectable activity did not appear until about 500 C, whereas the stabilization of Cr(VI) begins at as low as 200 C[1] and the actual esterification has been reported to occur between 150 and 300 C.[5] Furthermore, other sources of chromium behaved similarly in activity to Figure 1 even though the particular mechanism and temperature of binding must vary somewhat. Therefore the activation step must achieve some other necessary effect in addition to formation of the surface chromate ester.

Notice in Figure 1 that the catalysts were not active immediately after introduction into the reactor, but underwent a dormant period or induction time. Afterward the rate then increased during the rest of the experiment. This is thought to be due to reduction of Cr(VI) to the active valence,[1,6] probably Cr(II),[7] by ethylene. Zakharov and Ermakov[8] and others [1,6] have suggested that the gradual rise in rate is due to an increase in the number of active centers with time, and not to a change in the propagation rate constant. However, even under the best of conditions only a small fraction of the chromium is likely to be active.[1,8] In Figure 1, increasing the activation temperature also increased the development of polymerization activity. It is possible that this is also due to an increased concentration of active sites.

Figure 2 shows more completely this relationship between activity and activation temperature. Here activity was defined as the inverse of the time needed to make 5000 grams of polymer per gram of catalyst. Activity increased with increasing activation temperature up to a maximum at around 925 C, and then declined as sintering set in. Thus although the chromium was fully attached by 300 C, the activity continued to increase right up to the point of sintering. During this range, all of the chromium remained hexavalent and no obvious changes were observed in the x-ray diffraction pattern or porosity of the silica, just a gradual dehydration of the surface as silanol groups condensed to release water.

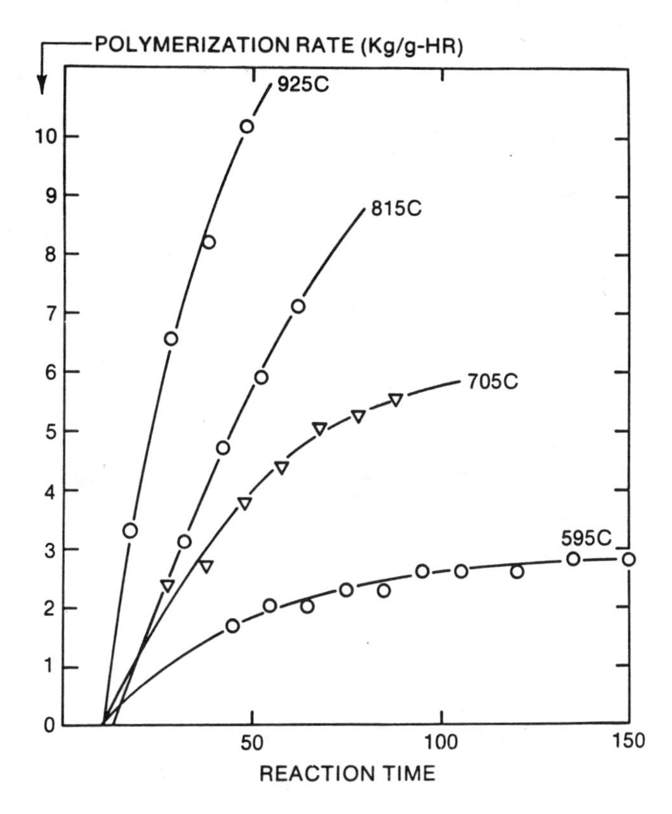

Figure 1. Dependence of polymerization kinetics on activation temperature.

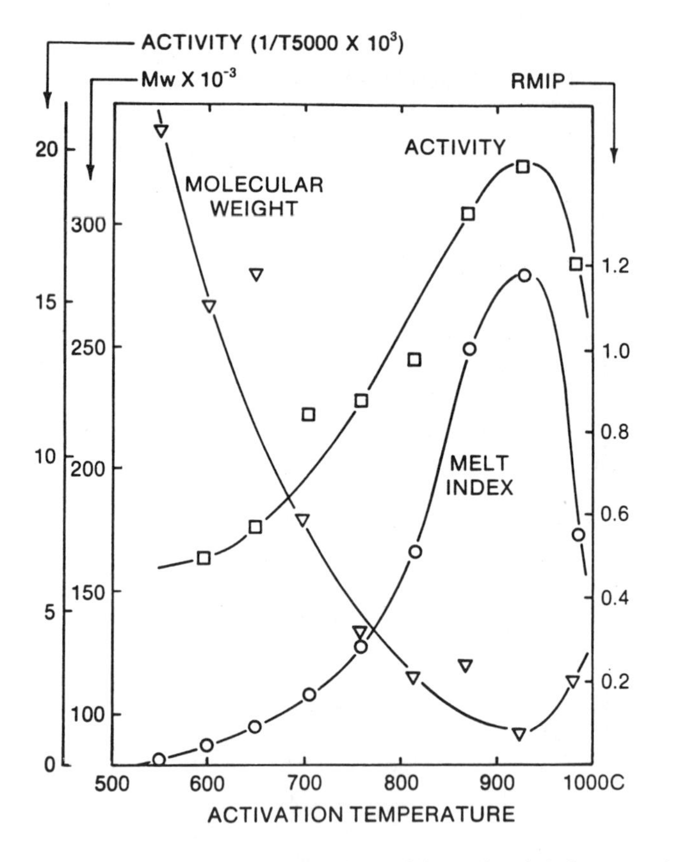

Figure 2. Dependence of activity, molecular weight, and melt index on activation temperature.

Melt Index. Also shown in Figure 2 is the molecular weight (MW) of the polyethylene produced by each catalyst. The weight average MW decreased with increasing activation temperature to a minimum near 925 C and then rose as the catalyst began to sinter. The MW can be taken as an indication of the termination rate relative to the propagation rate. And if, as has been suggested,[8] the propagation rate constant does not change much anyway, then the MW may even be taken as an indication of the absolute termination rate. If the propagation rate constant changes at all, it probably increases with temperature as does the overall activity. Thus a low MW suggests a high termination rate and we see in Figure 2 that the termination rate increased with activation temperature up to 925 C.

Molecular weight is most often judged by the viscosity of the polymer melt, to which it is inversely related. One convenient measure of this parameter is melt index. Manufacturers and customers of polyethylene are usually more concerned with melt index than with the actual molecular weight because melt index gives a direct indication of the flow of the polymer during processing. The molecular weight distribution of the polymer also affects the melt index, but in these experiments that parameter was held relatively constant. From a commercial point of view, the control of melt index is usually more important and difficult than control of the activity.

Figure 2 also plots the melt index (RMIP = relative melt index potential) of these same polymers. Since a high RMIP indicates a high termination rate, it also (like the activity) increased with increasing activation temperature up to the point of sintering. Other measures of the termination rate, such as the vinyl content of the polymer, also displayed this same pattern.

Hydroxyl Population. All of these facts indicate a connection between the hydroxyl population on the silica surface and the catalyst's activity and relative termination rate. Figure 3 plots this decrease in the hydroxyl population. Silica, containing no chromium , was calcined at various temperatures and then reacted with CH_3MgI solution.[9] The amount of methane released was taken as an indication of the surface hydroxyl content. As the activation temperature was increased, the hydroxyl population decreased from over 4 OH/nm^2 at 200 C to less than 1 OH/nm^2 at 900 C. However, it never actually reached zero even at the highest temperatures studied, but was always significant compared to the coverage by chromium.

Thus as hydroxyls became scarce the activity and termination-initiation rate increased. This argues against mechanism (2), that each site contains a hydroxyl as the initiator, but does not rule it out because even at 900 C there are sufficient hydroxyls to act in this capacity.

A further decrease in the hydroxyl population was obtained when the silica was calcined, not in air or nitrogen, but in carbon monoxide.[10] This is also shown in Figure 3. The curve representing the OH population in carbon monoxide split away from that in air at 600 C, and the separation increased up to sintering at 925 C. We believe that this is primarily due to a water gas shift mechanism in which surface moisture is removed by conversion to CO_2 and H_2. The CO could act through a direct attack on hydroxyls (4), or perhaps indirectly by eliminating moisture in the gas stream and thereby shifting the hydration equilibrium (5). The formation of a surface silane (6) is also conceivable.

$$2{\equiv}Si\text{-}OH + CO \longrightarrow H_2 + CO_2 + {\equiv}Si\text{-}O\text{-}Si{\equiv} \tag{4}$$

$$2{\equiv}Si\text{-}OH \quad {\equiv}Si\text{-}O\text{-}Si{\equiv} + H_2O \xrightarrow{CO} H_2 + CO_2 \tag{5}$$

$${\equiv}Si\text{-}OH + CO \longrightarrow CO_2 + {\equiv}Si\text{-}H \tag{6}$$

Even more effective than carbon monoxide for removal of hydroxyl groups, was sulfur in the presence of carbon monoxide.[10] Figure 3 also plots the hydroxyl population on silicas treated with CS_2 vapor and carbon monoxide. The OH level separates from that of the air treated samples at as low as 400 C, and by 950 C very few hydroxyls remained. The exact mode of action of the sulfur in these experiments is unknown, but one possibility is that it somehow affects the water gas shift equilibrium. Sulfur was not detected on these samples.

Sintering in CO. Not only did calcining the silica support in CO rather than air result in a lower surface hydroxyl population, but it also retarded the usual sintering process. This is shown in Figure 4 where the surface area of catalyst A is plotted after some samples were calcined in dry air and others in CO. As the temperature was raised, the onset of sintering became evident by the rapid drop in surface area. The effect was more pronounced in air than CO. This protection by CO could also be seen in the pore volume of the catalyst, which likewise decreased rapidly during sintering. Just why catalysts should be less susceptible to sintering in CO is not clear, unless it results from decreased moisture in the gas stream. The effect was found whether or not chromium or titania promoter were present.

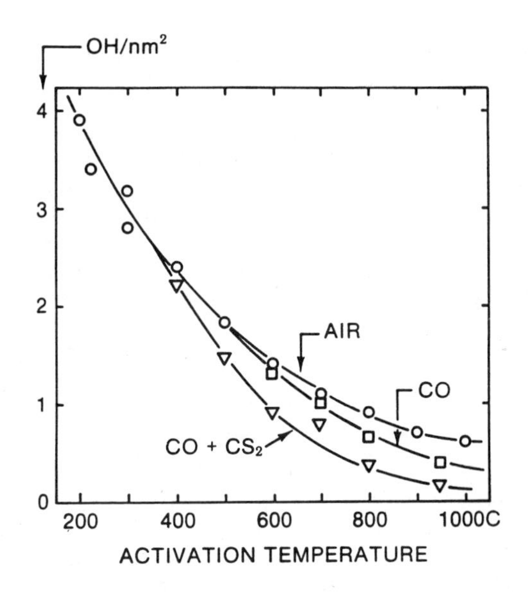

Figure 3. Dependence of the surface hydroxyl population on activation temperature in different gases.

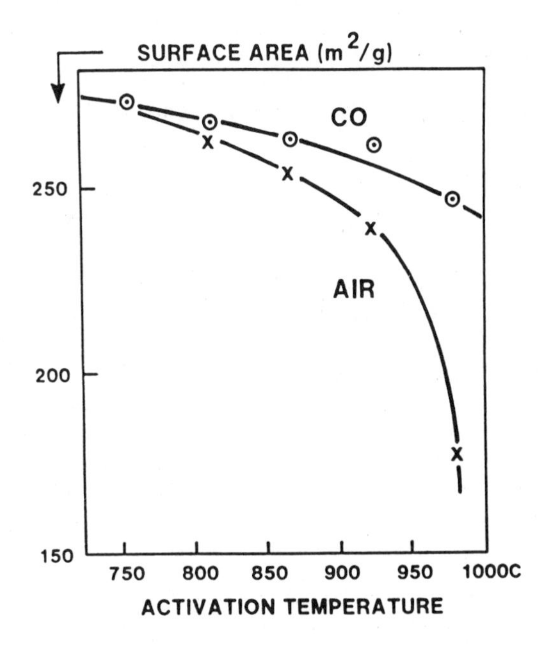

Figure 4. Surface area measurements show that Cr/silica is less prone to sinter in CO than in dry air.

R/R Activation. Figure 5 shows that the enhanced dehydroxylation by carbon monoxide also had a pronounced effect on the termination rate during polymerization. In these experiments, two series of Cr/silica catalyst samples were activated and allowed to polymerize ethylene to a yield of about 5000g PE/g. In one series the catalyst samples were simply calcined five hours in air as usual at the temperatures shown. The relative melt index potential (RMIP) has been plotted against activation temperature and the expected increase up to the point of sintering was observed. In the other series, however, samples were instead calcined in carbon monoxide for three hours, followed by air for another two hours to reoxidize the chromium. Thus in both series the chromium was hexavalent, but those samples treated in carbon monoxide exhibited considerably higher termination rates. The two curves first split apart at about 600 C, where also the enhanced dehydroxylation by CO first became visible in Figure 3. Similarly when a sulfur compound was added, the enhancement of the termination rate became even more pronounced, and was noticed even below 600 C.

This procedure, sometimes called reduction and reoxidation, or R/R activation, provides a commercially valuable means of increasing the melt index potential of a catalyst. Any number of reducing agents can be used for special applications. However, the reoxidation temperature is usually chosen to be lower than the temperature of reduction. In this way one can achieve considerable dehydroxylation during reduction at 800-900 C, but without the sintering which would normally result in air at that temperature (see Figure 4). There is also evidence, which will not be presented here, that a special type of bonding to the silica is achieved at low reoxidation temperatures, perhaps owing to decreased surface mobility. If the reoxidation step is omitted altogether, however, little activity is observed because the chromium tends to lump during reduction at 800-900 C.

Two-Step Activation. Other experiments also implied a negative correlation between the hydroxyl population and the termination rate. For example, silica samples containing no chromium were first dehydrated at 870 C, and then impregnated with a chromium compound anhydrously to avoid rehydrating the silica. A short secondary activation in dry air at 315-650 C removed the solvent and converted the chromium to the surface-attached hexavalent species. This two-step activation produced a very active catalyst having a high RMIP even though only the silica, and not the chromium, had experienced the high temperature.

This result was independent of the type of chromium compound impregnated, provided the impregnation was done anhydrously. Some examples are dicumene chromium (0) in hexane, chromium (III) octoate in toluene, CrO_3 in CH_3CN, di-t-butyl chromate in hexane, or

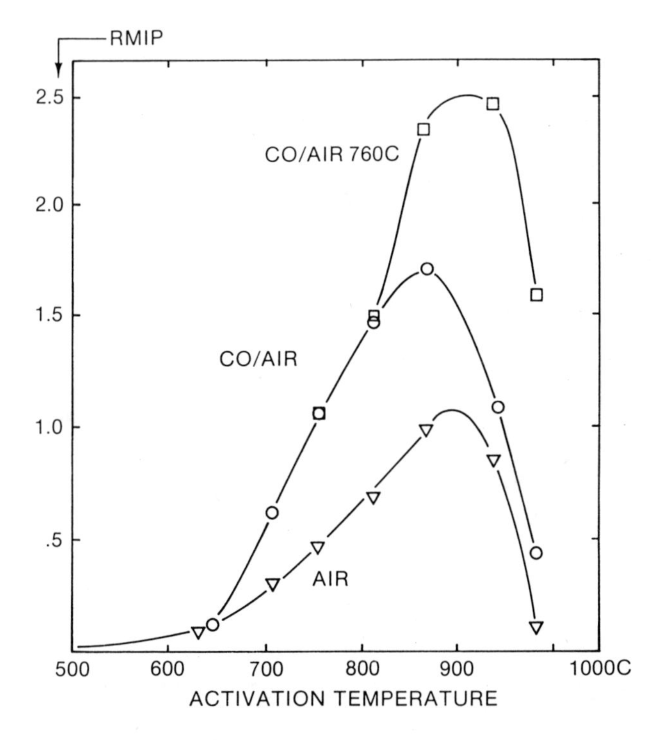

Figure 5. RMIP of catalyst B activated (1) in air 5 h at temperature shown, (2) in CO 3 h then air 2 h both at temperature shown, (3) in CO 3 h at temperature shown then air 2 h at 760 °C.

even CrO_2Cl_2 vapor. However, when the precalcined silica was impregnated with CrO_3 from an aqueous solution, then the activity and RMIP reflected only the second activation temperature, and not the first. Thus we again conclude that the activity and termination rate are controlled by (or at least related to) the level of surface dehydroxylation.

A summary of these two-step experiments is shown in Table I. Here the support was calcined at 870 C in the composition shown, then impregnated with 1/2% Cr as dicumene chromium (0) which was oxidized in O_2 between 315 C and 650 C to achieve maximum RMIP. Notice that when the support was calcined in carbon monoxide rather than air, a further elevation in RMIP was noticed. Since in these experiments the chromium was added only after the CO treatment, we must conclude that the beneficial action of CO is on the silica and not the chromium.

The same reasoning also holds for the action of sulfur compounds but the effect was far more pronounced. Sulfur compounds containing no hydrogen seemed to work best, such as CS_2 or COS, but even organic sulfur compounds like CH_3SH, Et-S-Et, or Et-S-S-ET, produced a large increase in the RMIP over that attainable by CO alone. All decomposed into elemental sulfur at the high temperatures, and in some cases a black sulfur-carbon residue was left. However, this residue played no part in the polymerization because in most cases it was burned off in O_2 at 870 C before the chromium was impregnated, with similar results. In fact, one of the cleanest compounds, COS, which left no detectable sulfur residue at all, also yielded among the highest increases in RMIP. Sulfur treatment in the absence of carbon or carbon monoxide, such as elemental sulfur vapor or SO_2 or H_2 did not yield the same beneficial effect. Selenium and tellurium compounds also exhibited this same effect.

TABLE I

THE EFFECT OF CHEMICAL DEHYDROXYLATION ON TERMINATION IN TWO-STEP ACTIVATION

TREATMENT*	RMIP
Air	2.5
Carbon Monoxide	6
Carbonyl Sulfide	150
$CO+I_2$, O_2	12
$CO+Br_2$, O_2	21

Halides. Another treatment which can lower the hydroxyl population, or even eliminate it altogether, is halogenation of the silica surface.[10,12-16] This removes hydroxyls, not by condensation as with CO and sulfur, but by replacement with halide, which prevents later attachment by Cr. The presence of halide probably also changes the electronic environment on the silica. Thus, fluoriding has long been used to increase activity but decrease RMIP.[14,15] Chloride also depresses RMIP, as well as the surface bromide and iodide of silica. However, these latter two have recently been studied,[13] and it was possible to burn off most of the iodide or bromide with oxygen above 600 C, leaving a partially dehydroxylated surface.

$$\equiv Si\text{-}OH + I_2 + CO \xrightarrow{800\ C} \equiv Si\text{-}I + CO_2 + HI \tag{7}$$

$$2\equiv Si\text{-}I + 1/2\ O_2 \xrightarrow{800\ C} \equiv Si\text{-}O\text{-}Si\equiv + I_2 \tag{8}$$

When this surface was then impregnated with chromium by the usual anhydrous impregnation (two-step process),[10] large increases in the melt index were obtained, as much as an order of magnitude over that obtained from samples activated similarly but in air only. An example is shown in Table I.

Although the surface chloride of silica cannot be burned off as easily, it will react with H_2 or CH_3OH and these can then be burned off. This treatment sometimes also produced an increase in RMIP.[17] However, the effect was minor by comparison to the bromide and iodide, probably because water was formed during the burn-off.

Another treatment which can completely remove all hydroxyls is to react them with $TiCl_4$.[18] In another series of experiments virgin silica was first calcined at 600 C to remove internal hydroxyls, then rehydrated in liquid water overnight. After being dried at 150°C the surface was dehydroxylated by reaction with $TiCl_4$ vapor at 150 C through 300 C. Chloride residue was partially burned away by O_2 at 800 C and afterward CrO_3 was impregnated from dry CH_3CN, leaving a dry orange powder. Finally samples were calcined in O_2 between 300 C and 800 C. These catalysts were highly active and exhibited a high melt index potential, despite the absence of hydroxyls. Titania is, of course, known to be a promoter.

(9)

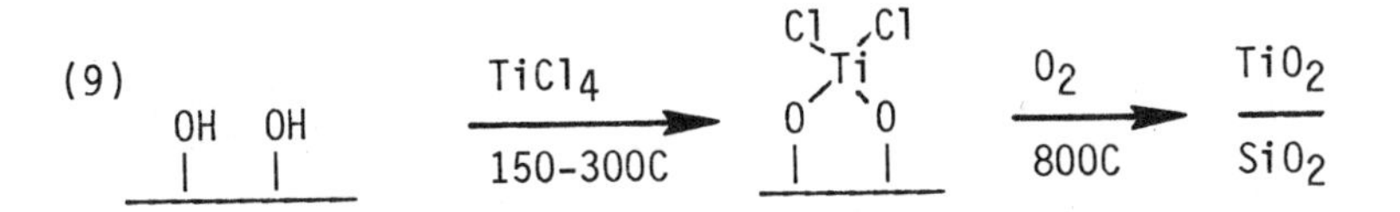

Although unlikely, one could still argue that some hydroxyls were replenished by oxidation of the organic solvent (unlikely because the solvent was removed at 200 C under flowing argon, there was no visible decomposition of the CrO_3, and other chromium compounds yielded the same results). To answer this objection one further experiment was done. Cr/silica was activated at 600 C in air, then reduced in CO at 300 C. $TiCl_4$ vapor at 200 C removed the hydroxyls, and a final treatment in O_2 at 800 C removed chloride and reoxidized the chromium (the reduction was necessary because Cr(VI) is stripped off by $TiCl_4$ as CrO_2Cl_2 vapor). This catalyst was again highly active (in fact more active than had the $TiCl_4$ not been employed, due to the promotional effect of titania) although no organics were used.

Discussion

These experiments indicate a strong negative correlation between the hydroxyl population on the Cr/silica catalyst, and its overall polymerization activity and termination rate. Far from being necessary to the active site, the hydroxyls seem to display an inhibiting effect. Dehydroxylation by thermal condensation, and also by several chemical means (including condensation and substitution) leads to enhanced activity and termination. Since some treatments removed all hydroxyls, and yet produced quite active catalysts, we must conclude that the first initiating hydrogen atom cannot come from a neighboring hydroxyl as in (2).

What then is the mechanism of initiation? One possibility is that the vinyl group is formed first, and the methyl lost:

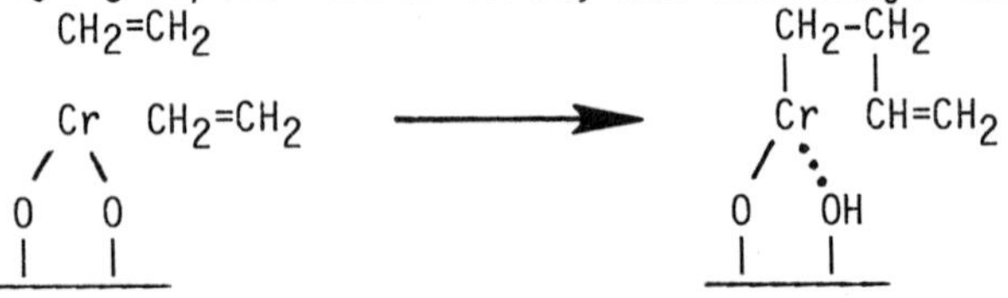

In which termination involves a proton jump from the newly formed hydroxyl:

Another possibility, as Hogan proposes,[1] is that the first chain is somehow different from those that follow. Hence the kinetics of polymerization shown in Figure 1 could result from a slow initiation step.

Whatever the mechanism, why should the behavior of the catalyst be so dependent on the activation temperature? From about 400 C to 900 C, where both activity and RMIP dramatically increase, the only significant change on the silica known to occur is the declining silanol population. Perhaps these hydroxyls coordinate to the active centers, blocking ethylene and thus poisoning the catalyst much as free water would do in the reactor? In some detailed experiments Krauss et.al.[19-21] have indeed found, by measuring the amount and ΔH of chemisorption by CO, N_2, and O_2, an inverse correlation between the coordinative unsaturation of Cr(II) centers and the surrounding hydroxyl population.

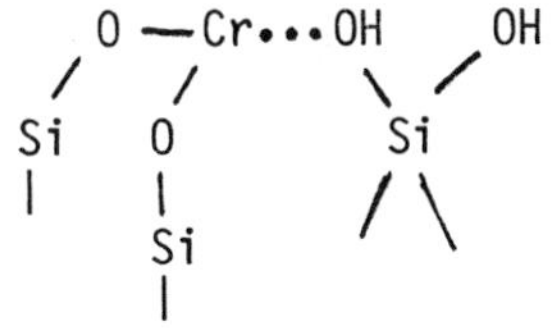

```
   O —Cr•••OH    OH
  /   /     \  /
Si   O       Si
|    |       /\
     Si
     |
```

Possible coordination between Cr and Silanol groups after reduction by ethylene from Cr(VI) to Cr(II)

However, it is hard to see from the common models of the silica surface just how the hydroxyls could come close enough to chromium to coordinate. For example in the model of Peri and Hensley,[22] with the chromium occupying vicinal pairs,[21] the nearest hydroxyl neighbor still seems to be 3 or 4A away. Nonetheless, the same problem is often encountered in explaining how neighboring silanols condense, so it is clear that our understanding of the silica surface is incomplete. Some type of distant coordination might even be possible between chromium and hydroxyls which does not completely block the active center but still influences its behavior.

A variant of this explanation would have the Cr(II) active center not coordinated to the hydroxyl, but oxidized by it to inactive Cr(III) and H_2. This might explain the dependence of the activity on activation temperature, and there is even evidence that this may happen when the catalyst is reduced in CO at 300 C or higher.[20,13] The conversion to Cr(II) is always more complete when the catalyst has first been highly dehydrated. Otherwise Cr(III) is also formed. However, both the surface protons and the chromium would probably have little mobility at 100 C, where reduction by ethylene occurs in the reactor, so again the hydroxyl would have to be in close proximity to the Cr.

Still another possibility is that the hydroxyls are not directly involved at all, but only reflect some other important change such as the strain introduced onto the surface by the condensation of hydroxyls. The RMIP always drops sharply when the first signs of sintering, which probably relaxes the strain, are observed. And we have seen that calcining the support in CO, which increases RMIP, also retards sintering. But without a more concrete knowledge of the silica surface it is difficult to examine this possibility in much detail.

Literature Cited

1. Hogan, J. P., J. Polym. Sci. 1970, A-1*(8)*, 2637.
2. Personal Communication with G.C A. Schuit.
3. McDaniel, M. P., J. Catal. 1981, *67*, 71.
4. McDaniel, M. P., "The State of Cr(VI) on the Phillips Polymerization Catalyst, II. The Reaction Between Silica and CrO_2Cl_2, III. The reaction with HCl, and IV. Saturation Coverage", J. Catal., In Press. Also summarized at the "Symposium on Transition Metal Catalyzed Polymerizations", Michigan Molecular Institute, (August, 1981).
5. Fubini, B., Ghiotti, G., Stradella, L. Garrone, E. and Morterra, C., J. Catal. 1980, *66*, 200.
6. Clark, A., Catal. Rev., 1969, *3(2)*, 145.
7. Merryfield, R., McDaniel, M. P., and Parks, G. D., "An XPS Study of the Phillips Cr/Silica Polymerization Catalyst", To be published, J. Cat.
8. Zakharov, V. A., and Ermakov, Y. I., J. Poly. Sci. 1971, A-1*(9)*, 3129.
9. Fripiat, J. J., and Uytterhoeven, J., J. Phys. Chem. 1962, *66*, 800.
10. McDaniel, M. P., and Welch, M. B., U. S. Pat. 4,248,735, (Feb. 1981).
11. Hawley, G. R., U. S. Pat. 4295998 (Oct. 1981).
12. Peri, J. B., and Hensley, A. L., J. Phys. Chem. 1966, *22(8)*, 2926.
13. McDaniel, M. P., J. Phys. Chem., 85, 532 (1981) and ibid. 537.
14. Hogan, J. P., and Banks, R. L., U.S. Patent 2,825,721 (Mar. 1953).
15. Kallenbach, L. M. U.S. Patent 3,445,367 (May, 1969).
16. Hockey, J. A., Chemistry and Industry, Jan. 9, 1965, pp 57.
17. McDaniel, M. P., Patent Pending, U.S. Appl. 151848.
18. Armistead, C. G., Tyler, A. J., Hambleton, F. H., Mitchell, S. A., and Hockey, J. A., J. Phys. Chem. 1969, *73(11)*, 3947.
19. Krauss, H. L., Z. Anorg. Allg. Chem., 1978, *446*, 23.
20. Krauss, H. L., Rebenstorf, B., and Westphal, U., Z. Anorg. Allg. Chem., 1975, *414(2)*, 97.
21. Hierl, V. G., and Krauss, H. L., Z. Anorg. Allg. Chem., 1975, *415*, 57.
22. Peri, J. B., and Hensley, A. L., J. Phys. Chem., 72,2926 (1968).
23. Groeneveld, C., Wittgen, P. P. M. M., van Kersbergen, A. M., Mestrom, P. L. M., Nuitjen, C. E., and Schuit, G. C. A., J. Catal. 1979, *59*, 153.

RECEIVED October 15, 1982

CELLULOSE, PAPER, AND TEXTILE CHEMISTRY

16

Novel Additives for Enhancing UV and Radiation Grafting of Monomers to Polymers and Use of These Copolymers as Ion Exchange Resins

CHYE H. ANG, JOHN L. GARNETT, RONALD LEVOT, and MERVYN A. LONG

The University of New South Wales, Kensington, N.S.W. 2033, Australia

Polyfunctional monomers (divinylbenzene and trimethylol propane triacrylate) are shown to be novel additives for enhancing the radiation copolymerisation of styrene in methanol to the polyolefins. These results are extrapolated to the radiation grafting of cellulose. The advantages of using these additives in preparing copolymers for ion exchange purposes is described. A comparison between mineral acid and these polyfunctional monomers in these reactions is discussed. Both additives increase the radiation grafting yields of styrene in methanol to polypropylene to approximately the same degree, however the mechanism of the enhancement is different for each additive and is important in determining the properties of the resulting copolymers and ion exchange resins. In the presence of the polyfunctional monomers, the ion exchange resins are cross-linked, leading to greater stability but slower migration of ions in and out of the resin. When compared with earlier work using similar radiation grafted ion exchange resins, the use of the present additives lowers the total radiation dose necessary to achieve a particular percentage graft by an order of magnitude. The value of this reduction in dose to the properties of the resulting copolymers and ion exchange resins is considered.

Grafting reactions are attractive one-step methods for modifying the properties of polymers to yield novel products which are of value in a wide range of applications. UV (1-4) and ionising radiation (5-11) are ideal initiators for such processes. The main limitations with the use of such radiation techniques are (i) the possible formation of competing homopolymer which affects the efficiency of the grafting process and (ii) the effects of

0097-6156/83/0212-0209$06.00/0

radiation, particularly energetic ionising types, on the stability of the backbone polymer being grafted.

Thus inclusion of any additive which will reduce the total radiation dose to achieve a particular percentage graft will be beneficial, particularly for radiation sensitive backbone polymers. Addition of mineral acid to monomer grafting solutions in the presence of ionising radiation has already been shown to significantly enhance copolymerisation yields for a variety of monomer/polymer systems (12) including the radiation grafting of styrene to synthetic backbone polymers such as the polyolefins (13, 14, 15), PVC (16) and polyesters (17) also the naturally occurring trunk polymers, cellulose (12), wool (18) and leather (19).

In the present paper, this acid effect will be compared with other recently discovered additives (19), in particular the polyfunctional acrylates, for the enhancement in radiation grafting. In preliminary work (19), divinylbenzene (DVB) has been reported to be a useful additive for enhancing the above grafting reactions. These early data (19) indicate that there are possible common mechanistic pathways between the acid effect and the DVB process. More detailed DVB studies are discussed in this paper for enhancing the radiation grafting yields of styrene in methanol to films of polyethylene and polypropylene. The work has been extended to include the use of other polyfunctional monomers such as trimethylol propane triacrylate (TMPTA) as additives. The possibility of being able to use these additives for copolymerisation of monomers to naturally occurring trunk polymers such as cellulose will also be considered.

Finally, using the styrene grafted polyolefin powders as representative polymers, the value of radiation grafting as a means of preparing novel ion exchange resins by sulfonation reactions will be discussed. The effect of grafting additives such as acid and DVB on the properties of the resulting ion exchange resins will be analysed.

Experimental Procedures

Of the backbone polymers used, polypropylene was isotactic, doubly oriented film of thickness 0.06 mm ex-Shell Chemicals (Aust) Pty. Limited, the polyethylene film was low density material of thickness 0.12 mm ex-Union Carbide (Aust) Pty Limited while the PVC film (0.20 mm) was supplied by ICI (Aust) Pty Limited. Styrene monomer (Monsanto Aust Limited), DVB and TMPTA (Polysciences) were purified immediately prior to irradiation by column chromatography on alumina.

Grafting Methods. The procedures used for the grafting runs were modifications of previously reported methods (11, 12, 13). All grafting experiments were performed in quadruplicate in pyrex tubes (15 x 2.5 cm) with styrene/ethanol/additive solutions at 20±1 ^{o}C. The backbone polymer films (4 x 2.5 cm) were fully immersed in the monomer solutions and the pyrex tubes stoppered for irradiation which was performed immediately after sample preparation. Two radiation sources were used for these studies, the first a 1200 Ci cobalt-60 installation and the second a spent fuel element facility at the Australian Atomic Energy Research Establishment, Sydney. At the completion of the reaction, the films were removed from solution, washed in an appropriate solvent, soxhlet extracted for 72 hours, dried in air, then at 45 ^{o}C to constant weight. With films at high graft of polystyrene (>50%), the films were soaked in chloroform at 50 ^{o}C for a further 3 hours after soxhlet extraction to ensure complete removal of homopolymer, then dried as above.

Sulfonation of Graft Copolymer. Styrene graft copolymers (10g) were sulfonated in concentrated sulfuric acid (90 ml) at various temperatures (60, 70, 80 and 90 ^{o}C) for 45 minutes. In separate experiments, the graft copolymers (10g) were sulfonated at 90 ^{o}C for different periods of time (20, 30, 45, 55 and 65 minutes), the mixtures being stirred continuously during reaction. At the conclusion of the sulfonation, the mixtures were cooled to 0 - 5 ^{o}C and distilled water added slowly until the concentration was 5M. The sulfonated product was filtered and washed successively with dilute sulfuric acid (2M, 500 ml), water (500 ml), sodium hydroxide (1M, 250 ml), water (500 ml), hydrochloric acid (1M, 250 ml) and finally water until the wash solution was neutral. The product was dried in a vacuum dessicator over silica gel to constant weight. After optimisation of sulfonation conditions, all graft copolymers for ion exchange experiments were sulfonated at 90 ^{o}C for 45 minutes.

Determination of pH Titration Curves and Capacities of Ion Exchange Resins. To the dry sulfonic acid exchange resin (1.00±0.01 g, 200-400 mesh) was added sodium chloride (0.1M, 125 ml), the mixture allowed to stand overnight then titrated with sodium hydroxide (0.1 M), the pH of the supernatant being measured with a glass electrode attached to a pH meter. A period of 5-10 minutes was allowed for establishment of equilibrium between exchange resin and the solution before each pH measurement. From the resulting pH titration curve, the capacity of the ion exchange resin was obtained from the inflexion point on the horizontal axis.

Stability of Sulfonic Acid Exchange Resins. Sulfonic acid exchange resin (2.00±0.01 g, 200-400 mesh) was allowed to swell overnight with water (30 ml) held in a glass column (40 x 1.5 cm) fitted with a sintered glass filter disc (porosity 3) near the outlet attached to a stopcock. After standing, the swollen resin was stirred with a glass rod and allowed to settle, while the stopcock was opened, and a volume of water (10 ml) maintained on top of the resin during preparation of the ion exchange column.

Sodium chloride (0.2 N, 100 ml) was poured gently, without disturbing the exchanger, into the column and sodium ion exchanged for hydrogen counter ions. The effluent was collected, the column freed from excess sodium chloride and interstitial hydrogen ions by washing with water, and washings being added to the effluent. The hydrogen ions released (i.e. the capacity of the ion exchange resin) were determined by titration with sodium hydroxide using methyl orange as indicator. The resin was regenerated with hydrochloric acid (0.2M, 100 ml), both the column and resin being subsequently washed with water until neutral when the resin was ready for further exchange. This whole process of converting the exchange resin to its acid form and regeneration was considered to be one complete cycle.

Rate of Exchange of Resins. The procedure was a modification of that reported by Reichenberg (20) and essentially involved the time required for 90% of the hydrogen ions to exchange with sodium ions.

Acid Effects in Radiation Grafting of Styrene to Cellulose and Polypropylene

Acid has previously been used as an additive to enhance the radiation grafting of monomers such as styrene to trunk polymers (12-19). The data in Table I compare the use of acid in such copolymerisation reactions with cellulose and polypropylene as backbone polymers. The results show that the acid enhancement in yield is generally larger with the synthetic polymer, polypropylene. Such synthetic materials can also tolerate relatively high acid concentrations (e.g. 3M in Table I) whereas the naturally occurring polymers like cellulose are readily degraded at acidities above 0.2M especially with nitric acid. Because of these acid results which show higher yields with the synthetic backbone polymers than with naturally occurring cellulose, particularly the polyolefins, these materials, especially polyethylene and polypropylene were chosen as model backbone polymers for the present preliminary studies using polyfunctional monomer additives in radiation grafting styrene in methanol.

Polyfunctional Monomers as Additives in Radiation Grafting

In Figure 1 are shown the data for the radiation copolymerisation of styrene in methanol to polyethylene film at a dose rate of 4.1×10^4 rad/hr to a total dose of 2.4×10^5 rad. It is observed that addition of divinylbenzene (DVB) in additive amounts (1% v/v) significantly increased the yield at virtually all concentrations studied. The enhancement is particularly evident at the monomer concentration associated with the Trommsdorff peak, namely 50% styrene in methanol. When DVB is replaced by trimethyl propane triacrylate (TMPTA) in 1% v/v concentration, a similar

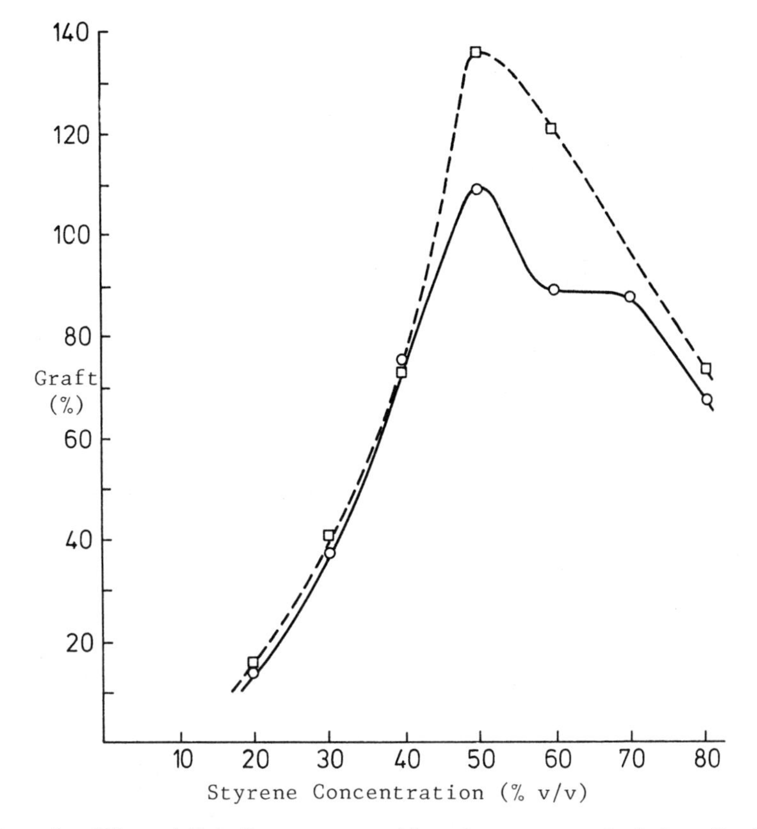

Figure 1. Effect of divinylbenzene on grafting of styrene to polyethylene film in methanol at dose rate of 4.1 × 10^4 rad/h to total dose of 2.5 × 10^5 rad. Key: ○, styrene-methanol; □, styrene-methanol-divinylbenzene (1% v/v).

Table I. Comparison of Cellulose with Polypropylene as Trunk Polymers for Acid Enhancement in Radiation Grafting of Styrene in Methanol

Styrene	Graft(%) with Cellulose[a]			Graft (%) with Polypropylene[b]			
(% v/v)	0.0	$5.4x10^{-2}$M	1.1M	0.0	$5.0x10^{-2}$M	M	3M
10	1.2	8.8	21	0	0	19	125
20	8.9	38	75	12	21	63	
30	14	48	27	51	72		
40	16	34	16	45	54		

[a] Irradiation in evacuated vessels at dose rate of $2.73x10^4$ rad/hr to $2.0x10^5$ rad.

[b] Irradiation in air at dose rate of $4.0x10^4$ rad/hr to $2.0x10^5$ rad.

enhancement pattern in grafting is observed (Figure 2), thus the functionality of the polyfunctional monomer does not appear to affect the degree of increase in grafting. If polypropylene film is used instead of polyethylene with DVB as additive similar patterns in grafting enhancement are obtained (Figure 3), however with polypropylene the position of the Trommsdorff peak has shifted slightly from 30% to 35% styrene in methanol in the presence of additive.

Comparison of Polyfunctional Monomers with Acid as Additives in Grafting

Inclusion of sulfuric acid (0.2M) in the styrene-methanol grafting solution gives an enhancement in radiation grafting with polyethylene film similar to that obtained with DVB (Table II). The optimum in the copolymerisation process occurs at 50% monomer concentration with both additives. The behaviour of sulfuric acid in these reactions is representative of the most reactive of the mineral acids (12). A comparison of TMPTA with H_2SO_4 (0.2 M) also indicates that similar trends in enhancement in radiation grafting to polyethylene film are observed with both additives (Table III). Again, when polypropylene film is used as backbone polymer (Table IV), acid and DVB show similar increases in grafting yield with styrene in methanol, however in this system, it is interesting to note that the concentration of monomer at the position of the Trommsdorff peak does not vary for neutral and acidified grafting solutions but changes from 30% to 35% when DVB is additive.

Comparison of Mechanisms of Acid and Polyfunctional Additive Enhancement Effects

The remarkable similarity in trends between acid and polyfunctional monomer additives for most of the present radiation

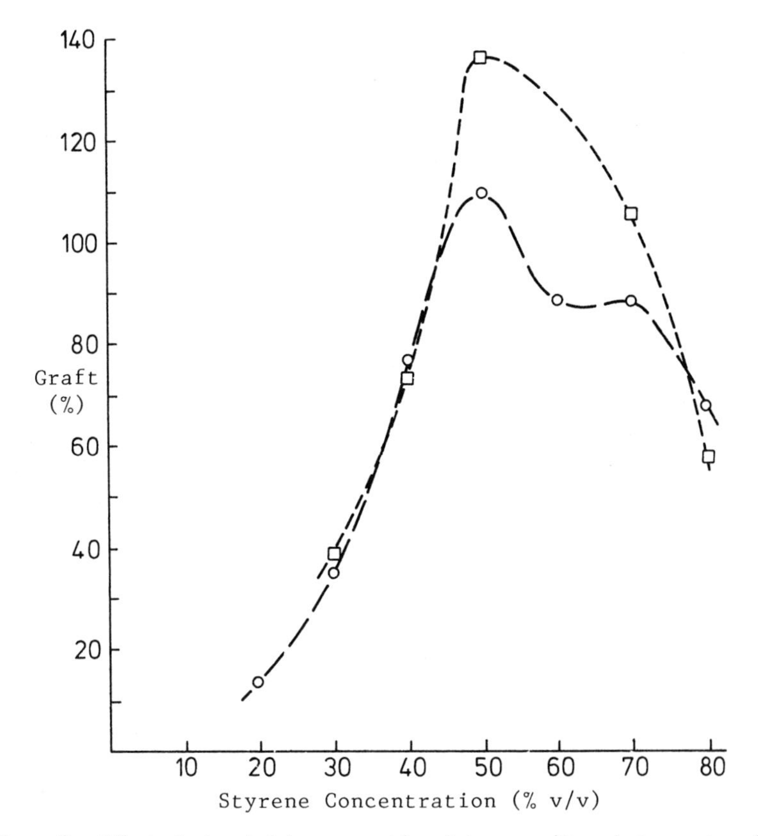

Figure 2. Effect of trimethylol propane triacrylate on grafting of styrene to polyethylene film in methanol at dose rate of 4.1×10^4 rad/h to 2.4×10^5 rad. Key: ○, styrene-methanol; □, styrene-methanol-trimethylol propane triacrylate (1% v/v).

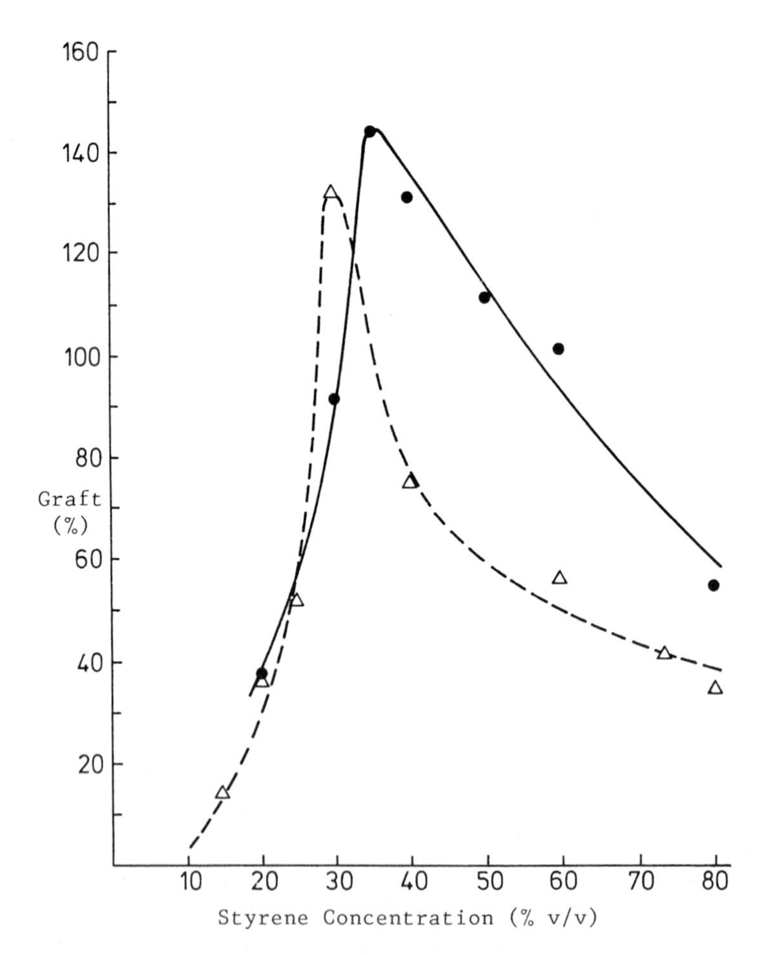

Figure 3. Effect of divinylbenzene on grafting of styrene to polypropylene film in methanol at dose rate of 4.1 × 10⁴ rad/h to 2.4 × 10⁵ rad. Key: △, styrene-methanol; ●, styrene-methanol-divinylbenzene (1% v/v).

Table II. Comparison of Divinylbenzene (DVB) with Mineral Acid as Additives in Styrene Grafting to Polyethylene[a]

Styrene (% v/v)	Graft (%) Neutral	H_2SO_4 (0.2M)	DVB (1% v/v)
20	14	19	15
30	37	51	41
40	76	81	74
50	109	134	136
60	89	119	121
70	89	73	-
80	68	62	74

[a] Styrene in methanol radiation grafted to polyethylene film (0.12 mm) at 4.1×10^4 rad/hr to dose of 2.4×10^5 rad in air.

conditions studied suggest that similar mechanistic pathways may be relevant for the enhancement reaction with both additives. This conclusion is supported by the observation that the functionality of the polyfunctional monomer additive does not appear to markedly influence the magnitude of the enhancement, indicating that cross-linking is not the major pathway for the increase in grafting observed with these additives.

The mechanismsof the acid effect has been extensively investigated (12-15, 21) whereas the current use of the polyfunctional monomers as enhancement additives in grafting is novel. The role of acid in these radiation grafting reactions is complicated and there is evidence that a number of pathways contribute to the overall enhancement effect. Thus mineral acid, at the levels used, should not affect the physical properties of the system such as swelling of the trunk polymer or precipitation of the grafted polystyrene chains. Instead evidence (12) indicates that the acid effect is due to a radiolytic increase in G(H) yields in the monomer-solvent system due to reactions similar to those depicted in Equations 1 and 2 for styrene-methanol.

$$CH_3OH + H^+ \rightarrow CH_3OH_2^+ \quad (1)$$

$$CH_3OH_2^+ + e \rightarrow CH_3OH + H\cdot \quad (2)$$

Such processes can lead to increased grafting yields from enhanced $H^\cdot$ scavenging by monomer and abstraction reactions with the backbone polymer. Evidence has also been presented (12) to indicate that the observed acid effect is due to an increase in monomer-solvent ($MR^\cdot$) radical intermediates in the grafting system. More recent studies (21) on the composition of the grafting solutions show that oligomer chains are of shorter length in the presence of

Table III. Comparison of Trimethylol Propane Triacrylate (TMPTA) with Acid as Additives in Styrene Grafting to Polyethylene[a]

Styrene (% v/v)	Graft (%) Neutral	H_2SO_4 (0.2 M)	TMPTA (1% v/v)
20	14	19	-
30	37	51	39
40	75	81	73
50	109	134	137
60	89	119	-
70	89	73	105
80	68	62	59

[a] Styrene in methanol, radiation grafted to polyethylene film (0.12 mm) at 4.1×10^4 rad/hr to dose of 2.4×10^5 rad in air.

acid, but the numbers of these chains are higher. Because of the smaller size of these chains, the radical intermediates associated with them can diffuse more readily into the swollen backbone polymer to give enhanced yields due to higher concentration and higher mobilities associated with smaller size.

The presence of acid can also lead to enhanced Trommsdorff peaks in radiation grafting (12), an observation that is important in the present work. This result can be attributed to the lower

Table IV. Comparison of Divinylbenzene (DVB) with Acid as Additives in Styrene Grafting to Polypropylene[a]

Styrene (% v/v)	Graft (%) Neutral	H_2SO_4 (0.2 M)	DVB (1% v/v)
20	37	41	38
30	132	156	92
35	-	136	144
40	75	97	132
50	-	-	111
60	56	51	101
70	41	39	-
80	35	31	55

[a] Styrene in methanol radiation grafted to polypropylene film (0.06 mm) at 4.1×10^4 rad/hr to dose of 2.4×10^5 rad in air.

$\bar{M}_n$ values and larger numbers of shorter chains in the presence of acid, since the lower molecular weights lead to a high solubility of oligomer radicals in monomer solution, leading to an increase in viscosity at the gel peak, a reduction in chain termination and a marked acceleration in grafting.

With respect to the inclusion of polyfunctional monomers (e.g. DVB) in the grafting solution, the magnitude of the increment in copolymerisation yield with these additives was comparable to the addition of acid. Preliminary studies show that the grafted polystyrene chains were cross-linked when DVB was in the grafting solution. In the presence of DVB, branching of the polystyrene can occur. This branching of the growing grafted chains takes place when one end of the DVB, immobilised during grafting, is bonded to the growing chain. The other end is unsaturated and free to initiate a new chain growth via scavenging reactions. The new branched polystyrene chain may eventually terminate, cross-linked by reacting with a neighboring polystyrene chain or an immobilised divinylbenzene radical. Grafting is thus enhanced mainly through branching of the grafted chain. The addition of TMPTA, a trifunctional monomer, gives similar results. The effect that a polyfunctional monomer has on $\bar{M}_n$ values of grafted chains is currently being investigated. It may well be that the result is similar to that found with acids and would explain the similarities in degree of grafting enhancement observed with both types of additives.

With radiation grafting, there is also an additional mechanism for enhancement unique to acid and not applicable to the polyfunctional monomer additives. This process is particularly relevant to irradiations performed in air and involves the acid induced decomposition of peroxy species formed radiolytically in the backbone polymer, thus generating further sites where copolymerisation may occur (Equation 3). Current evidence (17) indicates that the contribution

$$P - \overset{|}{\underset{|}{C}} - OOH + H^+ \rightarrow P - \overset{|}{\underset{|}{C}} - O^{\bullet} + H_2O \qquad (3)$$

of this pathway to the overall acid enhancement in grafting is small compared with the previous processes.

Sulfonic Acid Exchange Resins from Graft Copolymers

Sulfuric acid is the most commonly used sulfonating agent for preparing ion exchange resins from chemically prepared styrene-divinylbenzene copolymers (22). Reaction at temperatures up to 100 °C indicates that monosulfonation only is occurring. Using these reaction conditions, the graft copolymers were sulfonated in an analogous manner, the results (Table V) showing that sulfonation efficiency was very low at less than 70 °C, gradually increasing with temperature as indicated by the increasing exchange capacities and reaching a maximum at 90 °C. Above this temperature the

Table V. Effect of Temperature on Sulfonation of Styrene-Polypropylene Graft Copolymers as Measured by Capacities of Resulting Ion Exchange Resins[a].

Copolymer (% Graft)	Capacities (m mole/g) with Temp (°C) 60	70	80	90	103
30	0.4	1.03	1.42	17.6	16.1
38	0.08	1.08(1.08)[c]	1.61(1.63)[b]	19.6(19.5)[b]	18.2
55	0.10	1.24	1.75(1.77)[c]	21.1(21.1)[c]	18.9

[a] Copolymers prepared by radiation grafting styrene in methanol to polypropylene powder. Sulfonation time, 45 min.

[b] Capacities of resins preswelled with chloroform prior to sulfonation.

[c] Capacities of resins preswelled with 1,2-dichloroethane prior to sulfonation.

product begins to char resulting in a reduction of exchange capacity. Time of sulfonation at 90 °C is also important and optimum resin performance was obtained after 45 minutes (Table VI), thus 45 minutes at 90 °C was used as standard sulfonation conditions for all subsequent experiments including those using DVB copolymers. The data in Tables V and VI also clearly show that the capacity of the exchange resin increases with yield of copolymer. The inclusion of chlorinated hydrocarbon solvents, previously used with chemically prepared resins (23), to swell the styrene graft copolymer prior to sulfonation did not improve the efficiency of the sulfonation. This suggests that little grafting occurred in the interior of the polypropylene powder, copolymerisation being predominantly on the surface of the powder.

Table VI. Effect of Time on Sulfonation of Styrene-Polypropylene Copolymers as Measured by Capacities of Ion Exchange Resins[a]

Copolymer (% Graft)	Capacities (m mole/g) with Time (min) 20	30	45	55	65
30	1.17	1.41	17.6	1.79	1.67
38	1.22	1.55	19.6	2.03	1.71
55	1.28	1.64	21.1	1.97	1.78

[a] Copolymers prepared by radiation grafting styrene in methanol to polypropylene powder. Sulfonation temperature, 90 °C.

Effect of DVB on Stability of Ion Exchange Resins

Generally the resins prepared from the graft copolymers of higher yield (30%) without DVB suffered more degradation during sulfonation than those of lower yield (18%) (Table VII), the latter resin retaining most of its original exchange capacity even after a large number of exchanges. Preliminary studies of the effect of sulfuric acid on the polypropylene powder indicated that oxidation readily occurred as well as noncommitant decomposition of the polystyrene chains. The excessive degradation of the graft copolymers of higher yield during sulfonation appears to be related to the nature of the graft copolymer. As grafting increased, the copolymer became enriched in polystyrene not only on

Table VII. Effect of DVB on Stability of Sulfonated Styrene-Polypropylene Graft Copolymer Ion Exchange Resins

	Capacities (mmole/g) for Graft (%)					
Exchange Cycles	18	30	38	55	61	409
1	1.31	1.76	1.96	2.11	12.28	1.99
3	1.21	1.46	1.62	1.70	1.80	1.90
6	1.20	1.38	1.44	1.50	1.56	1.85
9	-	-	1.38	1.43	-	1.89
11	-	-	1.34	1.40	-	1.79
14	1.17	1.24	-	-	-	-
15	-	-	1.25	1.29	1.16	1.74
17	-	-	1.24	1.20	1.12	1.72

[a] Divinylbenzene (1% v/v with respect to styrene) added during synthesis of graft copolymer.

the surface but also within the bulk of the powder. During sulfonation, polystyrene on the surface of the copolymer was sulfonated, the polarity introduced by the acid group enabling the copolymer to swell in sulfuric acid, thus contact of the sulfonating reagent with the aromatic sites deeper within the graft copolymer became possible, leading to the possibility of direct oxidation of the parts of the trunk as sulfonation proceeded. Where the degradation was excessive, fragments of graft copolymer were gradually washed away in the exchange process, resulting in a reduction in exchange capacity. With graft copolymer of lower yield (18%), sulfonation was essentially confined to the surface of the graft, thus intimate contact of trunk polymer and sulfonating reagent was minimised, excessive oxidation avoided and higher exchange stability attained.

Inclusion of DVB into the copolymer also minimised the degradation effect during sulfonation and the resins produced were more stable (Table VII). However the migration of ions in and out of the grafted resins were slower when DVB was included (Table VIII). The results indicate that the DVB resins are structurally

Table VIII. Effect of DVB on Rates of Exchange of Sulfonated Styrene-Polypropylene Graft Copolymer Ion Exchange Resins

Graft %	DVB[a] (%)	Rate of Exchange (Sec)
30	0	<4
38	0	<4
41	1	11-12
40	2	11-12
42	4	15-16
55	0	<5
53	1	11-12

[a] Percentage (v/v with respect to styrene) in graft solution.

stronger and able to withstand the harsh sulfonation conditions although the cross-linked structure tends to hinder the movements of the ions into the resin. The resins without DVB obviously consist of polystyrene chains, immobilised to the trunk polymer, but free to move about with little restriction in accommodating the movements of the exchanging ions. With the DVB resins, there is a network of polystyrene chains randomly cross-linked which gives structural stability to the resin at the expense of some mobility. Obviously a compromise in the use of acid and DVB additives is required to give optimum properties to the resins. In this respect, one of the essential advantages of the current work is that the radiation doses required to achieve a particular percentage graft for ion exchange resin preparation, in the presence of additives, are an order of magnitude lower than those previously used for the same purpose (24), thus indicating the advantages of mineral acid and polyfunctional monomers in this work. The present polyolefin work also acts as a model for analogous studies with other backbone polymers including the naturally occurring materials such as cellulose. The preliminary results in Table I confirm this conclusion since they show that additive effects similar to those discussed in detail in this paper for radiation grafting to the polyolefins are also applicable to cellulose.

Acknowledgment

The authors wish to thank the Australian Institute of Nuclear Science and Engineering, the Australian Atomic Energy Commission and the Australian Research Grants Committee for the support of this research.

Literature Cited

1. Geacintov, N.; Stannett, V.; Abrahamson, E.W.; Hermans, J.J. J.Appl. Polym. Sci. 1960, 3, 54.
2. Reine, A.H.; Arthur, J.C., Jr. Text. Res. J. 1972, 42, 155.
3. Tazuke, S.; Matoba, T.; Kimura, H.; Okado, T. A.C.S. Symp. Ser. 1980, 121, 217.
4. Davis, N.P.; Garnett, J.L.; Urquhart, R. J. Polym. Sci., Polym. Lett. Ed.]976, 14, 537.
5. Chapiro, A. "Radiation Chemistry of Polymeric Systems", Interscience, New York, 1976.
6. Charlesby, A. "Atomic Radiation and Polymers", Pergamon, Oxford, 1960.
7. Krassig, H.A.; Stannett, V.T. Adv. Polym. Sci. 1965, 4, 111.
8. Arthur, J.C. Jr., Adv. Chem. Ser. 1971, 99, 321.
9. Hebeisch, A.; Guthrie, J.T. "The Chemistry and Technology of Cellulosic Copolymers", Springer-Verlag, Berlin, 1980.
10. Nakamura. Y.; Schimada, M. A.C.S. Symp. Ser. 1977, 48, 298.
11. Dilli, S.; Garnett, J.L.; Martin, E.C.; Phuoc, D.H. J. Polym. Sci. C. 1972, 37, 57.
12. Garnett, J.L. J. Rad. Phys. Chem. 1979, 14, 847.
13. Garnett, J.L.; Yen, N.T. A.C.S. Symp. Ser. 1980, 121, 243.
14. Garnett, J.L.; Yen, N.T. Aust. J. Chem. 1979, 32, 585.
15. Chappas, W.J.; Silverman, J. J. Rad. Phys. Chem. 1979, 14, 847.
16. Barker, H.; Garnett, J.L.; Levot, R.; Long, M.A. J. Macromol. Sci.-Chem. 1978, A12(2), 261.
17. Ang, C.H.; Garnett, J.L.; Levot, R. unpublished work.
18. Garnett, J.L.; Leeder, J.D. A.C.S. Symp. Ser. 1977, 49, 197.
19. Davis, N.P.; Garnett, J.L.; Ang, C.H.; Geldard, L. A.C.S. Symp. Ser. 1982, in press.
20. Reichenberg, D. J. Am. Chem. Soc., 1953, 75, 589.
21. Davis, N.P.; Garnett, J.L.; Jankiewicz, S.V.; Sangster, D. F. Proc. 1st PRI Conference Radiation Processing for Plastics and Rubber, Brighton, England, Paper 271, 1980, in press.
22. Pepper, K.W.; Reichenberg, D. Z Electrochem. 1953, 57, 183.
23. Hoover, M.F.; Thompson, R.N.; U.S. Patent 3, 128, 257, 1964.
24. Machi, S.; Sugo, T.; Miyaka, T.; Yonemochi, J. JAERI-M-6244 (Japan).

RECEIVED August 24, 1982

17

Initiation of Polymerizations Involving Polysaccharides and Radio Frequency Cold Plasmas

O. HINOJOSA, T. L. WARD, and R. R. BENERITO

U.S. Department of Agriculture, Agricultural Research Service,
Southern Regional Research Center, New Orleans, LA 70179

The ESR signals of cotton after exposure to both charged particles and uv light of rf cold plasma were identical but of greater intensity than those obtained when cotton was shielded from plasma particles by quartz or CaF_2. Chemiluminescence (CL) was oxygen dependent and greatest with unshielded cotton. CL of shielded and unshielded cottons was more prolonged than that of CL produced by irradiation with a Hg lamp. Plasma uv to 100 nm initiated graft polymerization of methacrylamide on cotton. A combination of depolymerization and polymerization was used with an equimolar N_2 and H_2 plasma to degrade chemically modified cottons in the electrode zone and to cause polymerization downstream to form products containing nitrogen and substituents of the chemically modified cottons.

The modification of cotton cellulose by treatment with low-temperature, low-pressure ammonia plasma created by passing ammonia gas through a radiofrequency (rf) electric field of 13.56 MHz has been reported (1). Earlier reports (2,3,4) were on the effects of rf plasma of argon, nitrogen or air on a group of polysaccharides that included cotton and purified cellulose.

In these earlier studies, the polysaccharides were in open sample holders within the rf reactor. Thus, samples were exposed to fast moving high temperature electrons, the slower moving positive and negative ions, and free radicals as well as to uv irradiations. In all types of rf plasmas investigated, changes in surface properties of the polysaccharides were analyzed by the techniques of electron spectroscopy for chemical analyses (ESCA), electron spin resonance (ESR), multiple internal reflectance infrared spectroscopy (MIR) and chemiluminescence (CL).

Free radicals indicative of breakage of the glucosidic bonds were detected as well as electronically excited carbonyl groups in all plasma irradiated polysaccharides. The products of CL on exposure of these irradiated polysaccharides to oxygen can be attributed to relaxation of excited carbonyls and to transfer of unpaired electrons to molecular oxygen. A linear regression analysis of ESR and CL from a set of six saccharides of varied molecular size and from another set of six celluloses from different plant origins established that CL intensity is highly correlated with ESR intensity (5). Thus, CL of an rf cold plasma irradiated polysaccharide can be predicted from the value of the ESR intensity of the irradiated sample.

The use of rf cold plasmas to cause polymerization of monomers on such non-conducting substrates as glasses (6) and other polymeric substances (7) is not new. However, the chemistry involved in rf cold plasmas is complicated, and within the reactor, depolymerizations of substrates and of newly formed polymers occur simultaneously with polymerization reactions. Clark (8) has reported on such phenomena, and it is known that the relative rates of polymerization to depolymerization vary within the reactor. Depolymerization, particularly of a polysaccharide substrate is greatest between the electrodes and almost insignificant downstream from the electrodes. Polymerization is the dominant reaction downstream from the electrodes, but within the glow area of the plasma.

This is a report of a study to use depolymerization or degradation and polymerization reactions within the rf reactor to advantage in the formation of new polymers and to investigate the relative contributions of rf produced particles (electrons and ions) and of rf produced uv light in the initiation of graft polymerizations on cotton substrates.

Cold Plasma Treatments

Details of the usual procedures followed with the 100-W, 13.56 MH_z rf generator and the capacitively and inductively coupled plasma reactor have been reported (2). In Figure (1) is a diagram of the reactor that indicates positions of the external electrodes, sample locations, and position of CaF_2 sample holder used in specified experiments. In brief, the reactor was evacuated to 20 mtorr before a selected gas used for the plasma was allowed to enter the reactor between the electrodes such that a flow rate of 0.1 standard cm^3/sec was obtained. After stabilization of the pressure in the reactor to about 400 mtorr, the rf generator was turned on, the power adjusted to 40W, and the sample irradiated for the desired time. After irradiation, the reactor pressure was returned to atmospheric with Ar and samples were removed and stored in Ar pending analyses.

Analyses

Chemiluminescence (CL) was measured on all samples immediately after plasma irradiation and before exposure to air or moisture and again after exposure to room air. CL in counts per min (CPM) was determined in a Packard 3255 liquid scintillation spectrometer equipped with low dark-noise photomultiplier tubes (RCA 4501/V4) and a Parkard 585 linear recorder.

Electron spin resonance (ESR) spectra were obtained on a Varian 4502-15 spectrometer system equipped with a variable temperature accessory and a dual sample cavity.

Electron spectroscopy for chemical analyses (ESCA) spectra were obtained on a Varian spectrometer Model VIEE-15 with a MgK_{α} x-ray source. Analyses were obtained on samples before and after plasma irradiations. Spectra represent electron binding energies (E_{BE}) in eV.

Syntheses via Degradation and Polymerizations

Others (9) have reported on the nature of solid products arising from rf plasmas of CO,N_2 and H_2 that were electrodelessly sustained at 13.56 MHz. While these investigators reported that hydrolysis of polymers showed presence of various amino acids, including those with aromatic groups, such as tyrosine, we (10) have formed proteinaceous material lacking in aromatic groups when the sole source of carbon within the reactor was cellulose. Aromatic groups were absent when the plasma gas was N_2, NH_3 or combinations of N_2 and H_2 (5). However, even with the same gaseous species in a given plasma, the walls of the container and more importantly a solid substrate such as a polysaccharide within the reactor can have a chemical or catalytic effect in determining the final complex organic product via gas-solid phase processes.

In attempts to form proteinaceous material with aromatic substituents, a cotton modified chemically to contain aromatic groups was placed in the depolymerization zone, that is, between the external electrodes and an untreated cotton placed downstream in the polymerization zone. The plasma was an equimolar mixture of N_2 and H_2 that had been shown to be optimal for the production of proteinaceous material when cotton was placed between the electrodes. According to ESCA analyses, the nitrogenous films formed downstream were identical whether benzoylated cellulose or benzylated cellulose was placed between the electrodes. Figure (2) illustrates the C_{1S} spectra obtained with a cotton control (lowest curve) and with a cotton that had been treated in a cold plasma of 1:1 mole ratio of N_2 and H_2 for one hour while a benzylated cotton placed between the electrodes acted as a source of carbon for syntheses of the proteinaceous product (uppermost curve). The middle curve is that for a cotton subjected to the N_2:H_2 plasma in the absence of the benzylated cotton. The C_{1S}

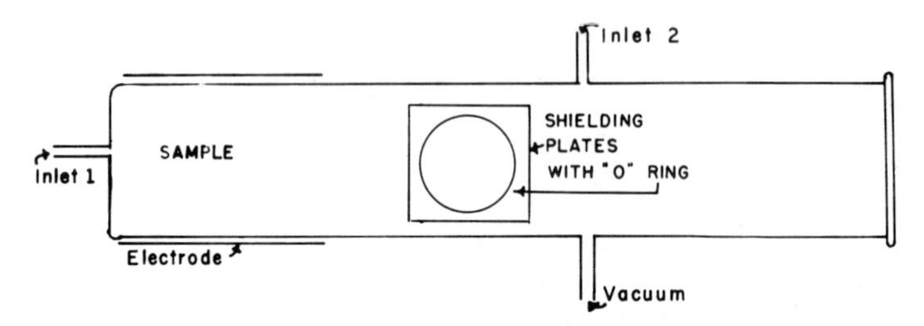

Figure 1. Plasma reactor showing location of external electrodes, sample location between electrodes, and location of sample holder for shielding substrate downstream from electrodes.

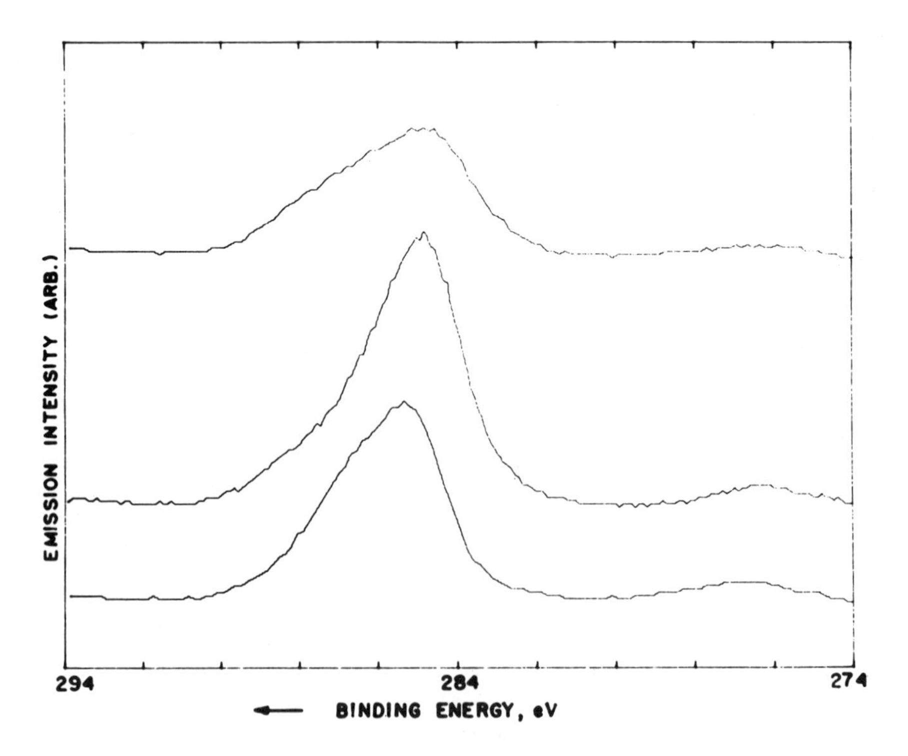

Figure 2. Electron emission spectra for C_{1S} electrons from cottons. Lowest curve is for an untreated cotton. Other curves are for cottons placed downstream in rf reactor with equimolar N_2:H_2 plasma. For uppermost curve, a benzylated cotton was placed in the depolymerization zone between external electrodes during irradiation of cotton.

peak at 284 eV with a full width at half height (FWHH) of 3.0 eV is typical of cellulose and other polysaccharides. The broadening of the C_{1S} peak in the product to a FWHH of 4.5 eV is indicative of formation of groups containing carbon atoms of higher and lower electron densities than those in the original cellulose. Aromatic groups contribute carbon atoms of E_{BE} about 284 and carbonyl groups contribute C atoms of E_{BE} about 286 eV. No significant differences were detected when a benzoylated cotton was substituted for a benzylated cotton as the source of aromatic groups.

Use of phytic acid placed between the electrodes when the N_2:H_2 plasma was used resulted in nitrogenous product containing the phytate group. These are examples of the use of a solid placed between the electrode as the source of a moiety to be used in a polymerization reaction downstream from the electrodes.

Photoinitiation

A cotton print cloth was treated with an aqueous solution of 0.5M methacrylamide (MA), which has a characteristic absorption at 224 nm. If such a fabric sample is air dried before being enclosed in a CaF_2 cell and subjected to an argon plasma for 30 minutes, only a small amount of polymer is grafted. A typical N_{1S} ESCA spectrum of a sample so treated is the upper curve in Figure (3). Attempts to do post polymerization of MA by first subjecting a cotton print cloth to 30 minutes of an Ar plasma and then immersing it in an aqueous solution of 0.5M MA also met with little success. The lower curve in Figure (3) is the N_{1S} spectrum of a sample irradiated in an open container and then subjected to MA. These ESCA spectra are the results of 100 scans.

Grafting was successful when a cotton print cloth treated with aqueous 0.5M MA was enclosed while still wet in the CaF_2 sample holder and then irradiated in an Ar plasma. The N_{1S} spectrum of a sample that had a 10% add-on is shown in the upper curve of Figure (4) and was obtained with only 10 scans. Note that spectra of Figure (3) are 100 scans. The control shown as the lower curve in Figure (4) was obtained on a sample that had been treated with 0.5M MA for one hour, then washed and air dried before being subjected to ESCA. The E_{BE} of each peak in Figures (3) and (4) is 399.9 eV, indicating amino type nitrogen. The FWHH for the samples of highest add-on (upper curve Fig. 4) was the least and was 2.2 eV. The C_{1S} ESCA spectrum for this optimally polymerized sample is the upper curve of Figure (5). The large FWHH of 3.6 eV is indicative of several types of C_{1S} electrons. The lower curve of Figure (5) is the C_{1S} spectrum of a cotton control showing a peak at 285 eV and a FWHH of 3.0 eV. The C_{1S} peak characteristic of a carbonyl group in polymerized MA would be about 287 eV and account for the shift of the C_{1S} spectrum to higher E_{BE} values. Other experiments have shown that MA can be grafted if treated in an open container within the

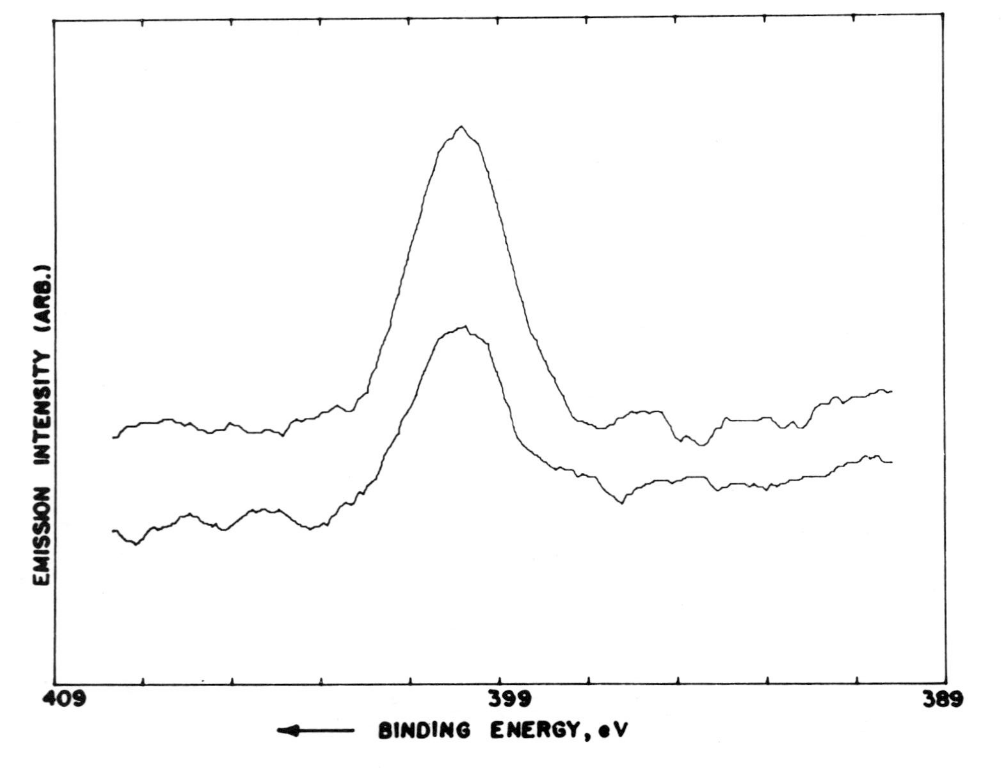

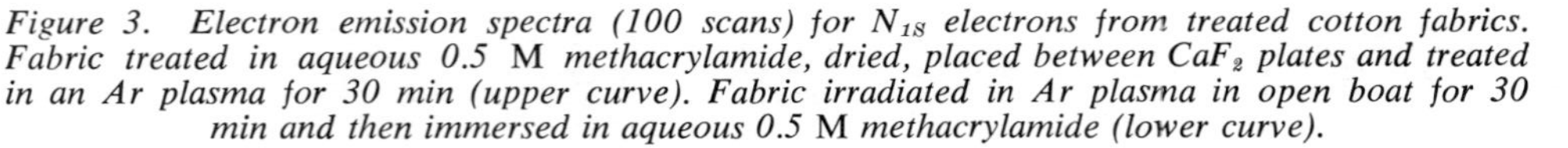

Figure 3. Electron emission spectra (100 scans) for N_{1S} electrons from treated cotton fabrics. Fabric treated in aqueous 0.5 M *methacrylamide, dried, placed between CaF_2 plates and treated in an Ar plasma for 30 min (upper curve). Fabric irradiated in Ar plasma in open boat for 30 min and then immersed in aqueous 0.5* M *methacrylamide (lower curve).*

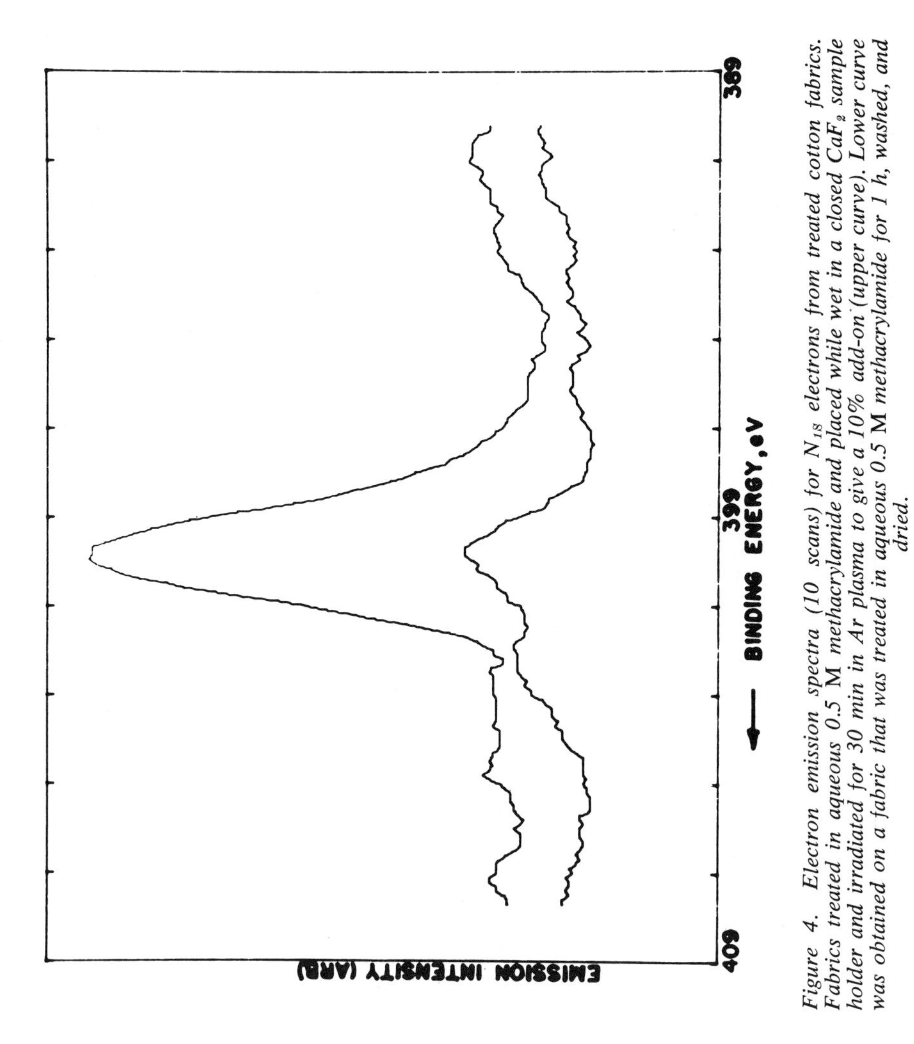

Figure 4. Electron emission spectra (10 scans) for N_{1S} electrons from treated cotton fabrics. Fabrics treated in aqueous 0.5 M *methacrylamide and placed while wet in a closed* CaF_2 *sample holder and irradiated for 30 min in Ar plasma to give a 10% add-on (upper curve). Lower curve was obtained on a fabric that was treated in aqueous 0.5* M *methacrylamide for 1 h, washed, and dried.*

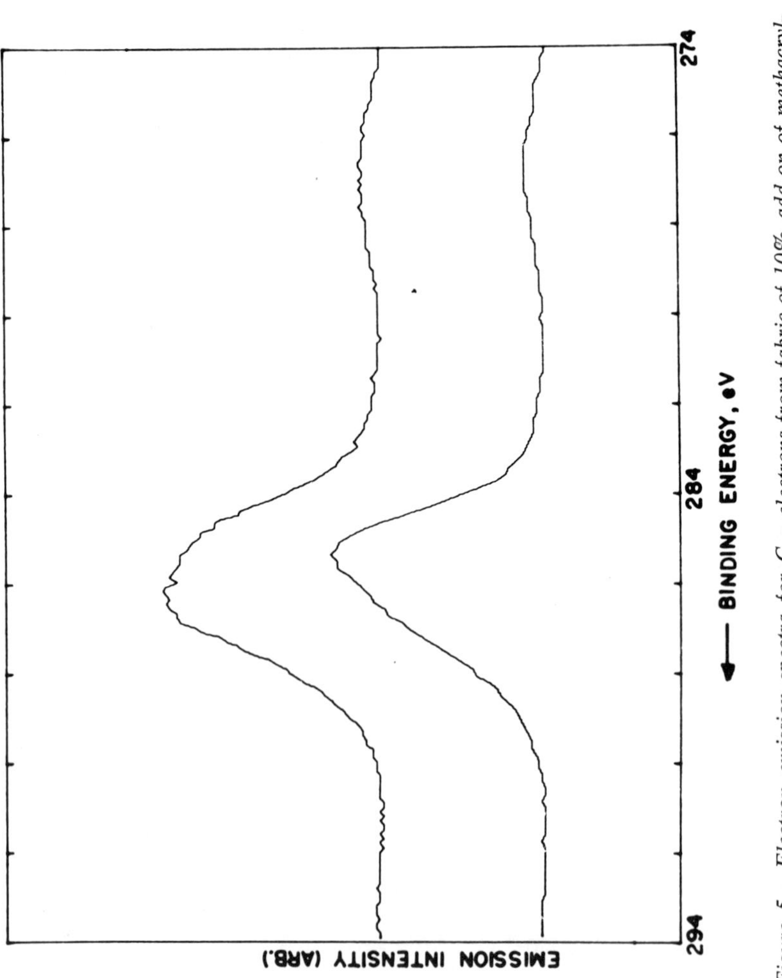

Figure 5. Electron emission spectra for C_{1S} electrons from fabric of 10% add-on of methacrylamide described in Figure 4 (upper curve), and of an untreated cotton control (lower curve).

reactor. Under such conditions, the sample is bombarded by electrcns, ions and radicals as well as by the short uv radiations. When the sample is enclosed in a quartz container, which allows for passage of light as low as 200 nm, some polymerization of MA is obtained. The use of a CaF_2 enclosure that passes light down to 100 nm allows for the photoinitiation of polymerization of MA.

ESR and CL

The production of paramagnetic species in cellulose can be attributed to interactions between particles, including electrons, ions, or free radicals, with cellulose or the effects of uv irradiations on cellulose. The ESR signals obtained from cotton samples placed in open boats within the reactor, within quartz sample holders, and within CaF_2 sample holders showed that the signal obtained in every case was similar but different in intensity. If the irradiated samples are kept in dry Ar, they are stable. In Figure (6) are the ESR signals obtained after 1 week storage in Ar of samples irradiated in open containers (magnified 1X), CaF_2 container (magnified 2X), and quartz container (magnified 4X). These ESR spectra are for samples that had been irradiated for identical times in identical zones of the plasma reactor, but with containers that provided for different shielding.

The CL of the variously irradiated samples were also investigated. If a cotton is irradiated with the uv light from a mercury lamp for 30 minutes, the CL produced is about 150,000 CPM. However, there is an extremely fast decay and the CL is insignificant within 3-4 minutes. The decay is fast even in dry N_2 or Ar gases. Those samples subjected to plasma show a slower decay in CL. Those samples subjected to Ar plasma in open boats, quartz or CaF_2 containers continued to chemiluminesce even after a week. The CL values immediately after plasma treatments were extremely high and were reduced to about 10,000 - 15,000 CPM after a week, as shown by the initial CPM in Figure (7). Figure (7) shows the increases in CL with time of exposure of the samples to air. All three samples show increases in CL on exposure. The increases for those samples shielded by quartz and CaF_2 are similar. Greatest increase in CL was observed with the sample treated in the open boat. It should be noted that the open boat treated sample had the greatest ESR signal (Figure 6). Apparently, the free radicals produced by the fast moving electrons or other particles are more effective in producing CL in the cellulose than the free radicals resulting from uv irradiation at the different wavelengths.

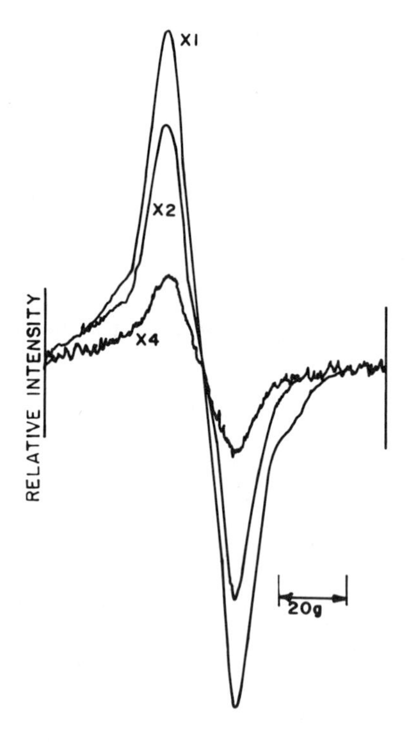

Figure 6. Electron spin resonance spectra for samples of cotton printcloth that had been treated for 30 min in an Ar plasma and then stored in Ar for 1 week. Sample in an open container (uppermost curve), sample shielded by CaF_2 (middle curve), and sample shielded by quartz (lowest curve). Intensities are of 1, 2, and 4 magnifications for comparative purposes.

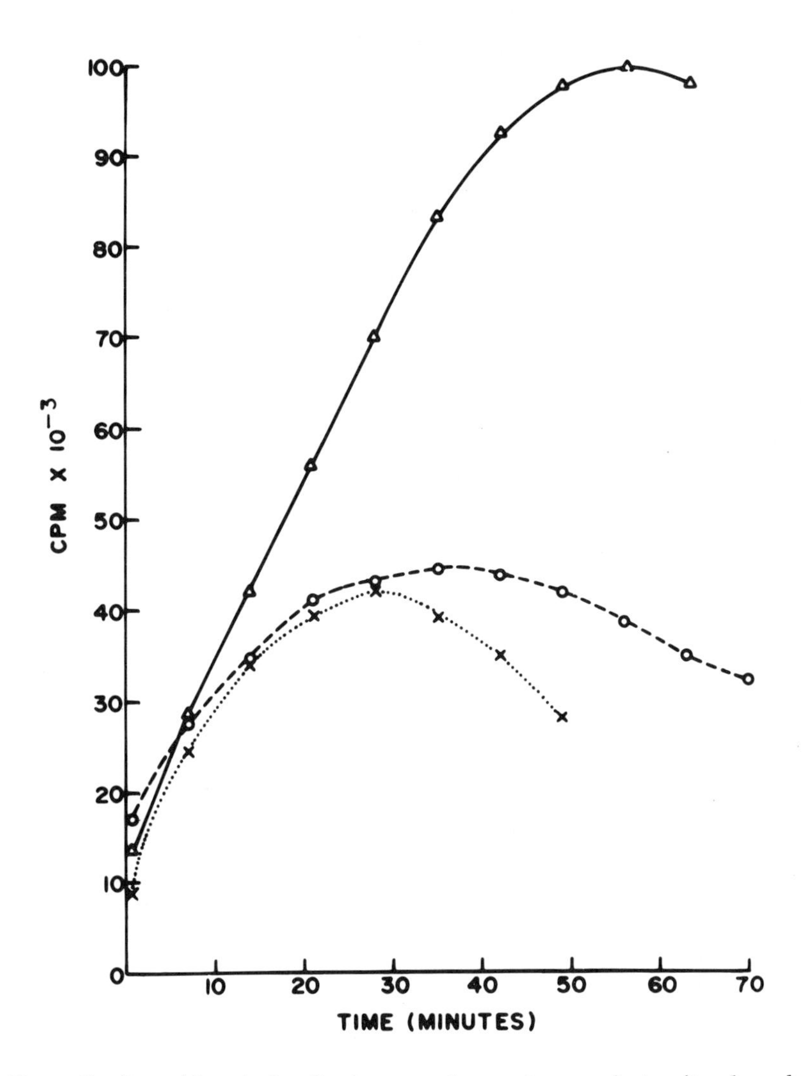

Figure 7. Intensities of chemiluminescence in counts per minute after 1 week ($t = 0$) of storage of samples in Ar and with time (t) after exposure to air for samples irradiated in Ar plasma for 30 min in an open container (uppermost curve), a CaF_2 container (middle curve), and a quartz container (lowest curve).

Literature Cited

1. Ward, T.L. and Benerito, R. R. Textile Res. J., (1982), 52, 256-262.
2. Jung, H. Z., Ward, T. L. and Benerito, R. R. Textile Res. J., (1977), 47, 217-222.
3. Ward, T. L., Jung, H. Z., Hinojosa, O. and Benerito, R. R., Appl. Polym. Sci., (1979), 23, 1987-2003.
4. Ward, T. L., Jung, H. Z., Hinojosa, O., and Benerito, R. R. J. Surface Sci., (1978), 76, 257-273.
5. Ward, T. L. and Benerito, R. R. Polym. Photochem. (1982), Accepted for Publication.
6. Goodman, J. J. Polym. Sci., (1960), 44, 551-552.
7. Coleman, J. H., U.S. Pat. 3,600,122, Aug. 1971.
8. Clark, D. Organic Coatings and Plastics Chem. (1980) 42, 460.
9. Hollahan, J. R. and Emanuel, C. F. Biochimica et Biophysica Acta, (1970), 208, 317-327.
10. Ward, T. L., Hinojosa, O., and Benerito, R. R. Organic Coatings and Applied Polymer Science Proceedings, (1981), 45, 382-385.

RECEIVED September 1, 1982

18

Thermal Decomposition of Cotton Modified by Grafting with Acrylamide and Bis(β-chloroethyl) Vinylphosphonate

MACHIKO SHIMADA and YOSHIO NAKAMURA

Gunma University, Faculty of Technology, Kiryu, Gunma 376, Japan

Cotton with flame-retardant property was made by grafting with acrylamide and bis-(beta-chloroethyl) vinylphosphonate, and subsequent treatment with stannic chloride. Thermal decomposition behaviors of unmodified and modified samples were studied under nitrogen atmosphere at 500°C. It was found that carbamoylethyl- and carboxyethylation, grafting and stannic chloride treatment increased the yield of char residue and water, but decreased that of tar. In flame-retardant cotton, the temperature of thermal decomposition onset and completion were lower than those in unmodified cotton. And the rate of thermal decomposition was slower than that in unmodified cotton. Phosphorus content in char residue did not change by thermal decomposition. But chlorine content introduced by grafting decreased by thermal decomposition.

Cotton is a natural fiber with many desirable properties as clothing. But it is flammable and leaves little residue after combustion in the air. Consequently many people are injured by fire. So we planned to make a flame-retardant cotton without loss of strength and of water imbibition.

Experimental

Materials. Scoured cotton cellulose fibers of the Acala variety supllied by Kanebo Co. Ltd. were purified by extracting with hot benzene-ethanol mixture (1:1 vol. ratio) for 24h. Then the cotton fibers were washed with methanol and distilled water and air-dried. Acrylamide (AM) was purified by recrystallization from benzene for several times. Bis-(beta-chloroethyl) vinylphosphonate (Fy) was purified by passing the monomer through a column filled up with activated alumina to remove inhibitors of polymerization. Other chemicals used were reagent grade, and were used without further purification.

0097-6156/83/0212-0237$06.00/0

Mercerization. The purified cotton fibers were mercerized in 20% sodium hydroxide aqueous solution for 5h at 20°C. Then the samples were washed with distilled water, immersed in 0.1% acetic acid, washed with distilled water until acetic acid was throughly removed and then air-dried.

Carbamoylethylation and Carboxyethylation. The purified cotton fibers were carbamoylethylated and carboxyethylated in aqueous solution containing 7% acrylamide and 20% sodium hydroxide for 10h at 20°C. Then the samples were washed with distilled water, 0.1% acetic acid, and distilled water and air-dried. Degree of carbamoylethylation and carboxyethylation per anhydroglucose residue of the samples was determined by micro Kjeldahl method and back-titration technique (1), respectively.

Grafting Reaction. After the samples were dried under reduced pressure at 50°C for 24h, they were irradiated with Co-60 gamma-rays at an exposure rate of 1.0×10^6 R/h for 1h at room temperature under nitrogen atmosphere. The irradiated samples were grafted with AM and Fy using emulsifying agent at 30°C under nitrogen atmosphere. The monomer concentration ratio was 1:1 in weight for AM and Fy. The extent of grafting was expressed as percent of weight increase by grafting. As the reactivity of mercerized cellulose (M) was more than that of carbamoylethylated and carboxyethylated cellulose (C), the total monomer concentration was determined as 6% and 14% for M and C, respectively.

Stannic Chloride Treatment. M and C were treated with stannic chloride aqueous solution. Grafted samples were also treated with stannic chloride solution. After the treatment, weight increase by stannic chloride treatment was calculated. As weight increase by stannic chloride treatment was much different from the samples, the concentration of the treatments were changed from the samples.

Isothermal Pyrolysis. Pyrolysis of about 0.5g sample was carried out at 500°C for 2min under nitrogen flow of 100ml/min.The products of pyrolysis were devided into four fractions, char, tar, water and volatile fractions. The char fraction was solid ash that remained in the sample tube and was cooled to room temperature under nitrogen atmosphere to prevent second decomposition by oxygen After drying, the weight of char was measured, and the yield of char was expressed as percent to original weight.The tar fraction was the liquid that condensed in the cool part of the sample tube and that was not volatile below 120°C. After drying at 120°C for 12h, the weight of tar was measured, and the yield of tar was expressed as percent to original weight. To determine the extent of water after pyrolysis, the sample tube was throughly washed with standard methanol. The yield of water was determined by Karl Fisher titration method, and expressed as percent to original weight. Volatile fraction was not analyzed in this experiment.

Flammability Test. Flammability of the samples was evaluated by two methods using cotton fabrics. Cotton fabrics were treated in the same way as cotton fibers. Oxygen index of the samples was measured according to the JIS K 7201 procedure. Another method was to estimate the charred area of the burned sample in air using micro burner.

Thermal Gravimetric Analysis(TGA). The TGA thermograms were obtained using Rigakudenki TGA standard type. Samples were cut with scissors to fine pieces and uniformly packed in sample pans. TGA curves were run under nitrogen atmosphere at a constant heating rate of $5^{\circ}C$/min. Alpha-almina was used as reference material.

Crystallinity Index. X-ray diffraction traces of the samples were measured with Rigakudenki X-ray diffractometer type D-9c. Samples partially and completely charred were taken from TGA experiments. Crystallinity indices of the samples were determined according to Segal's method (2).

Elementary Analysis. Samples were burned in combustion flask filled up with oxgen to determine the phosphorus and chlorine contents. Phosphorus content was colorimetrically determined by phosphovanadomolybdatate method (3). Chlorine content was determined by titration with mercuric nitrate.

Analysis of Levoglucosan. Tar was trimethylsilylated in the mixture of pyridine, trimethylchlorosilane, and hexamethylenedisilasane at room temperature for 1h (4). The trimethylsilylated sample was analyzed by gas chromatography (Hitachi GC type 163). Identification of levoglucosan was made up with mass spectrometry (JEOL JMS 07) (5).

Results and Discussion

Effects of Carbamoylethyl and Carboxyethyl Groups on Pyrolysis Products. Three samples with different properties were pyrolyzed at $500^{\circ}C$ for 2 min under nitrogen atmosphere. The yields of products are shown in Table I. Sample C-HCl was made by treating sample C with 3N-HCl at $50^{\circ}C$ for 1h, and it was assumed that carbamoylethyl group was converted to carboxyethyl group. In M, tar fraction was very much compared with char and water fractions. In C and C-HCl, char and water fractions were more and tar fraction was less than those in M. The tendency was more remarkable in C thanC-HCl. Though all samples have the crystalline structure of cellulose II, the crystalline region in C and C-HCl is rough compared with that in M, because substituents are introduced in amorphous and crystalline regions in C and C-HCl (6). So the difference in pyrolysis products is considered to be caused by carbamoylethyl and carboxyethyl groups, and stiffness of crystalline region. To clarify this point, carbamoylethyl and carboxyethyl groups were introduced only in amorphous region of M, and pyrolysis of the sample was carried out qualitatively. The result showed that carbamoylethyl group make increase char and water fractions and decrease tar fraction. Carboxyethyl group also contributes to make increase char and water fractions, but ability is weak compared with carbamoylethyl group. And it was also clarified that cellulose with rough structure left more char than cellulose with stiff structure.

Effects of Grafting on Pyrolysis Products. The relations between extent of grafting and yield of char, tar and water frac-

Table I. Pyrolysis Products of M, C and C-HCl at 500°C for 2min under nitrogen atmosphere.

Sample	Char	Tar	Water	Volatile
M	9.1	37.6	16.2	37.1
C	18.9	24.5	24.8	31.8
C-HCl	11.2	27.4	20.9	40.5

tions by pyrolysis of M and C at 500°C for 2min under nitrogen flow are shown in Figure 1. In the case of M, the yields of char and water increased remarkably as extent of grafting increased. The yield of tar decreased abruptly as extent of grafting increased up to 3.5% and then gradually decreased. Char and water fractions increased linearly and tar fraction decreased as extent of grafting increased in the case of C. So it can be said that graft polymer of AM and Py is effective to increase char and water fractions and to decrease tar fraction by pyrolysis in both samples. Char fraction of M is less than that of C until extent of grafting increased up to 3.5%. As main chain of polyacrylamide and carbamoylethyl group have similar structure, the effects to increase char by pyrolysis is assumed to be the same. After extent of grafting exceeded 3.5%, char fraction of M was more than that of C. So the effects of carbamoylethyl and carboxyethyl groups to increase char residue are ignored at high extent of grafting. So the effects of graft polymer to increase char are more in M than in C. The difference is assumed to be due to the difference of structure between two samples. Tar and water fraction showed similar behavior in two samples. So it is considered that amount of graft polymer influences the yields, and that structure of trunk polymer does not influence the yields.

Effects of Stannic Chloride Treatment on Pyrolysis Products. The relations between weight increase by stannic chloride treatment and yields of pyrolysis products of M and C are shown in Figure 2. And those of grafted samples are shown in Figure 3. To compare with C, C was treated with 3N-HCl and yields of pyrolysis products are also shown in Figure 2b (C-HCl). In the case of M, char and water fractions increased as weight increase by stannic chloride treatment increased. The tendency was remarkable for char fraction. Tar fraction decreased as weight increase by stannic chloride treatment increased. In the case of M, char and water fractions increased as weight increase by stannic chloride treatment increased. The tendency was remarkable for char fraction.Tar fraction decreased as weight increase by stannic chloride treatment increased. In the case of C, char fraction did not change as weight increase by stannic chloride treatment increased. Stannic chloride treatment was not useful for C to increase yield of water.

Yield of char and tar showed very similar behavior in two samples as weight increase by stannic chloride treatment increased. But yield of water was different in two samples. So it is assumed

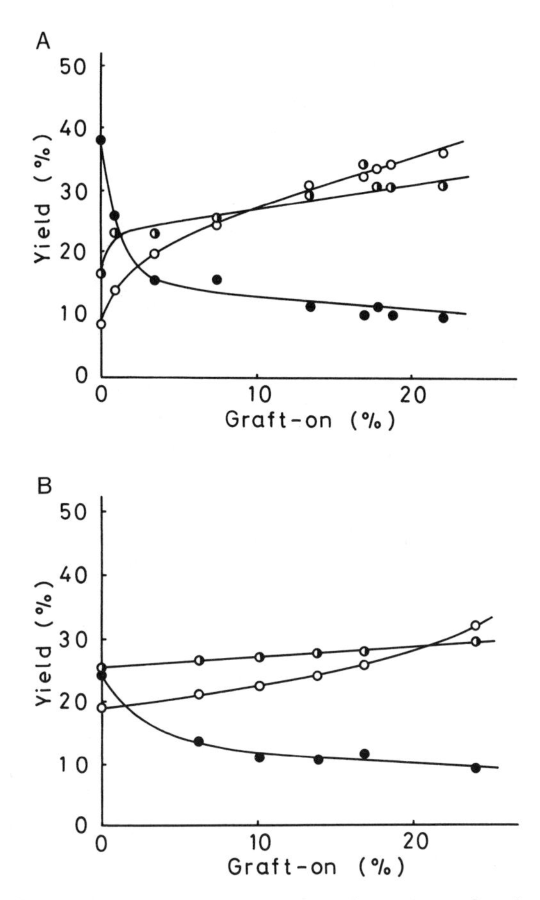

Figure 1. Effects of grafting on pyrolysis products. Key: ○, char; ●, tar; ◑, water. A, mercerized cellulose; B, carbamoylethylated and carboxyethylated cellulose.

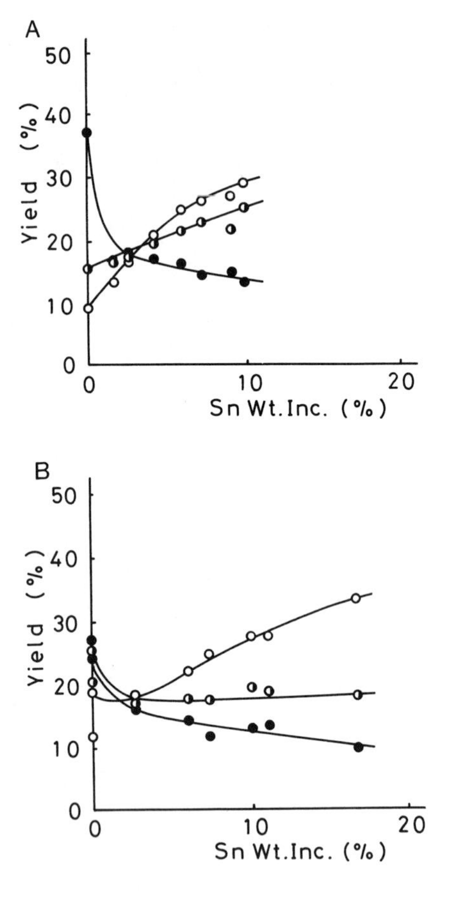

Figure 2. Effects of stannic chloride treatment on pyrolysis products. Key: ○, char; ●, tar; ◑, water. A, mercerized cellulose; B, carbamoylethylated and carboxyethylated cellulose.

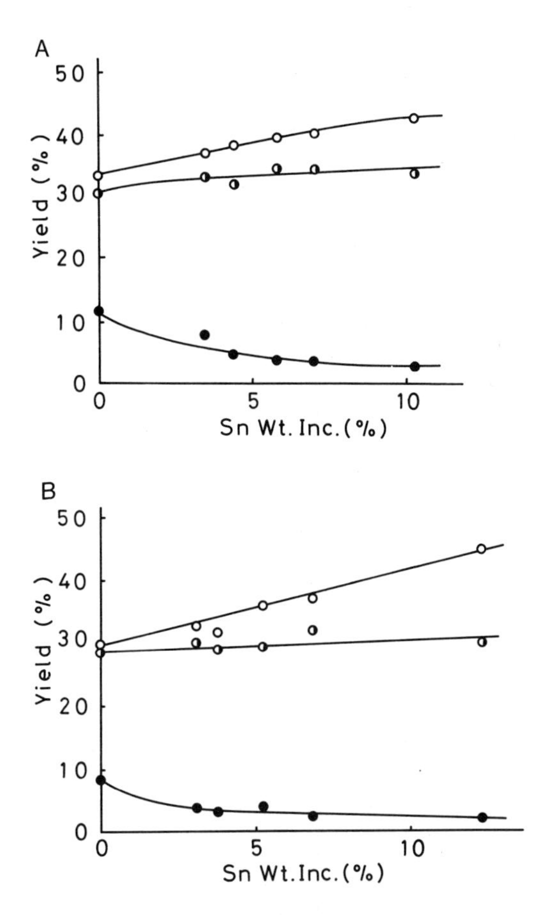

Figure 3. Effects of stannic chloride treatment on pyrolysis products of grafted samples. Key: ○, char; ●, tar; ◑, water. A, mercerized cellulose (extent of grafting is 18.1%); B, carbamoylethylated and carboxyethylated cellulose (extent of grafting is 23.8%).

that char and tar are influenced only by the amount of tin compounds introduced. But water is influenced by structural difference between two samples. As mentioned before, weight increase by stannic chloride treatment was different in two samples. It is easy to increase weight by stannic chloride treatment in the case of C. So the strength of interaction between cellulose and tin compound is different from the samples. In the case of grafted sample of M, char fraction increased, water fraction increased very little, and tar fraction decreased as weight increase by stannic chloride treatment increased. The char fraction increased, tar fraction decreased, and water fraction hardly increased as weight increase by stannic chloride treatment increased in the case of grafted sample of C. Yields of three fractions showed similar behavior in grafted samples. Probably effects of carbamoylethyl and carboxyethyl groups and crystalline roughness of C are shielded by grafting. Char and water fractions are influenced only by the amount of tin compounds introduced. But water fraction was hardly influenced by stannic chloride treatment. So tin compounds introduced do not influence the dehydration of copolymer of AM and Fy.

LOI and Charred Area of Burned Fabric. Limited oxgen indices (LOI) of the samples and the result of flammability test are shown in Figure 4. M burned completely by combustion in the air. C burned but left stiff char residue with figure of fabric after combustion in the air. Grafted samples and stannic chloride treated samples burned little. But stannic chloride treated samples of grafted samples were charred only at the point close to flame. LOI of the sample showed that M-G-Sn and C-G-Sn are self-extinguishable in the air.

TGA. TGA curves of the samples are shown in Figure 5. M decomposed abruptly at 300-350°C. C began and finished to decompose at lower temperature than those of M. Char residue at 400°C was 25% and 37% for M and C, respectively. The difference between two samples is considered to be caused by the change of crystalline structure and substituents. The TGA curves of the grafted samples are similar to each other (Figure 5B). The samples started to decompose at 200°C and ended at 310-330°C. Char residue at 400°C was 46% and 43% for grafted samples of M and C, respectively. The effects of graft polymer on M were greater than those on C. So it is clear that the difference of trunk polymer does not affect the pyrolysis behavior in grafted sample. Graft polymer affects the decomposition behavior. The TGA curves of stannic chloride treated samples showed that decomposition started at 250°C and ended at 320°C in the case of M. But in the case of C, decomposition started at 200°C and decomposed gradually and gradually. Char residue at 400°C was 39% and 41% for stannic chloride treated samples of M and C, respectively. The effects of stannic chloride treatment are considered to be different between M and C. The TGA curves of stannic chloride treated samples after grafting indicated that decomposition started at 250°C and did not

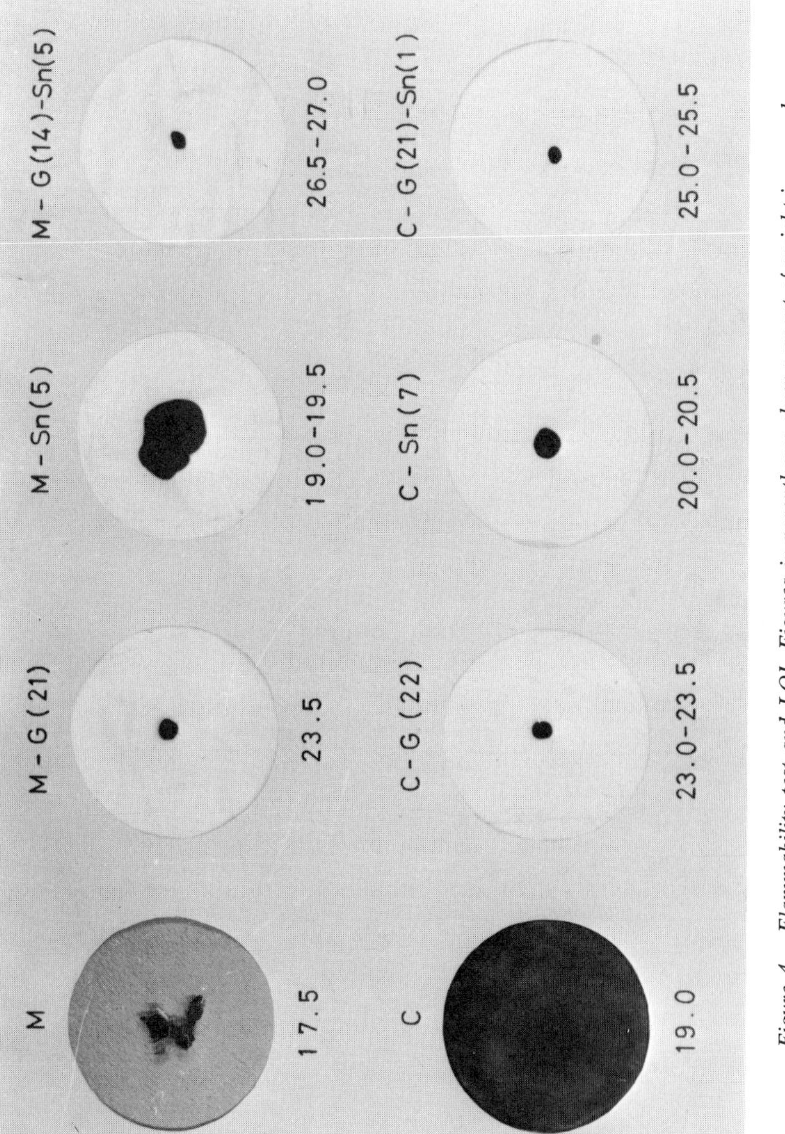

Figure 4. Flammability test and LOI. Figures in parentheses show percent of weight increase by grafting and stannic chloride treatment. Figures not in parentheses show LOI.

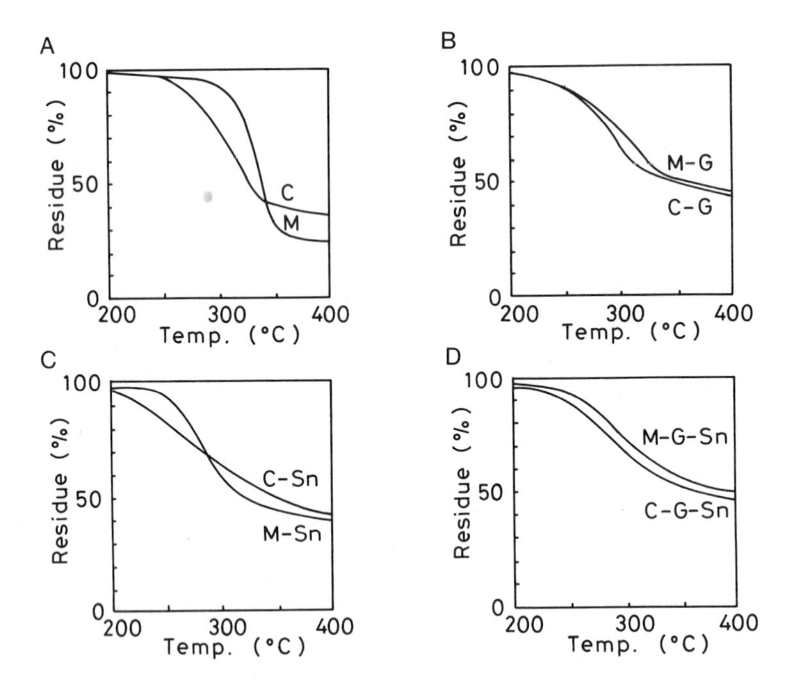

Figure 5. TGA curves of unmodified and modified cellulose. A, mercerized cellulose (M) and carbamoylethylated and carboxyethylated cellulose (C); B, grafted samples of M and C; C, stannic chloride treated cellulose of M and C; D, stannic chloride treated cellulose of grafted samples.

finish even at 400°C in both samples. Char residue at 400°C was 50% and 46% for M and C, respectively. As mentioned above, it is clear that carbamoylethylation and carboxyethylation, stannic chloride treatment, and grafting expedite thermal decomposition temperature, delay decomposition rate, and leave much char.

Crystallinity Index. Crystallinity index of the samples charred partially and completely were measured by X-ray diffraction and shown in Table II. M kept crystal structure even at 320 °C, but decomposed almost completely at 340°C. The decomposition was very abruptly. The result is consistent with that obtained by TGA. C decomposed gradually and gradually. The decomposition initiated already at 280°C and finished completely at 320°C. Grafted samples decomposed at lower temperatures than M. Stannic chloride treated samples decomposed at lower temperatures. This is most remarkable in C-G-Sn. It finished decomposition at 300°C. Like this, the decomposition of cellulose crystal structure initiated and finished at lower temperatures in C, grafted cellulose, and stannic chloride treated cellulose than those in M.

Table II. Crystallinity Index.

Sample	G-on (%)	Sn wt.inc. (%)	Crystallinity Index (%)					
			R.T.	260°C	280°C	300°C	320°C	340°C
M	-	-	51	-	51	53	52	6
C	-	-	47	-	35	24	0	0
M-G	20	-	58	-	39	25	0	-
C-G	22	-	53	-	38	10	0	-
M-Sn	-	9	57	-	46	45	0	-
C-Sn	-	11	47	36	38	24	0	-
M-G-Sn	20	10	57	-	30	18	0	-
C-G-Sn	22	10	42	38	32	0	0	-

Levoglucosan. The amount of tar and levoglucosan at 500°C is shown in Table III. The amount of tar and levoglucosan is shown as relative value to that of M. The amount of tar was very much in M, but that in other samples was very little. Same phenomena are recognized in the amount of levoglucosan. So it can be said that flame-retardant cotton yields little amount of tar and levoglucosan.

Phosphorus and Chlorine Contents. Phosphorus and chlorine contents of the samples in char residue at 400°C are shown in Table IV. Phosphorus content did not change by thermal decomposition in all samples. But chlorine content decreased by pyrolysis. Chlorine introduced by stannic chloride treatment hardly decreased, but that by grafting decreased easily by pyrolysis. Stannic chloride is not introduced in cotton in the form of stannic chloride, because the chlorine content of stannic chloride treated cellulose is very lower than that calculated from weight increase. X-ray diffraction trace of burned sample stannic chloride treated showed that tin compounds introduced in cotton was similar to stannic oxide. But all tin compounds are not in the form.

Table III. Amount of Tar and Levoglucosan by Pyrolysis.

Sample	G-on (%)	Sn wt.inc. (%)	Tar (M=100)	Levoglucosan (M=100)
M	-	-	100	100
C	-	-	65	32
M-G	18	-	30	17
C-G	24	-	22	16
M-Sn	-	10	33	20
C-Sn	-	10	36	22
M-G-Sn	18	10	10	5
C-G-Sn	24	12	6	5

Table IV. Phosphorus and Chlorine Contents in Char.

Sample	G-on (%)	Sn wt.inc. (%)	Phosphorus* (%)	Chlorine* (%)
M	-	-	-	-
C	-	-	-	-
M-G	20	-	100	31
C-G	22	-	100	39
M-Sn	-	9	-	93
C-Sn	-	11	-	83
M-G-Sn	20	10	100	59
C-G-Sn	22	10	100	72

* percent of element remained in char.

Literature Cited

1. Reinhardt, R. M.; Fenner, T. W.; Reid, J. D. Text. Res. J. 1957, 27, 873.
2. Segal, L.; Creely, J. J.; Martin, E. Jr.; Conrad, C. M. Text. Res. J. 1959, 29, 786
3. Dirscherl, A.; Erne, F. Microchem. Acta, 1960, 775.
4. Sweeley, C. C.; Bentley, R.; Makita, M.; Wells, W. W. J. Am. Chem. Soc. 1963, 85, 2497.
5. Heyns, K.; Scharman, H. Chem. Ber. 1966, 99, 3461.
6. Shimada, M.; Kuribara, H.; Matsumoto, I.; Nakamura, Y. J. Appl. Polymer Sci. 1979, 24, 1017.

RECEIVED August 24, 1982

19

Photo-Induced Grafting of Monomers on Prewetted Fiber Substrates

HOWARD L. NEEDLES

University of California, Division of Textiles and Clothing, Davis, CA 95616

Photo-induced graft polymerization of acrylic monomers on fiber surfaces pre-wetted with water or aprotic polar solvents is reviewed. The role of solvent-induced fiber swelling and monomer penetration in the grafting process is examined. The role of dyes, biacetyl, metal oxides, polar solvents, and hydrogen donors as accelerants in photo-grafting is explored. Grafting of monomers introduced as vapors is compared with grafting of monomers applied from wetting solution, and the nature of polymer deposition is considered for both systems. The physical and chemical properties of the resultant fiber grafts are examined and compared with untreated fibers. Grafting is shown to have a marked effect on the dyeing and color characteristics of the fibers. Photo-induced vapor phase grafting of acrylic monomers on pre-wetted fibers is shown to proceed rapidly with little homopolymer formation or interfiber bonding to give modified fabrics that retain major aesthetic characteristics and yet have improved properties dependent on the polymer grafted.

Dye-Sensitized Photografting on Fibers Immersed in Monomer Solutions

Dye-sensitized photopolymerizations of moderate to high concentrations (>10%) of acrylic monomers have been studied extensively (1-10); however, until 1967 little was known concerning dye-sensitized graft photopolymerizations onto fibers. In general, dye-sensitized photopolymerizations at low monomer concentrations required the presence of small quantities of oxygen and a mild reducing agent (added or as part of the dye-moiety) to give photosensitized initiation of polymerization by dyes. The rate of dye-sensitized photopolymerizations was proportional to the square of the monomer concentration (2, 7, 8), so photopolymerizations of low monomer concentrations (>5%) proceeded very slowly if at all and essentially no polymerization of monomer was noted after 4

0097-6156/83/0212-0249$06.00/0

hours irradiation with light. We found that suspension of wool and certain other fibers into 1-5% monomer solutions at 40.0 ± 0.1°C and subsequent irradiation of the sealed solution for 2 hours resulted in extensive photopolymerization of monomer (11-15). A portion of the resulting photopolymer was grafted onto the fiber (0.5-23%), and the remainder of the polymer existed as ungrafted homopolymer. Photopolymerization proceeded after an induction period of 30-60 min in which time excess oxygen in solution was consumed. Riboflavin (11-14), fluorescein (12-14), and anthraquinone (15) photosensitizers in conjunction with an RS-type mercury sunlamp source were found effective. The quantity of polymer grafted onto wool and related fibers and the extent of accompanying homopolymerization were dependent on the concentration and chemical nature of both monomer and dye. In general, concentrations of 3% monomer and 10^{-4}M or 10^{-5}M dye in solution have been found satisfactory for maximum photografting onto wool with minimum homopolymerization. Higher monomer concentrations led to extensive homopolymerization with no increase in polymer grafted onto wool, while higher dye concentrations led to longer induction periods and often little or no photopolymerization within a 2 hour irradiation period.

In general less homopolymerization and lower molecular weight homopolymers were found with anthraquinone dye-sensitizers (15) than with riboflavin (11-14) and fluorescein (12-14) sensitizers. Monomers which form water-insoluble polymers gave a minimum of homopolymer, presumably through precipitation of the growing homopolymer chain and removal of the radical from further initiation. Higher photografts on wool occurred with monomers containing electron-donating functional groups than with monomers containing electron-withdrawing functions, and the quantity of photopolymer grafted was dependent on the dye-monomer combination used.

As mentioned earlier oxygen was necessary in the photoinitiation redox mechanism; however, excess oxygen was also responsible for long induction periods before photopolymerization (11-13). After excess oxygen was consumed through reaction with the photo-activated dye to form hydrogen peroxide, initiation of photopolymerization proceeded until oxygen remaining in solution inhibited polymerization of remaining monomer. This would explain why complete and rapid polymerization for a short period of time following the induction period was observed. In addition, rigorous exclusion of oxygen tended to inhibit photopolymerization (11-13), the effect being more pronounced with riboflavin and fluorescein than with anthraquinone dyes. Also, the oxygen scavenger tetrakis(hydroxymethyl)phosphonium chloride lowered the amount of polymer grafted and increased the extent of homopolymerization (16).

Addition of hydrogen peroxide to solution, a photolysis product from photoreaction of the dye, lowered polymer grafting on wool (14). The apparent molecular weight of homopolymer formed in the presence of hydrogen peroxide depended on the dye-sensitizer; however, the amount of homopolymerization observed was generally higher. In addition, graft photopolymerizations in the presence of hydrogen peroxide were less sensitive to oxygen in solution, as oxygen exclusion had little effect on the amount of polymer grafted on wool. Hydrogen peroxide apparently entered into the redox photo-initiation rather than oxygen and thereby cancelling the effect of oxygen exclusion. Lower apparent molecular weights of homopolymers in the presence of peroxide were probably caused by more rapid termination of growing polymer chains by peroxide.

Introduction of the hydrogen-donating accelerators N,N,N',N'-tetramethylethylenediamine and triethanolamine (0.05-0.5%) had little effect on the rate at which polymer was photografted onto wool (16). The extent of homopolymerization, however, was dramatically lowered, and these compounds can be considered homopolymerization retardants.

In a further study, we examined (20) graft photopolymerization of acrylamide and related monomers onto wool and certain other fibers sensitized by anthraquinones. Modified wools containing 1.5-9.6% grafted polymer were obtained, accompanied by homopolymer formation. The substituent on the anthraquinone, the nature of substitution on acrylamide, the solvent, oxygen exclusion, and the fiber-type all were found to affect the degree of photografting and homopolymerization. The tensile properties of the grafted wools showed increased breaking strengths and increased energies to break when compared with ungrafted wool. The grafted wools were found to dye by an acid dye at different rates than ungrafted wool, dependent on the availability of amide groups within the polymer structure.

Stannett and coworkers (17, 18) have shown that anthraquinone-sensitized photopolymerizations on celluloses and nylon involve the triplet state of the dye and semiquinone intermediates in photoinitiation. In addition, anthraquinones (17, 18) were also known to photosensitize degradation of polymers. Although the stability of free radical intermediates was only slightly affected by substituents in our study, substitution at the 2-position of an electron-donating function increased grafting and decreased the degree of homopolymerization, whereas 2-substitution of mildly to strongly electron-withdrawing functions had little effect on polymer grafting but markedly decreased the extent of homopolymerization. Substitution in the 1,5-positions of anthraquinone caused the sensitizer to be incapable of initiation of polymerization, and 1-substituted anthraquinones are known to be less effective than 2-substituted anthraquinones in photosensitization (19) presumably due to quenching of the excited anthraquinone by the adjacent 1-substituent. The loss in weight of the wool samples was thought to be due to phototendering (17-19), similar to phototendering of cellulose and nylon sensitized by anthraquinone dyes.

Another study involved irradiation of wool by short-wavelength ultraviolet light in the presence of acrylate or methacrylate esters in dimethyl sulfoxide (21). Intermediate grafts (2%-10%) of polymer on the surface of the wool were observed. The amount of grafting and homopolymerization and the evenness and regularity of photografted polymer were dependent upon the monomer used. With methyl acrylate, the reaction time, monomer concentration, and introduction of water as cosolvent affected the degree of fiber grafting and the amount and molecular weight of accompanying homopolymer. Polymers grafted onto wool by this technique changed the tensile properties and the water desorption characteristics of the wool.

We also have examined polyester fabric pretreated at 120°C with N,N-dimethylformamide (DMF) alone or with acrylic acid (10-40%) followed by photo-irradiation in the presence or absence of biacetyl vapors for periods up to one hour (22). Graft uptakes of 2.7 to 5.1% acrylic acid on the polyester were found. The effect of DMF-treatment, irradiation, and graft uptake on selected properties of the polyester were examined. The area shrinkage of the fabric caused by DMF treatment was found to be affected by the presence of acrylic acid, whereas irradiation alone or in the presence of acrylic acid caused a decrease in tensile properties. Moisture regain of the treated fabrics was only slightly affected by irradiation and grafting, whereas the surface wetting properties showed large effects. Irradiation and grafting significantly affected the dyeing characteristics of the fabric.

Ogiwara, Kubota, and Yasunaga (23) have examined the effect of fiber swelling, photosensitizer, and solvent on photo-induced graft polymerization of methyl methacrylate on vinyl, nylon 6, and polyester fibers. Solvent-induced swelling of fiber as well as the presence of certain sensitizers were found to increase photo-induced grafting through enhancement of radical formation on the fibers.

Photo-Induced Vapor Phase Grafting on Fibers

In the above studies, it has been shown that photo-induced grafting on fiber substrates can be carried out at low monomer concentrations over short reaction times and that homopolymer formation can be minimized by selection of appropriate reaction conditions, monomers, and photo-initiation systems. Fiber swelling was found to have a marked effect on the degree and location of polymer grafting with solvents causing the greatest swelling giving the greatest degree of grafting.

Since homopolymer had a marked effect on the aesthetic properties of grafted textiles, we considered methods whereby homopolymer formation could be minimized. Earlier Howard, Kim, and Peters had successfully photografted methacrylic acid vapors on nylon 6 film using benzophenone as photosensitizer (24). They found that the technique was efficient only at low concentrations of monomer vapor and that little homopolymer was formed during grafting. The irradiation time necessary for 10% to 20% grafting was short (2-10 min); however, the techniques initially used to introduce monomer and benzophenone vapors onto the film required long periods (up to 24 hr). In addition, if longer irradiation periods were used, a drop in polymer grafting accompanied by film degradation was noted with the short-wavelength (254 nm) ultraviolet source used.

Since a number of polymer substrates are adversely affected by short-wavelength ultraviolet irradiation, a rapid, higher wavelength (>300 nm), sensitized, vapor-phase grafting technique for polymer substrates would be desirable and might lead to an economically viable photografting process. In the course of our research into photografting on fiber surfaces, we obtained extremely even grafts of up to 20% polymer on pre-wetted surfaces when near UV light and biacetyl vapors were used to initiate graft polymerization using vapor-phase monomers (25). Inexpensive equipment that operates near atmospheric pressure with minimum radiation shielding was used in this photografting process.

Biacetyl-Induced Photografting of Monomer Vapors on Fibers

This novel vapor-phase process was used to graft relatively volatile acrylic monomers onto various polymeric substrates, using photo-initiation by near ultraviolet irradiation in the presence of biacetyl vapors. With it, very even graft polymerizations on the substrates, with minimum amounts of homopolymerization, were found. Furthermore, there were essentially no changes in the tensile or aesthetic properties of the treated surfaces. The degree of photografting was dependent upon the chemical composition and porosity of the substrate, the volatility and reactivity of the monomers, pre-wetting of the substrate with a suitable wetting agent, and the conditions of irradiation used. The effects of various reaction parameters on the photo-induced grafting of methyl acrylate, methyl methacrylate, and acrylonitrile on wool keratin were studied in detail. Increasing biacetyl and monomer flow rates and flow times, irradiation times, and moisture content of the wool all caused progressive increases in the amount of polymer grafted to the wool, up to limiting values dependent on the reaction parameters involved and monomer used. In all instances, the amount of homopolymer found on the fiber was limited and remained essentially constant over the range of conditions studied. A series of acrylic monomers of different volatilities and reactivities including methyl acrylate, methyl methacrylate, butyl acrylate, acrylic acid, acrylamide, acrylonitrile, N,N-dimethylaminoethyl methacrylate, and 2,2,2-trifluoroethyl methacrylate was successfully grafted onto several hydrophilic and hydrophobic textile fibers (wool, cotton, rayon, nylon, acrylic, polyester, and polypropylene) and other polymeric surfaces such as filter paper, cellophane, and acetate film by this process. The wetting agents used included water, methyl and n-propyl alcohol, N,N-dimethylformamide, dimethylsulfoxide, benzene, and chlorinated hydrocarbon solvents.

From this study we were able to show that pre-wetting of the surface was a critical feature of the photo-initiated vapor-phase grafting reaction. The degree of grafting on the surface increased with the degree of wetting and wet-pickup of the fabric, up to a limiting value. Liquids that tended to wet and swell the fiber surface increased the amount of grafting. Use of a wetting solvent with an affinity for biacetyl and monomer vapors favored grafting. Since vapor-phase grafting and accompanying homopolymerization occurred at or under the fabric surface, and since the degree of grafting was dependent on both the biacetyl and monomer flow times and rates, the wetted fiber was intimately involved in the grafting process, with the concentrations of biacetyl and monomer on or slightly penetrated into the surface affecting both the rate and degree of grafting. The generally linear dependence of the degree of grafting on irradiation time when monomer was present in a constant concentration indicated the critical importance and constant effect of light quanta falling on the surface that was being grafted. The low degree of homopolymer formation that occurred with the vapor-phase graft process attested to the efficiency of the photo-initiation system with regards to grafting. Initiation of grafting occurred through facile hydrogen abstraction from the fabric surface, and the termination mechanisms in operation did not lead to extensive initiation of homopolymerization. Although formation of homopolymer entrapped within the fiber could not be excluded as a possibility in this process, several factors tended to minimize the input of such a mechanism. The excited biacetyl would be more likely expected to favor abstraction of accessible hydrogens from polymer backbones within the fiber than abstraction of hydrogens from monomer dissolved within the solvent wetting the fiber. Furthermore, since it did not appear critical that the monomer be soluble in the solvent wetting the fabric, free-radical abstraction of hydrogen from entrapped monomer seemed less likely than free-radical addition of monomer diffusing to the surface of fibers where substrate radicals are present.

Biacetyl diffused onto the pre-wetted fiber initially. Then, as monomer vapors were introduced and the surface was irradiated with ultraviolet-visible light, the excited biacetyl dissociated to yield acetyl radicals or decayed to its more stable triple state through intersystem crossing. These radical species abstract accessible hydrogens from the substrate or near the surface of the fiber substrate, which in turn react with monomer in the proximity or possibly within the solvent wetting the substrate. Growing polymer chains were terminated in a manner whereby only limited initiation of homopolymerization occurred.

The biacetyl-sensitized vapor-phase grafting technique was

shown to provide a novel and rapid method for even deposition of polymers on substrates at atmospheric pressure and at moderate temperatures, using a simple photoreactor system. Due to its simplicity, this technique possessed distinct advantages over high energy and glow discharge processes.

Unsensitized Photografting of Monomer Vapors on Fibers

Following the biacetyl-sensitized grafting study, we explored the effect of certain physical and chemical variables on unsensitized photografting of monomer vapors on pre-wetted wool (26, 27) and on pre-wetted polyamide and polyester fabrics (28). One study (26) showed that the rate and degree of unsensitized grafting of monomer vapor on wool was dependent upon the monomer, wetting agent, hydrogen donor present, and the time and intensity of photo-irradiation. The resulting grafted wools possessed improved properties without loss of the aesthetics inherent to wool. A further study examined the unique role of tetrakis-(hydroxymethyl)phosphonium chloride (THPC) in grafting monomer vapors on pre-wetted wool (27). Wetting with 1% aqueous THPC increased the amount of polymer grafted on the wool but had only a slight effect on the color, tensile properties, stiffness, and wrinkle-recovery properties of the grafted wools compared to grafted wools pre-wetted with water only. The THPC pretreatment had a marked effect on the dyeing and resultant color properties of the grafted wools which dyed at faster rates, to higher exhaustion, and to different shades than the corresponding grafted wools pre-wetted with water only. No significant differences between the THPC-wetted and water-wetted grafted wools were found by scanning electron microscopy.

The role of wetting solvent on photo-grafting of methyl acrylate vapors on nylon and polyester was examined. Photo-induced polymerization of methyl acrylate vapors on polyamide and polyester fibers occurred when these fibers were wetted with the polar solvents methanol, N,N-dimethylformamide, or dimethyl sulfoxide (neat or in aqueous solution). Polymer grafting was accompanied by homopolymer formation, with the amount of grafting and ratio of grafting to homopolymerization being dependent on the fiber type, the solvent used to wet the fiber, and the amount of water present in the wetting solution. Overall deposition of polymer was higher on polyamide than on polyester under all wetting conditions. Water present in the wetting agent had a limited effect on polymerization of poly(methyl acrylate) on polyamide, whereas water caused a rapid decrease in polymerization on polyester. The nature of polymer grafting on the fibers under various wetting conditions was examined by scanning electron microscopy and tensile property measurments. The presence of polar solvents was shown to permit greater penetration of methyl acrylate vapor into these fibers.

Metal Oxide-Induced Grafting of Methyl Acrylate on Fibers

A method was developed to cause surface grafting of methyl acrylate vapor on fibers (29). Textile fabrics of cotton, wool, nylon, polyester, acrylic, and polyolefin pretreated with aqueous dispersions of photosensitive metal oxides (antimony, tin, titanium, and zinc oxide) were exposed to methyl acrylate vapors with simultaneous ultraviolet irradiation (>3100 Å) for up to 2 hours. The metal oxides acted either as effective photosensitizers, causing increased polymer grafting on the fiber surface, or as photo-absorbers causing a net decrease in grafting compared to unsensitized photografting. Metal oxide-induced grafting occurred more readily on hydrophilic fibers and was accompanied by less homopolymer formation in comparison to grafting on more hydrophobic fibers. Antimony and tin oxides were more effective on hydrophilic fibers, while zinc oxide was more effective on hydrophobic fibers. Titanium dioxide was essentially ineffective as a photosensitizer.

There was a strong possibility that initiation occurred via photosensitized formation of hydrogen peroxide on the metal oxide surface followed by desorption and subsequent photolysis of hydrogen peroxide to form free radicals. Radicals formed by this means would lead to abstraction of hydrogen from the fiber surface and subsequent grafting of poly(methyl acrylate) at these sites.

Summary

Photo-induced grafting of fibers on pre-wetted fiber substrates occurs by a variety of methods. Introduction of monomer as a vapor in the grafting process on pre-wetted fibers gives rapid photografting with a minimum of homopolymer formation. Biacetyl vapor greatly increases the rate of photografting of monomer vapors on pre-wetted fibers and provides a method of fiber modification without loss of fiber aesthetics.

Literature Cited

1. Oster, G. Nature 1954, 173, 300.
2. Oster, G. K.; Oster, G.; Prati, G. J. Am. Chem. Soc. 1957, 79, 595.
3. Oster, G. Phot. Sci. Eng. 1960, 4, 237.
4. Oster, G. U. S. Pat. 2,875,047 Feb. 24, 1959.
5. Oster, G. U. S. Pat. 2,850445 Sept. 8, 1958.
6. Oster, G. Ger. Pat. 1,009,914 Jan. 18, 1956 to June 6, 1957.
7. Nishijima, Y. Kyoto Daigaku Nippon Kogakuseni Kenkyusho Koenshy 1959, 16, 141; Chem. Abstr. 1960, 54, 8616f.
8. Delzenne, G.; Dewinter, W.; Toppet, S.; Smets, G. J. Polym. Sci. A 1964, 2, 1069.
9. Oster, G.; Bellin, J. S.: Holmstrom, B. Experientia 1962, 18, 249.

10. Chen, C. S. H. J. Polym. Sci. A 1965, 3, 1107, 1127, 1137, 1155.
11. Needles, H. L. J. Polym. Sci. B 1967, 5, 595.
12. Needles, H. L. J. Appl. Polym. Sci. 1967, 11, 719.
13. Needles, H. L.; Wasley, W. L. Text. Res. J. 1969, 39, 97.
14. Needles, H. L. Polymer Preprints, Am. Chem. Soc. 1969, 10, 302.
15. Needles, H. L. Text. Res. J. 1970, 40, 860.
16. Needles, H. L. Text. Res. J. 1970, 40, 579.
17. Geacintov, N.; Stannett, V.; Abrahamson, E. W.; Hermans, J. J. J. Appl. Polym. Sci 1960, 3, 54.
18. Geacintov, N.; Stannett, V.; Abrahamson, E. A. Makromol. Chem. 1960, 36, 52.
19. Baugh, P. J.; Phillips, G. O.; Worthington, N. W. J. Soc. Dyers and Colourists 1969, 85, 241.
20. Needles, H. L.; Sarsfield, L. J. Appl. Polym. Symp. 1971, 18, 569.
21. Needles, H. L. J. Appl. Polym. Sci. 1971, 15, 2259.
22. Siddiqui, S. A.; McGee, K.; Lu, W.-C.; Alger, K.; Needles, H. L. Am. Dyestuff Reptr. 1981, 70, No. 7, 20.
23. Ogiwara, Y.; Kubota, H.; Yasunaga, T. J. Appl. Polym. Sci 1975, 19, 887.
24. Howard, G. J.; Kim, S. K.; Peters, R. H. J. Soc. Dyers and Colourists 1969, 85, 468.
25. Seiber, R. P.; Needles, H. L. J. Appl. Polym. Sci, 1975, 19, 2187.
26. Needles, H. L.; Alger, K. W.Schrifenreiche, Special Issue, III 1976.
27. Needles, H. L.; Alger, K. W.; Tai, A. Textile Res. J. 1978, 48, 123.
28. Needles, H. L.; Alger, K. W. J. Appl. Polym. Sci. 1978, 22, 3405.
29. Needles, H. L.; Alger, K. W. J. Appl. Polym. Sci. 1975, 19, 2207.

RECEIVED August 24, 1982

20

Mechanochemically Initiated Copolymerization Reactions in Cotton Cellulose

DAVID N.-S. HON

Virginia Polytechnic Institute and State University, Department of Forest Products, Blacksburg, VA 24061

The potentiality of using mechanical stress to initiate graft copolymerization onto cotton cellulose was investigated. A Norton ball mill and a Wiley mill were used. The absorption of mechanical energy by cellulose molecules during milling and its consequence on cellulose properties were taken into consideration. Decreases in degree of polymerization and crystallinity, and increases in accessibility and copper number of milled cellulose were observed. Free radicals formed in the interim were detected by electron spin resonance (ESR) techniques. Three types of mechanoradicals contributing singlet, doublet and triplet ESR signals were identified. ESR studies also revealed that cellulose mechanoradicals were capable of initiating graft polymerization. Methylmethacrylate propagating radicals were identified when the monomer was in contact with cellulose mechanoradicals.. High grafting efficiency was obtained for ball milled and cut fibers, but a higher degree of grafting was obtained from the ball milled fiber.

Cotton is a major world fiber and cellulose resource, contributing to the health, safety, and well being of all people. And even more significant, it is a renewable organic raw material by the fixation of solar energy by green plants. Cotton cellulose fiber (*gossypium* spp.) is the seed fiber of cotton plant which normally has a higher purity and a higher molecular weight than other celluloses such as those isolated from wood (1). Cotton cellulose provides high strength, durability and thermal stability, ability to absorb moisture, easy dyeability and wearing comfort. In defiance of these serviceable properties, cotton cellulose disallows itself for wider commercial applications due to poor solubility in common inexpensive solvents, lack of ther-

0097-6156/83/0212-0259$06.25/0

moplasticity, low dimensional stability and poor crease resistance. Many scientists with restless and probing minds have sought ways to improve cotton cellulose properties for a long time. Many techniques have been developed in the past decades; of these, graft copolymerization reactions appear to be a propitious one to synthesize cellulose copolymers with unique and useful properties (2, 3).

Graft copolymers can be synthesized by various initiation methods such as using ultraviolet and ionizing radiations, and thermal and chemical reactions (2, 4). Despite considerable research in the field, unfortunately, due to the low grafting efficiency, high degree of homopolymer formation, cellulose copolymers are unable to reach commercial successes (2). In order to establish a grafting technique to achieve cellulose copolymers with high grafting efficiency, it has been recognized that mechanically generated free radicals, i.e., mechanoradicals, are capable of initiating graft copolymerization with high grafting efficiency (5). Although the synthesis of block- and graft-copolymers by mechanical forces has been studied extensively for synthetic polymers (6, 7), very little work has been performed on cellulosic materials. Whistler and Goatley (8) had investigated the possibility of using free radicals generated by ball-milling of corn starch for acrylamide polymerization. Deter and Huang (9) had attempted to graft acrylonitrile, methyl methacrylate and vinyl acetate by a vibromill. Various degrees of grafting were obtained, but vinyl chloride did not graft well onto cellulose. Hon (5, 10, 11) had demonstrated that mechanoradicals generated in wood, high yield pulps, cellulose and lignin by a glass-bead mill are capable of initiating graft copolymerization, and a higher degree of grafting efficiency was obtained from mechanically initiated grafting systems than from those initiated by ultraviolet light irradiation. Ordinarily, the disadvantages of using mechanical stress for chemical reactions are a relatively high energy consumption and equipment complexity. Fortunately, it has been recognized that grafting reactions can be carried out directly during polymer processing and in standard equipment, such as in cutters and grinders, without adding extra energy and production cost. Details are reported in this paper.

The manufacture of cotton cellulose products involves complex conversion methods. In order to convert cotton cellulose fiber into useful consumer and industrial products, mechanical processings, such as grinding, crushing, cutting, etc., are inevitably carried out at gins, textile and paper mills, as to render the cotton fiber processible and commercially useful. As a consequence, it is opportune if the existing mechanical operation energy can be utilized to initiate grafting reaction. Accordingly, the possibility of utilization of this energy for initiating grafting reaction was experimented using a Wiley mill and a Norton ball mill as to simulate a cutting and a milling operation in cotton mills. In addition to the evaluation of graft copolymers, the absorption of mechanical energy generated by these mills by cellulose molecules and its consequence on cellulose properties were taken into consideration. Experimental results revealed that a high degree of grafting efficiency can be obtained during cutting or grinding process. Only a very short cutting time is required for the former process to achieve high grafting efficiency, whereas a longer milling time is required for the latter process.

Experimental

Materials. Purified acetate-grade cotton fiber in a sheet form was used. Methyl methacrylate was used as the monomer after purification by alkaline extraction and followed by distillation under reduced pressure.

Procedures. Cotton cellulose was either cut in a Wiley mill or milled in a Norton ball mill (Roalox porcelain jar) of 1.3 gallon capacity. The charge always consists of 3500 grams of high carbon chrome steel balls (3/8" diameter) and 25 grams of cotton fiber. The mill was rotated at a constant speed of 60 rpm The change in degree of polymerization before and after mechanical treatments was determined from limiting viscosity number obtained by using a capillary viscometer. The measurements were carried out in a thermostat at 298.00 ± 0.05°K in cupriethylenediamine solution and converted to degree of polymerization using the following equation (12):

$$DP = 190\ [\eta]$$

Crystallinity of cotton cellulose was measured using a density method by means of a density gradient column (Techne Inc., Model DC-2). Xylene and carbon tetrachloride were used to make up the solution. Based on the density data, crystallinity of cellulose can be calculated from the following equation (13):

$$\text{Crystallinity} = \frac{Va - V}{Va - Vc};\quad \text{Density} = \frac{1}{V}$$

where Va, Vc and V are specific volume of amorphous portion, crystalline portion and unknown sample, respectively. Due to Kast (14), the values of Va and Vc are 0.680 and 0.628, respectively.

Accessibility of cellulose was determined by an iodine absorption method described by Hessler and Power (15) using the following equation:

$$\frac{\text{mg of iodine}}{\text{g of sample}} = \frac{(a-b) \times 2.04 \times 2.54}{0.3}$$

where a is the volume of 0.02N thiosulfate for the blank and b is the corresponding volume for determination, a ratio of the milligrams of iodine absorbed per gram of cellulose to 412 (the mg of iodine adsorbed per gram of methocel) gives a value for the amorphous fraction. The percentage of crystallinity is thus equalled to 100 minus percentage of amorphous portion.

Copper number, a method of determination of reducing end groups in cellulose, was measured using a standard method described by Earland and Raven (16).

The average number of chain scission per chain unit ($\bar{S}$) and the degree of degradation ($\bar{\alpha}$) were calculated based on the following equations:

$$\bar{S} = \frac{\text{DP initial}}{\text{DP cut}} - 1$$

$$\bar{\alpha} = \frac{1}{\text{DP cut}} - \frac{1}{\text{DP initial}}$$

Graft Copolymerization by Norton Ball Milling: Cotton fiber sheets were torn into lengths of 1-2 cm and soaked with methyl methacrylate monomer and water (9:1 ratio in volume) for 24 hrs prior to milling in nitrogen atmosphere.

Graft Copolymerization by Wiley Mill Cutting: Cotton fiber sheets (2 inches in width) were soaked with methyl methacrylate and water (9:1 ratio in volume) for 24 hrs prior to cutting in nitrogen atmosphere.

The polymerization was finally terminated by the addition of hydroquinone. The copolymerization products were collected and extracted with benzene to remove the homopolymers. The ungrafted fiber under identical milling or cutting conditions were also extracted with benzene in consideration of the possible loss of fiber bundles during extraction. Degree of grafting and grafting efficiency were calculated as follows:

$$\text{Degree of grafting (\%)} = \left(\frac{A-B}{B}\right) \times 100$$

$$\text{Grafting efficiency (\%)} = \left(\frac{A-B}{C-B}\right) \times 100$$

where A is the weight of cellulose after copolymerization and extraction, B is the weight of original cellulose, and C is the total weight of products after copolymerization.

Electron Spin Resonance (ESR) Studies: ESR spectra were measured with an X-band ESR spectrometer (Varian E-12, 100 KHz field modulation). To avoid distortion of the spectra by a power saturation, the ESR measurements were carried out at a microwave of 3mW. The g-value was measured by comparison with the strong pitch provided by Varian Associates. In all cases, ESR spectra were recorded at 77°K.

For ESR measurements, the ball milled or cut fibers were transferred to a Dewar flask filled with liquid nitrogen immediately after mechanical treatments in order to avoid significant decay of unstable free radicals. The fibers were then transferred slowly to ESR sample tubes together with liquid nitrogen, which was removed by a vacuum pump afterward. Subsequently, the ESR tube was sealed in vacuum for ESR measurements.

Results and Discussion

Mechanical Effect on Cotton Fiber Properties. It has been reported earlier (5, 17, 18) that in the mechanical processing, wood, cellulose, and lignin are inescapably absorbing mechanical stress, i.e., shear forces. The consequence of this energy uptake normally leads to the slippage of secondary bonds due to the intra- and intermolecular hydrogen bonds, and the rupture of covalent bonds, which leads to the shortening of fiber length. Mechanoradicals are formed in the interim.

The reduction of degree of polymerization of cotton fiber was studied after cutting and milling. Results are shown in Figures 1 and 2. When cotton sheets were cut through a 6 mm sieve of the Wiley mill for several consecutive cycles, the cotton sheet was disaggregated into fibrillar bundles as mechanical processing progressed. The fiber lost its DP significantly at the initial state of cutting, i.e., the first cutting cycle, followed by a slow drop of DP. Calculation of chain scission showed that 0.19 cleavage takes place per molecular chain at the first cutting cycle. No significant increment of the reducing end group of cellulose, i.e., aldehyde group, as determined by copper number, was observed (Figure 1). When cotton fiber was milled in the Norton ball mill, however, the drastic change in DP was observed (Figure 2). It was noticed that after 50 hrs of milling, a blend of two fractions of cellulose was formed: one fraction of deformed fiber retained most of its fibrous structure, and one form of cellulose powder has lost its fibrous structure completely. For the fiber fraction, the loss of DP was less severe than the powder fraction. The loss of DP was 17, 31.5, 45.5, 49.5, and 50.5% of their original value after 50, 100, 198, 336 and 400 hours of milling, respectively. For the powder fraction, the loss of DP was significant. Cotton cellulose lost 50.1, 66.4, 72.3, 77.3 and 77.6% of its original DP value after the same periods of milling, indicating that mechanochemical chain scission was extensive. The rate of chain scission and degree of degradation as a function of milling time for the fiber and powder fractions are shown in Table I. It is clearly evident that severe degradation of fiber took place during milling.

The increment of reducing end group due to the cleavage of main chains was also observed in terms of copper number study. Results are also shown in Figure 2. It is obvious that the rate of producing reducing end group in milled powder was much faster

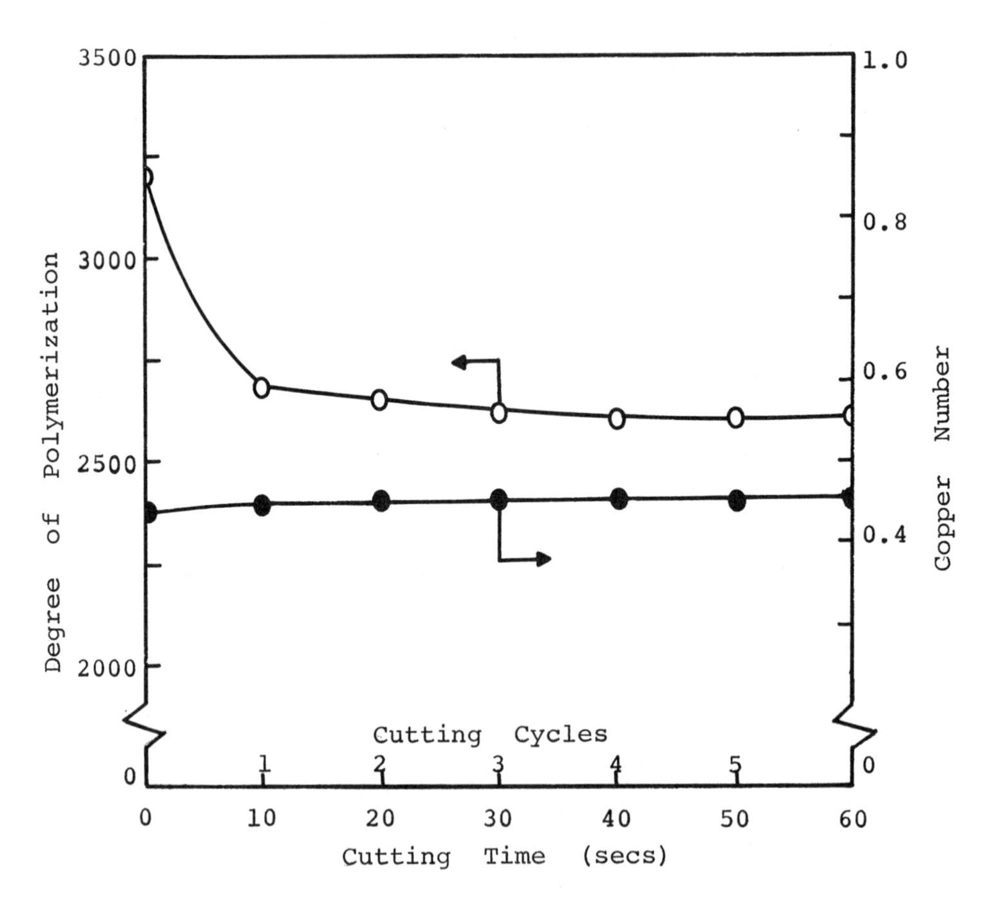

Figure 1. Changes in degree of polymerization and copper number of cotton cellulose during cutting with a Wiley mill.

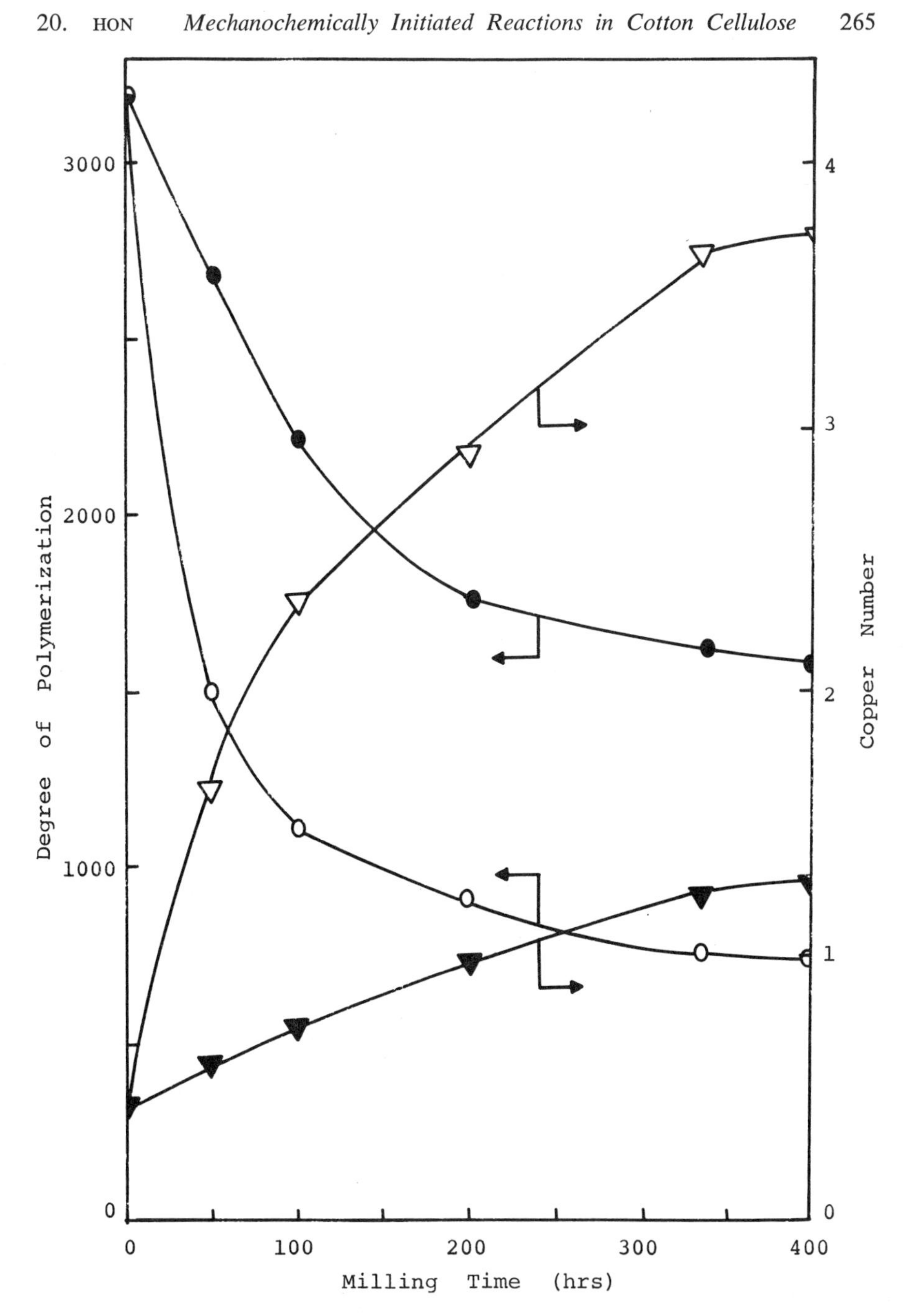

Figure 2. Changes in degree of polymerization and copper number of cotton cellulose during milling with a Norton ball mill. Key: ▼, ●, *milled fiber;* ▽, ○, *milled powder.*

TABLE I

Chain Scission and Degree of Degradation of Cotton Fiber After Milling in a Norton Ball Mill

Milling Time (hrs)	Fiber Portion			Powder Portion		
	DP	$\bar{S}$	$\bar{\alpha}(10^{-4})$	DP	$\bar{S}$	$\bar{\alpha}(10^{-4})$
Control	3211	0	0	3211	0	0
50	2673	0.20	0.63	1475	1.18	3.67
100	2209	0.45	1.41	1083	1.96	6.12
198	1750	0.83	2.60	892	2.60	8.10
336	1622	0.98	3.05	729	3.40	10.06
400	1593	1.02	3.16	720	3.46	10.77

DP: Degree of polymerization.

$\bar{S}$: Average number of chain scissions

$\bar{\alpha}$: Degree of degradation

than in the milled fiber. It is in agreement with the severe reduction of DP for milled powder.

Moreover, the direct evidence of the rupture of primary bonds was originated from ESR studies. When cotton fiber was cut in the Wiley mill for 60 sec at 298°K in nitrogen, a singlet signal with a line width of 18 gauss, and a g-value of 2.003, due to the alkoxy radicals (5), was observed (Figure 3a). This indicated that mechanoradicals were generated in the cut fiber due to a chain scission reaction. When the cotton fiber was milled in the Norton ball mill, an ill-defined five-line signal was detected (Figure 3b). This spectrum is very similar to those observed from wood cellulose milled at 77°K (5), which was a superposition of a singlet, a doublet and a triplet signal (5). Accordingly, three types of mechanoradicals were produced in cellulose during ball-milling. The assignment of these radical structures contributing to the signals has been discussed elsewhere (5). Comparison between the mechanically cut and the milled fibers clearly indicated that ball-milled fiber exhibited more intense ESR signal than the cut fiber, implying that a higher amount of mechanoradicals was generated in ball-milled fiber. It also suggested that the degree of degradation of the ball-milled fiber was much more critical than that of the cut fiber.

Much of the chemical behavior of cellulose fiber can be attributed to cellulose structure. Since cellulose is a highly crystalline polymer, it can absorb mechanical energy efficiently for mechanical stress reaction (5, 19). The mechanically activated thermal energy, in addition to rupture of main chains, may alter morphology or microstructure of cotton cellulose. Accordingly, the crystallinity and accessibility of cotton fiber may be influenced.

When cotton fiber was cut through the Wiley mill through a 6 mm sieve for different cycles, i.e., for different cutting times, the change of crystallinity was not observed (Figure 4). When cotton fiber was milled for 10 hrs, the change of crystallinity was hardly recognized, although the increment of accessibility was observed (see Figure 5). The noticeable change in crystallinity was observed only after 20 hrs of milling. It should be noted here that the destruction of crystallinity heavily depended upon the milling equipment and operation conditions. Howsmon and Marchessault (19) destroyed the crystallinity of wood cellulose in a relatively short ball-milling period. Forzaiti et al. (20) had reported that by using a vibratory ball mill, the cotton cellulose was converted almost completely to the amorphous form in 30 mins.

As mentioned earlier, a fiber fraction and a powder fraction of cotton cellulose were formed during milling; the changes in crystallinity for these two fractions are different. For the fiber fraction, the loss of crystallinity was critical for the initial 100 hrs of milling, as shown in Figure 4. The rate of change in crystallinity was then levelled off after 100 hrs of

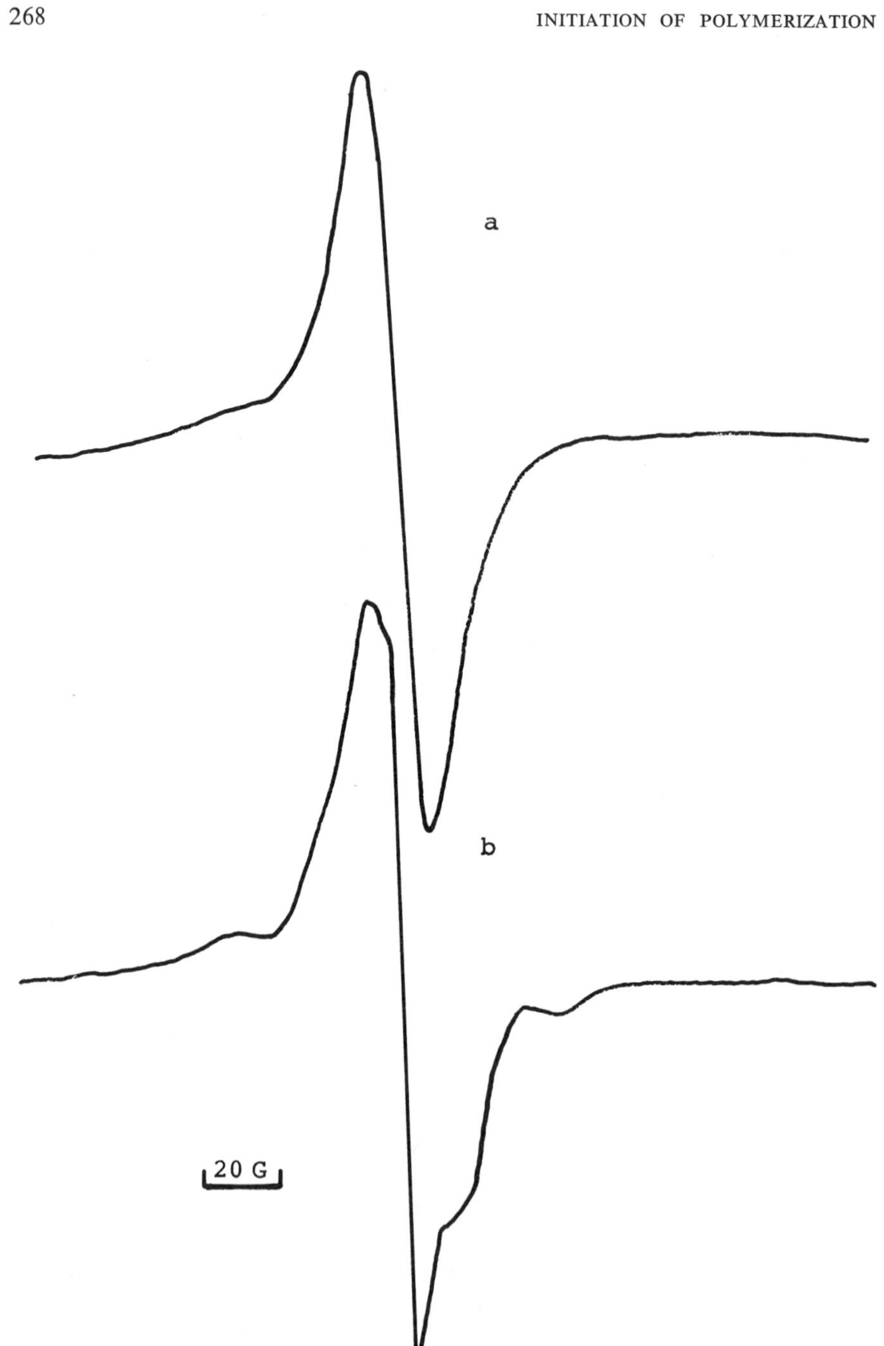

Figure 3. ESR spectra of cotton cellulose cut in a Wiley mill for 60 s at 298 K in nitrogen (a) and cotton cellulose milled in a Norton ball mill for 4 h at 298 K in nitrogen (b).

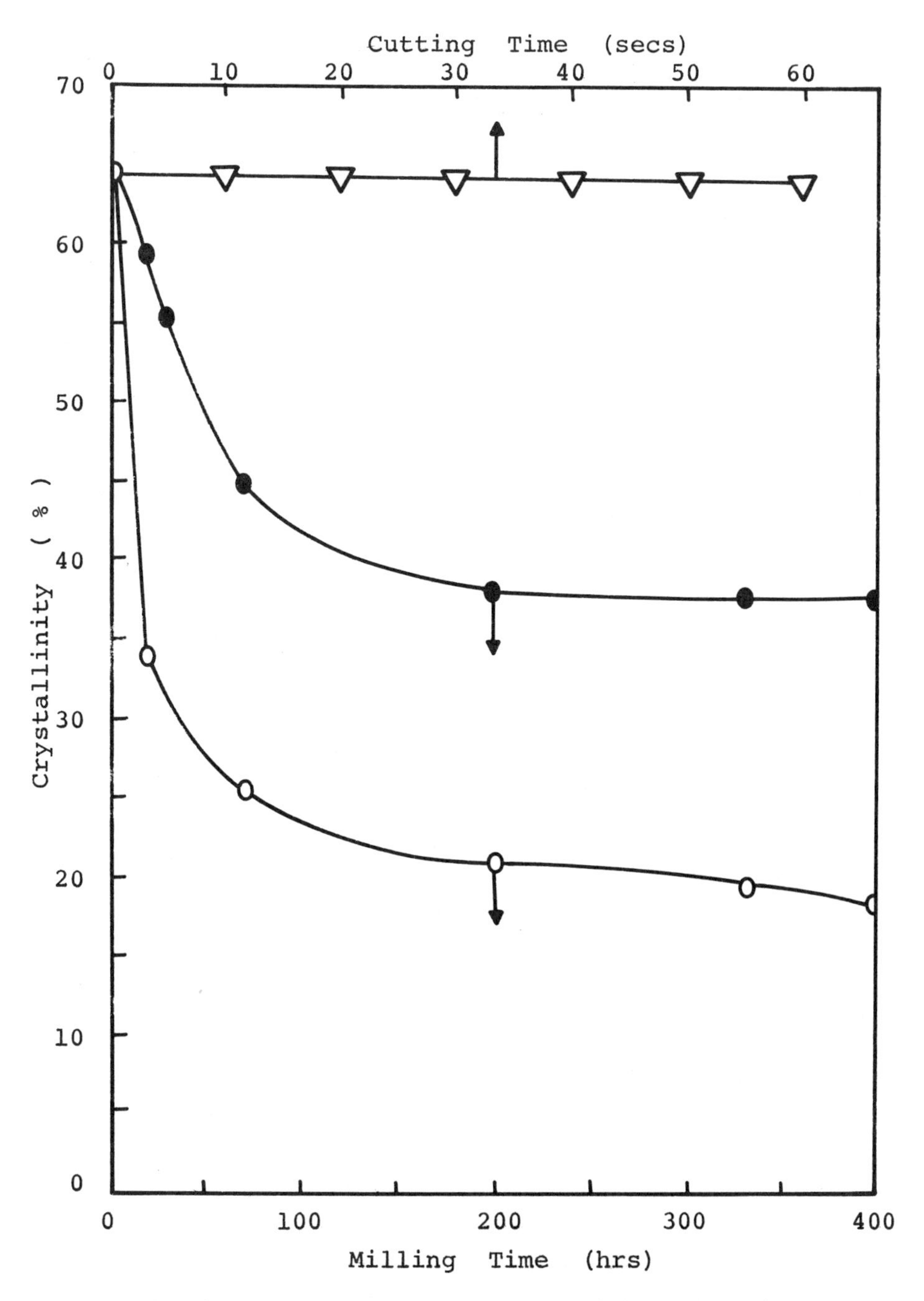

Figure 4. Change in crystallinity of cotton cellulose during cutting and milling. Key: ▽, cut fiber; ●, milled fiber; ○, milled powder.

milling. After 72 and 200 hrs of milling, the loss of crystallinity was 29.7 and 40.7% of their original crystallinity values, respectively. Additional several percentage of crystallinity may be lost if the milling time is prolonged. But it is believed that if the milling time is prolonged, the powder fraction of the cotton would probably be increased, which actually has a lower crystallinity than the fiber fraction. For the powder fraction, at the initial 22 hrs of milling, the crystallinity was only 34%, which was 46.9% reduction of crystallinity of its original value, i.e., 64.04%. After 200 hrs of milling, about 67% of the crystallinity was lost. The reduction of crystallinity was continuously observed if the milling time was prolonged.

Since the consequences of mechanical energy uptake were disaggregation of fiber bundles, shortening of fiber length and reduction of DP, it is plausible to consider that new surface areas were also created. Accordingly, the change of accessibility of mechanically treated fibers may be observed. The accessibility of mechanically treated fibers was evaluated based on the iodine absorption techniques. Results are shown in Figure 5.

Although the crystallinity of cotton fiber was not influenced in the cutting process, a slight increase in accessibility was observed for the cut fiber. As shown in Figure 5, the accessibility increased from 10.06% to 17.51% after 60 sec of cutting. The increase in accessibility was enhanced when cotton fiber was ball-milled. During the first 72 hrs of milling, the accessibility of fiber fraction increased to 32.12% (from 10.06%), whereas for the powder fraction, it even increased to 50%. The rate of increment of accessibility was slowed down for the fiber fraction even if the milling time was prolonged, but for the powder fraction, the accessibility was continuously increased. After 400 hrs of milling, 72% of accessibility was achieved.

Based on the experimental data of crystallinity and accessibility, it is revealed that the mechanical shear forces involved in the cutting process were able to open up the unaccessible regions in cellulose by creating new surfaces without damaging the crystalline structure. It is likely that the surfaces of crystallites are made more accessible during cutting. The mechanical shear forces involved in the milling process are able to increase accessibility perspicuously in accompanying with destruction of crystalline regions. The increment of accessibility usually set forward the penetration of monomer into channels and pores of fiber to achieve high degree of reaction. However, it should be borne in mind that the loss of crystallinity also is an indication of the loss of physical properties of cotton fiber. Hence, the proper control of the mechanical process to achieve high accessibility with less destruction of crystalline structures of fiber is beneficial to accomplish the purpose for which a chemical modification is designed.

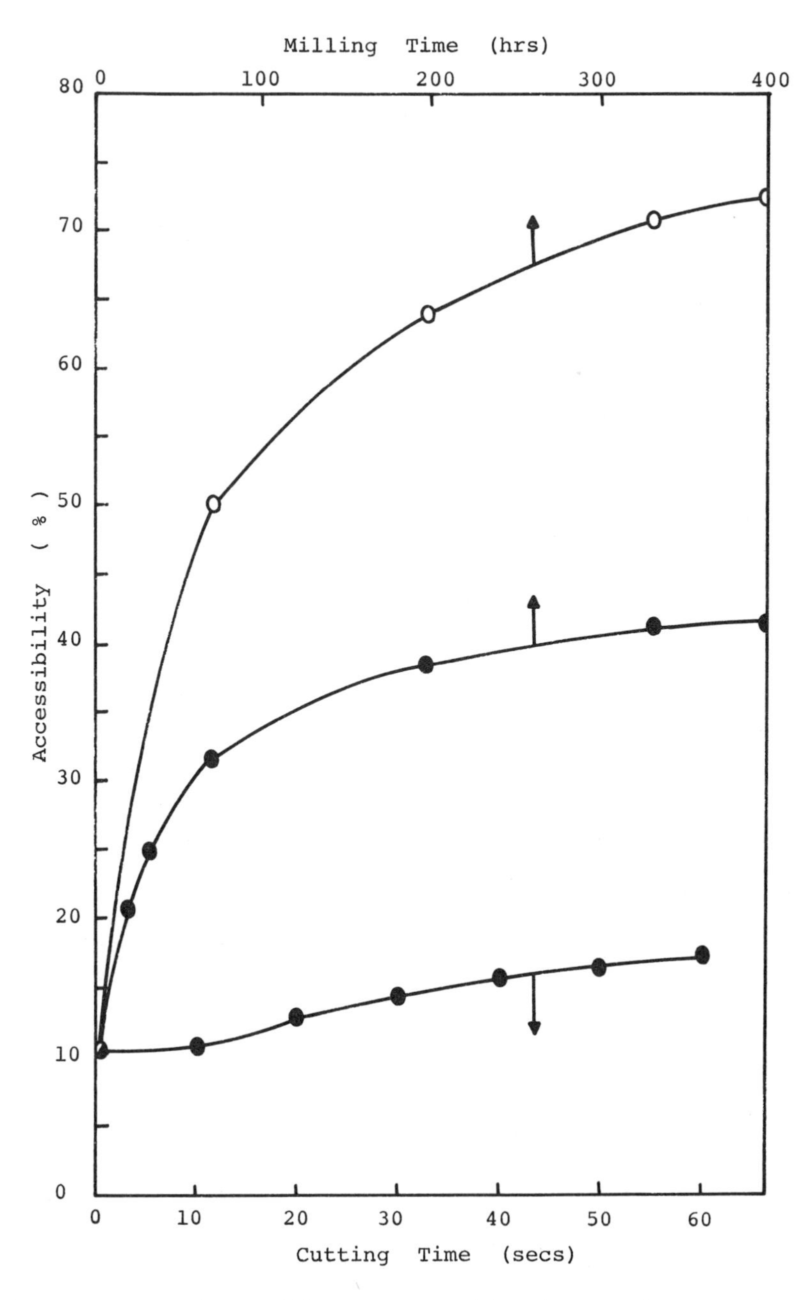

Figure 5. Change in accessibility of cotton cellulose during cutting and milling. Key: ○, powder; ●, fiber.

Mechanochemically Initiated Graft Copolymerization. According to the ESR study, it is clear that mechanoradicals were generated in cellulose either by means of mechanical cutting or ball milling. These mechanoradicals may be utilized as reaction sites for the initiation of vinyl polymerization which would result in graft copolymerization of cellulose. Based on this principle, the ability of cellulose mechanoradicals to initiate copolymerization was pursued.

When cotton fiber was ball-milled for 4 hrs at 298°K and transferred immediately to a sample tube for ESR measurement at 77°K, an ill-defined five-line spectrum with a g-value of 2.003 was detected (Figure 6a). When cotton fiber was milled in the presence of MMA, only a singlet signal with a line width of 18 gauss was observed (Figure 6b). This implied that other free radicals which generated signals other than the singlet component were interacted with MMA during milling. No MMA-propagating radicals were observed. Moreover, immediately following the milling, MMA was introduced into the mechanically treated cellulose for 2 mins at 298°K and recorded its ESR signal at 77°K, the five-line ESR signal of cellulosic mechanoradicals was converted to an asymmetrical multiplet spectrum (Figure 7b), and these signals were further intensified when the sample was warmed at 298°K for 10 mins, the ESR spectrum observed at 77°K was a nine-line spectrum, as shown in Figure 7d. The nine-line spectrum originated from the characteristic propagating radicals of MMA for polymerization (21). Similar results were observed when wood cellulose was milled with glass beads at 77°K (5). MMA-propagating radicals were also detected from cut fiber under identical warm-up treatment, with the exception that the signal intensity was very weak. This could be due to the low free radical concentration being generated by cutting. The propagating radicals of MMA were not detected when monomer was added to the untreated cellulose fibers in identical experiments. This clearly indicated that propagating radicals of MMA were created by contact of MMA with cellulosic mechanoradicals. Subsequently, it is evident that cellulosic mechanoradicals are capable of initiating vinyl polymerization. In light of these findings, graft copolymerization of MMA onto cellulose fiber was carried out at ambient temperature by impregnating monomer into fiber prior to cutting or milling. Results are shown in Figures 8 and 9.

For the vinyl graft copolymerization initiated by the cutting process, the degree of grafting was approached at 50% after 60 sec of cutting. The grafting efficiency reached its maximum of 89% between 20 and 30 sec of cutting. The initiating reaction for the ball mill system took place steadily. A longer induction period seemed to be needed to reach the maximum degree of grafting of 196% for 5 hrs of milling, and the grafting efficiency reached its maximum after 2-3 hrs of milling. When the grafting reaction time was prolonged, the degree of grafting as well as grafting

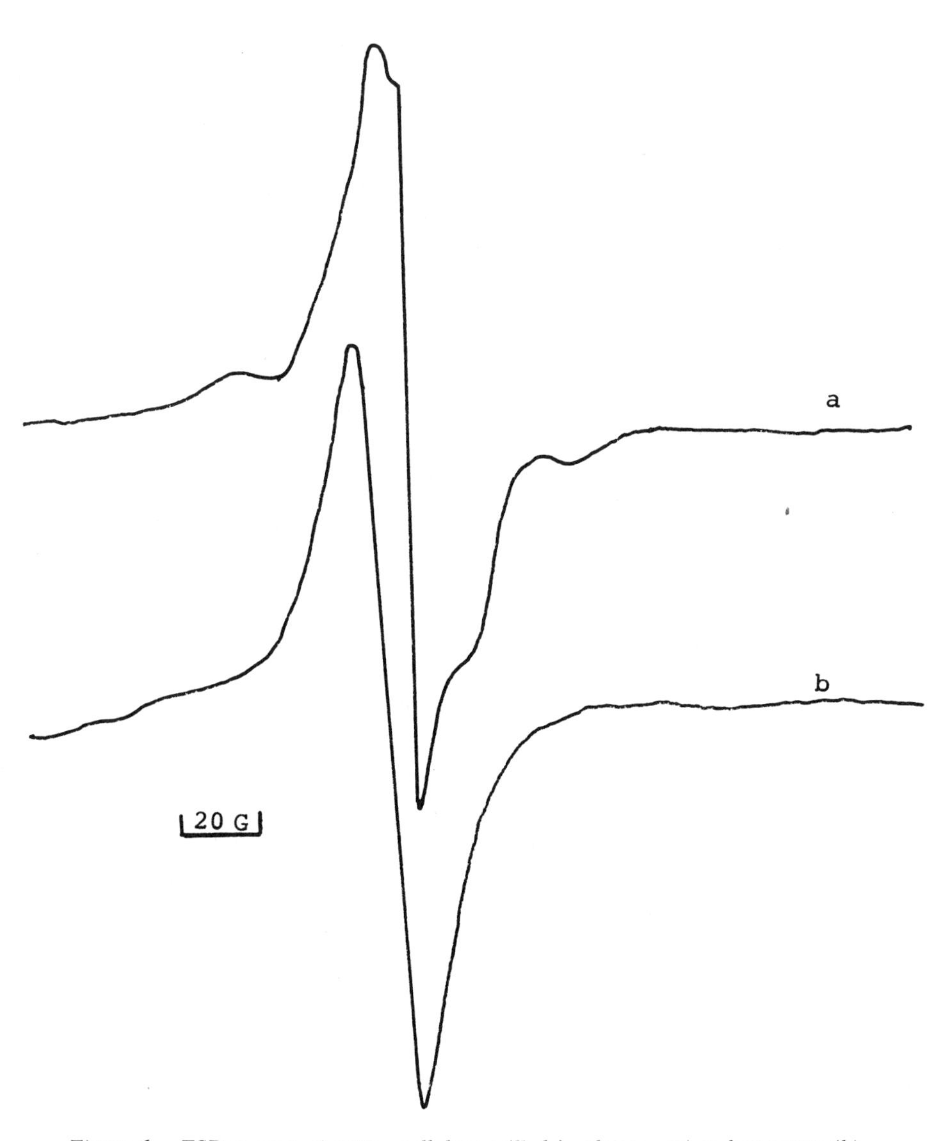

Figure 6. ESR spectra of cotton cellulose milled in absence (a) and presence (b) of methyl methacrylate for 4 h at 298 K. ESR spectra were recorded at 77 K.

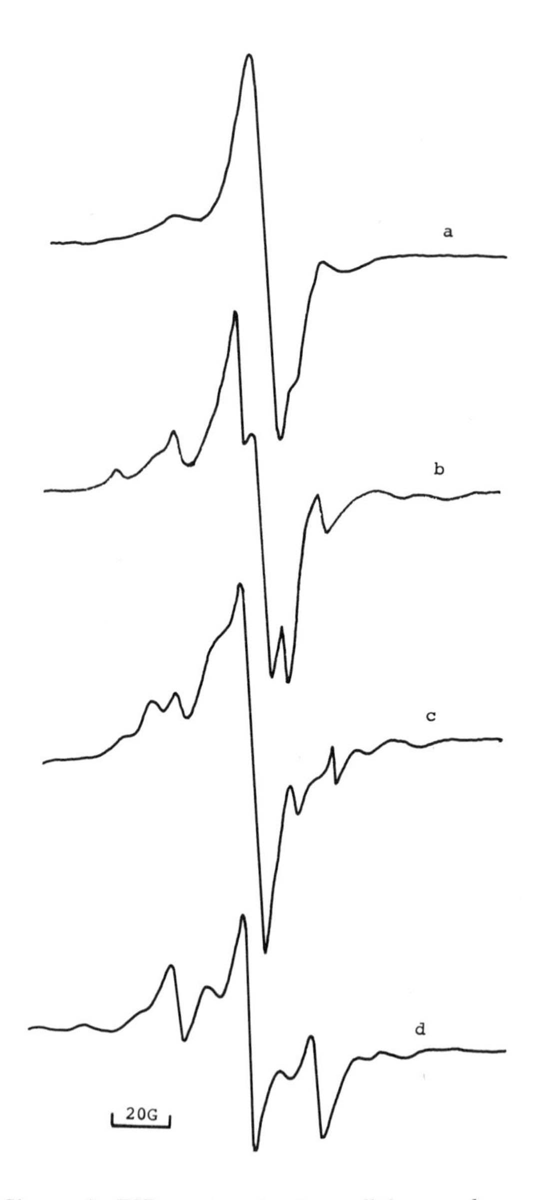

Figure 7. Changes in ESR spectra of cotton cellulose mechanoradicals. Key: a, initial spectrum observed at 77 K immediately after milling for 4 h at 298 K; b, cotton cellulose after milling was contacted with methyl methacrylate and warmed at 298 K for 2 min; c, 5 min; d, 10 min; recorded at 77 K.

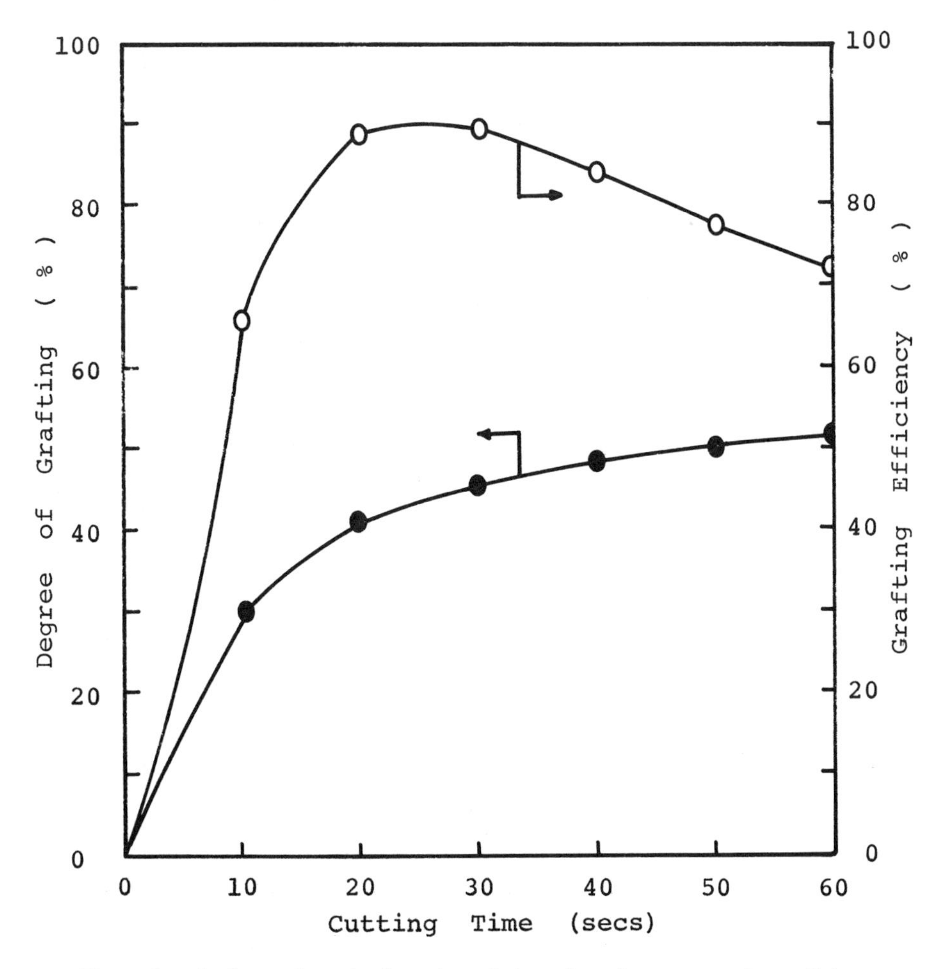

Figure 8. Graft copolymerization of methyl methacrylate onto cotton cellulose induced by mechanical cutting with a Wiley mill.

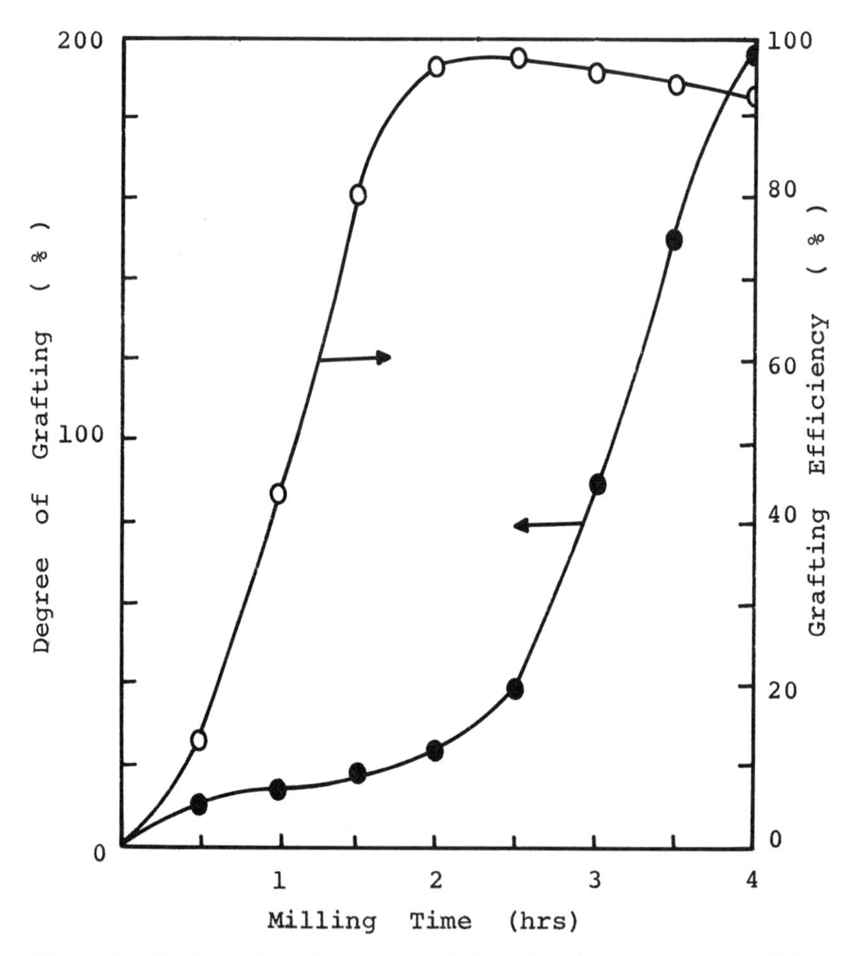

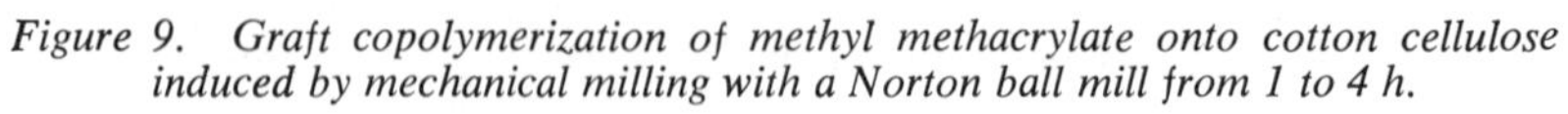
Figure 9. Graft copolymerization of methyl methacrylate onto cotton cellulose induced by mechanical milling with a Norton ball mill from 1 to 4 h.

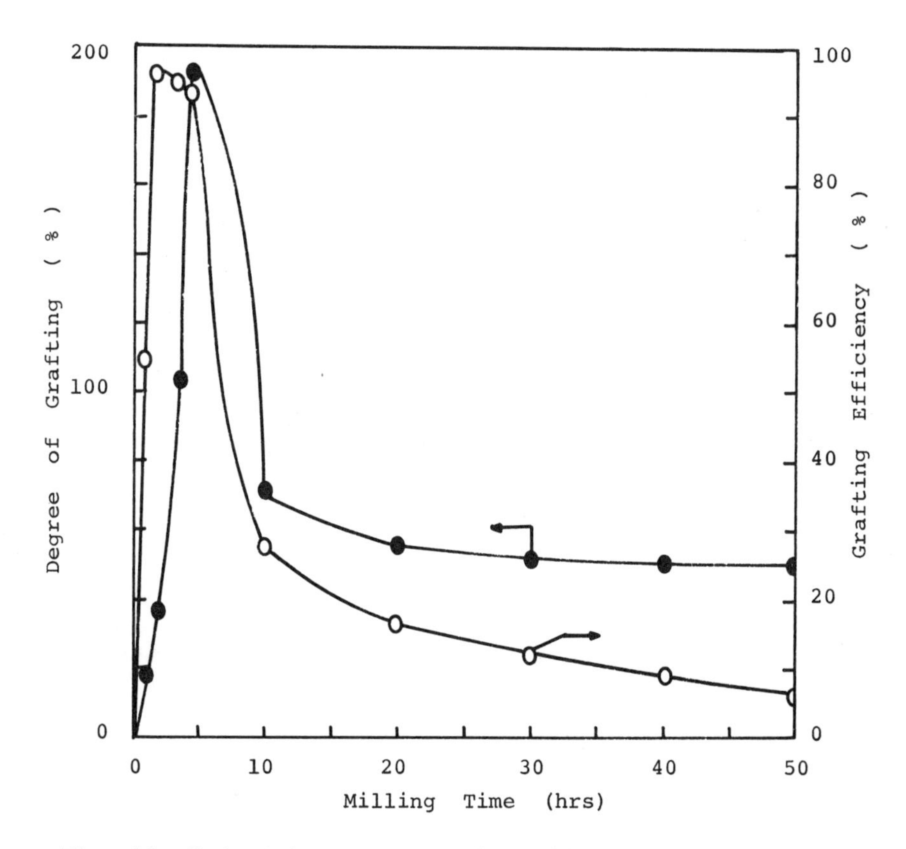

Figure 10. Graft copolymerization of methyl methacrylate onto cotton cellulose induced by mechanical milling with a Norton ball mill from 1 to 5 h.

efficiency reduced sharply after 10 hrs of milling (Figure 10). It is possible that if the milling was continued, the grafted PMMA chains might well have also suffered cleavage of covalent bonds by mechanical stress which led to low degree of grafting.

In comparison, a higher degree of grafting was achieved for the milling system than for the cutting system. It is plausible that higher mechanical energy was supplied for the former system, in which cellulose absorbed more energy to produce mechanoradicals for grafting reaction. The increment of accessibility and destroying of crystalline structure of cotton fiber during the milling process could also be the factors contributing to the high degree of grafting. On the other hand, the change in physical and chemical properties of milled cotton fiber were more critical than those of the cut cotton fiber.

Conclusions

On the basis of the experimental findings, the following conclusions may be drawn:

1. Cotton cellulose is susceptible to degradation by mechanical shear forces supplied by either a Wiley mill or a Norton ball mill. Mechanoradicals are produced in the interim.
2. The consequences of the energy absorption by cotton cellulose molecules are reduction of degree of polymerization and increment of accessibility and reducing end group. Cotton fiber suffered less degradation and its crystalline structure was not influenced when it was treated with a Wiley mill. When cotton fiber was treated with a Norton ball mill, a deformed fiber fraction and a powder cellulose fraction were formed. Drastic changes in crystallinity and accessibility were observed from both fractions. Accordingly, it is important to control the mechanical processing properly in order to avoid losing physical and chemical properties of cotton fibers.
3. Mechanoradicals are capable of initiating graft copolymerization. A high degree of grafting efficiency and a low degree of homopolymer formation were obtained from the fibers treated with the Wiley mill and the Norton ball mill, but a higher degree of grafting was obtained from the ball-milled fiber.

These auspicious findings will certainly shed light on the development of graft copolymerization techniques with high grafting efficiency. And the utilization of existing mechanical processing energy for graft copolymerization reaction is certainly an attractive mode of operation for commercial applications.

Literature Cited

1. Hon, D. N.-S. "Yellowing of Modern Papers" in "Preservation of Paper and Textiles of Historic and Artistic Value II", (Williams, J. C., ed.), Advances in Chemistry Series, 1981, 193, 119.

2. Stannett, V. "Some Challenges in Grafting to Cellulose and Cellulose Derivatives" in "Graft Copolymerization of Lignocellulosic Fibers" (Hon, D. N.-S., ed.), ACS Symposium Series, in press.
3. Hebeish, A., and Guthrie, J. T. "The Chemistry and Technology of Cellulosic Copolymers", Springer-Verlag, Berlin-Heidelberg-New York, 1981, p 351.
4. Hon, D. N.-S. "Graft Copolymerization of Lignocellulosic Fibers", ACS Symposium Series, in press.
5. Hon, D. N.-S. J. Appl. Polym. Sci., 1979, 23, 1487.
6. Casale, A., and Porter, R. S. "Polymer Stress Reactions", Academic Press, New York, Volume 1, 1978, Chapter V.
7. Ceresa, R. J. J. Polym. Sci., 1961, 53, 9.
8. Whistler, R. L., and Goatley, J. L. J. Polym. Sci., 1962, 62, 5123.
9. Deter, W., and Huang, D. C. Faserforsch. Textil. Tech., 1963, 14, 58.
10. Hon, D. N.-S. J. Polym. Sci. Polym. Chem. Ed., 1980, 18, 1957.
11. Hon, D. N.-S. "Modification of Lignocellulosic Fibers." Paper presented at the Second Chemical Congress of the North Americal Continent, San Francisco, CA., Aug., 24-29, 1980, Abstracts of Papers, Part 1, cell-39.
12. Browning, B. L. "Methods of Wood Chenistry", Academic Press, New York, 1967, Vol. 2, p 519-529.
13. Migita, N., Yonezawa, H., and Kondo, T. "Wood Chemistry", Kyoritsu, Tokyo, 1968, Vol. 1, p 102.
14. Kast, K. Z. Elektrochemie, 1953, 57, 525.
15. Kessler, L. E., and Power, R. E. Textile Res. J., 1954, 18 (9), p 822-827.
16. Earland, C., and Raven, D. J. "Experiments in Textile and Fiber Chemistry", Butterworth, London, 1971, p 134.
17. Hon, D. N.-S., and Glasser, W. G. TAPPI, 1979, 62, 107.
18. Hon, D. N.-S., and Srinivasan, K. S. V. J. Appl. Polym. Sci., in press.
19. Howsmon, J. A., and Marchessault, R. H. J. Appl. Polym. Sci., 1959, 1, 313.
20. Forziati, F. H., Stone, W. K., Rowen, J. W., and Appel, W. D. J. Res. of Natl. Bur. of Stand., 1950, 45, 2116.
21. Campbell, D. Macromol. Rev., 1970, 4, 91.

RECEIVED November 5, 1982

RUBBER CHEMISTRY

21

Factors Affecting the Isomeric Chain Unit Structure in Organolithium Polymerization of Butadiene and Isoprene

MAURICE MORTON and J. R. RUPERT[1]

The University of Akron, Institute of Polymer Science, Akron, OH 44325

Organolithium initiators are used extensively in the polymerization of butadiene and isoprene, because of their solubility in a variety of solvents. It is well known that non-polar media lead to polymers having a high 1,4 chain unit structure, while polar solvents can exert a dramatic effect in producing a high proportion of 1,2 or 3,4 structures. However, recent work, using the most accurate available NMR spectroscopy, has also shown that, even in non-polar media, the concentration of initiator and the presence of solvent can have a very significant effect on the cis-1,4/trans-1,4 ratio, without noticeably influencing the proportion of side-vinyl units. Thus, at very low initiator concentrations ($\sim 10^{-5}$ M.), and in the absence of any solvents, it is possible to attain a cis-1,4 content of 96% for polyisoprene and 86% for polybutadiene. These chain structures have not been found to be noticeably affected by temperature or extent of conversion.

The intensive investigations that followed the discovery (1) that lithium and its compounds can lead to the polymerization of isoprene to a very high cis-1,4 configuration, close to that of natural rubber, showed that both the type of alkali metal and the nature of the solvent can have a profound effect on the chain structure of polydienes (2-8). The conclusions reached from these investigations are as follows:

1. In non-polar media, the 1,4 structure is highest for lithium and decreases with increasing electropositivity of the alkali metals.
2. Polar solvents (e.g., ethers, amines, etc.) lead to a higher side-vinyl content (1,2 or 3,4), especially

[1] Current address: Monsanto Company, Akron, OH 44313.

0097-6156/83/0212-0283$06.00/0

in the case of lithium, apparently by solvating the metal cation and thus increasing the ionic character of the carbon-metal bond.

With regard to the effects of solvents on the chain microstructure of polydienes, the most thoroughly investigated systems have been those involving organolithium initiators, since these operate in a variety of solvents, both polar and non-polar, and can be controlled to undergo little or no side reactions. Furthermore, in non-polar media, such systems yield polybutadienes and polyisoprenes having a very high (>90%) 1,4 unit content, and these are important synthetic rubbers.

Most of the data referred to above were obtained in earlier work, and were based on infrared spectroscopy. In recent years, more reliable data were obtained by means of NMR spectroscopy, using both H^1 and C^{13} resonances (9-14). Some of these investigations suggested that, aside from the dramatic effects of polar solvents on the chain structure in organolithium systems, there were some subtle effects even in non-polar media, e.g., caused by initiator concentration and type of non-polar solvent. Sinn and coworkers (15), for example, used infrared spectroscopy to show an effect of butyl lithium concentration on the chain structure of polyisoprene and polybutadiene. Hence an extensive study was carried out recently (13, 14) on the influence of reaction parameters on the chain structure of polybutadiene and polyisoprene prepared in non-polar media.

Experimental

All spectra were obtained with the Varian HR-300 NMR Spectrometer, using H^1 in normal mode, with occasional use of Fourier transform for very high molecular weight samples. Hexachlorobutadiene was used as solvent, with 1% hexamethyldisiloxane as reference. The temperatures used were 110°C. for polyisoprene and 125°C. for polybutadiene. The polymer samples were prepared with sec-butyllithium as initiator, using the high vacuum techniques described elsewhere (13).

NMR peak assignments were as follows. For polyisoprene, the 3,4-unit content was determined from the olefinic methylene protons at 4.61 and 4.67 ppm, while the trans-1,4 units were determined from the methyl proton resonance at 1.54 ppm. The cis-1,4 content was then calculated by difference. No 1,2 olefinic methylene protons could be seen at 5.4 ppm, where they would be expected. For polybutadiene, the 1,2-unit content was calculated from a comparison of the 1,2-olefinic methylene protons at 4.8 ppm with the 1,4-methine protons at 5.4 ppm. The cis/trans ratio was then computed from the 1,4-methylene protons at 1.98 and 2.03 ppm. These methods made it possible to estimate the chain unit structures within about 1%.

Results and Discussion

The reaction variables investigated were: a) initiator concentration, b) monomer concentration, c) type of solvent (i.e., benzene, n-hexane or cyclohexane), d) temperature, and e) degree of conversion. Polymerizations were carried out generally at room temperature to complete (or near complete) conversion except when the effects of these two variables were being studied.

Polyisoprene

Effect of Initiator and Monomer Concentration. The effect of initiator concentration (s-butyllithium) and solvents on the chain structure of polyisoprene is shown in Table I. The following conclusions can be drawn from these results:

1. The 3,4 content is affected very little, if at all, by the initiator or monomer concentration, or by the type of solvent present.
2. The presence of solvent decreases the cis-1,4 content, an aromatic solvent like benzene having a greater effect than an aliphatic solvent like n-hexane.
3. A decrease in initiator concentration increases the cis-1,4 content, in the absence of solvent, or even in the presence of n-hexane.

Table I

Effect of Initiator Concentration and Solvents on Chain Structure of Lithium Polyisoprene (Polymerization temperature = 20°C)

Solvent	[s-C_4H_9Li]	Microstructure - mol.%		
		cis-1,4	trans-1,4	3,4
Benzene*	$9x10^{-3}$	69	25	6
Benzene*	$4x10^{-5}$	70	24	6
n-Hexane*	$1x10^{-2}$	70	25	5
n-Hexane*	$1x10^{-5}$	86	11	3
None	$3x10^{-3}$	77	18	5
None	$8x10^{-6}$	96	0	4

* [Monomer] = 0.5 M.

Thus it appears that low concentrations of lithium, in general, enhance the formation of the cis-1,4 structure, at the expense of the trans-1,4 structure, and that this effect is most noticeable in the polymerization of the undiluted monomer. However, in the presence of benzene, no such effect can be seen, presumably because the "solvent effect" of this aromatic compound can counter the effect of the lithium concentration. This is discussed further in a later section.

Effect of Conversion. The effect of increasing degrees of conversion on the chain structure in polymerization of undiluted isoprene is shown clearly in Table II. As can be seen, the extent

Table II

Effect of Conversion on Chain Structure of Lithium Polyisoprene

(No solvent, polymerization temp. = 20°C. [s-C_4H_9Li] = 10^{-5} M.)

Conversion (%)	Microstructure - mol. %		
	Cis-1,4	Trans-1,4	3,4
13	93	2	5
29	92	3	5
40	92	2	6
46	92	3	5
48	95	1	4
86	95	1	4

of conversion of monomer to polymer has no noticeable effect on the chain structure. Since this is a non-terminating chain growth reaction, this means that each chain maintains a constant chain structure as it grows.

Effect of Temperature. Table III shows the effect of polymerization temperature on the chain structure of lithium polyisoprene, both in the case of undiluted monomer and in the presence of n-hexane as solvent. Within the ranges shown, there does not appear to be any influence of temperature on the placement of the various isomeric chain unit structures.

Table III

Effect of Polymerization Temperature on Chain Structure of Lithium Polyisoprene

Temp. (°C)	[s-C_4H_9Li]	Solvent	Microstructure - mol. %		
			Cis-1,4	Trans-1,4	3,4
46	$1x10^{-5}$	None	95	1	4
20	$1x10^{-5}$	None	95	1	4
0	$1x10^{-5}$	None	95	2	3
-25	$1x10^{-5}$	None	93	3	4
40	$1x10^{-3}$	n-hexane*	76	18	6
-25	$1x10^{-3}$	n-hexane*	78	18	4

* [Monomer] = 2.5 M.

Polybutadiene

Effect of Initiator Concentration and Solvents. In this case, three different solvents were used to carry out the polymerization, in addition to the use of undiluted monomer, as shown in Table IV. It is well known that the organolithium

Table IV

Effect of Initiator Concentration and Solvents on Chain Structure of Lithium Polybutadiene

(Polymerization Temp. = 20°C)

Solvent	[$s\text{-}C_4H_9Li$]	Microstructure - mol. %		
		Cis-1,4	Trans-1,4	1,2
Benzene*	$8x10^{-6}$	52	36	12
Cyclohexane*	$1x10^{-5}$	68	28	4
n-Hexane*	$2x10^{-5}$	56	37	7
n-Hexane*	$3x10^{-2}$	30	62	8
None	$7x10^{-6}$	86	9	5
None	$3x10^{-3}$	39	52	9

* [Monomer] = 0.5 M.

polymerization of butadiene does not yield as high a content of cis-1,4 units as in the case of isoprene. In fact, under practical polymerization conditions in industry ([RLi] $\sim 10^{-3}$), the trans-1,4 content is generally greater than 50%. However, the data in Table IV show exactly how these reaction parameters affect the chain structure, as follows:

1. Again, as in the case of isoprene, the side-vinyl content (1,2-units) is not greatly affected either by the initiator concentration or the presence of solvent. It does seem to be increased slightly by higher initiator levels, and apparently increases noticeably in the presence of benzene.
2. Compared to the case of isoprene, the cis-1,4 content changes dramatically as the initiator concentration is decreased, especially in the absence of solvents, where the cis-1,4 content is shown to rise from 39% to 86%. This also happens in the presence of solvents, although not to as great an extent.

It is interesting to note that a review of the literature shows no reference to the successful synthesis of polybutadiene, by organolithium initiators, with a cis-1,4 content as high as 86%

although the trend toward higher values than 50% has been shown. This is perhaps not surprising, since the attainment of 86% cis-1,4 was only found possible in this work at extremely low initiator concentrations (<10^{-5} M) and with undiluted monomer, conditions which are quite impractical since they result in very slow rates and very high molecular weights (10^7-10^8).

Effect of Polymerization Temperature. The effect of temperature on the polymerization of butadiene is shown in Table V. It can be seen at once that, within the small temperature range shown, there is no noticeable effect of temperature on the chain structure of the polymerizing monomer.

Table V

Effect of Polymerization Temperature on Structure

(No solvent. [s-C_4H_9Li] = $7x10^{-6}$ M.)

Temp. °C	Microstructure - mol. %		
	Cis-1,4	Trans-1,4	1,2
35	85	9	6
20	86	9	5
0	86	9	5

General Discussion

It appears from this work that the chain unit structure in the organolithium polymerization of butadiene and isoprene is sensitive to certain reaction parameters even in non-polar media. This is especially true for the effect of initiator concentration and the amount and type of hydrocarbon solvent present. These effects apparently influence mainly the cis/trans ratio of 1,4 units in the chain, being largely ineffective in changing the side-vinyl content.

The present state of knowledge about the true mechanism of these polymerization reactions is not sufficiently advanced to permit a satisfactory rationalization of these effects. It should be remembered that the growing chains in these systems have been convincingly demonstrated (13) to be associated in pairs at the site of the carbon-lithium bond. Hence it appears that the incoming monomer must react with the associated complex, which apparently can affect the mode of entry. This undoubtedly can explain the greater extent of cis-1,4 addition in the case of isoprene compared to butadiene. Furthermore, such factors as lithium concentration and presence of different solvents can be assumed to have an effect on the structure and reactivity of the associated carbon-lithium bond at the active chain end. This would certainly be expected for the highly polar carbon-lithium bond in non-polar media. Thus even the difference between an aromatic and an aliphatic solvent can affect the chain structure.

In this connection, a recently proposed theory (16) to explain the effect of lithium concentration on polyisoprene chain structure deserves mention. This theory is based on a proposed competition between the rates of chain propagation and isomerization of the chain end, which presumably changes from the cis-1,4 to the trans-1,4 configuration. Although this theory may have some merit, it cannot account for the results demonstrated in Tables II, III and V above, i.e., the absence of any effect of temperature or degree of conversion, both of which would strongly affect the propagation rate, but would not be expected to influence the chain-end isomerization rate. It is far more likely, therefore, that the effects on chain structure described above are due to subtle effects of these reaction parameters on the structure and reactivity of the carbon-lithium bond complex at the active chain end.

Literature Cited

1. Stavely, F. W. and co-workers. Ind. Eng. Chem., 1956, 48, 778.
2. Foster, F. C.; Binder, J. L. Adv. Chem. Ser., 1957, 17, 7.
3. Morita, H.; Tobolsky, A. V. J. Am. Chem. Soc., 1957, 79, 5853.
4. Tobolsky, A. V.; Rogers, C. E. J. Polym. Sci., 1959, 40, 73.
5. Stearns, R. S.; Forman, L. E. J. Polym. Sci., 1959, 41, 381.
6. Morton, M.; Fetters, L. J. J. Polym. Sci., 1964, A2, 3311.
7. Uraneck, C. A. J. Polym. Sci., 1971, Part A1, 9, 2273.
8. Antkowiak, T. A.; Oberster, A. E.; Halasa, A. F.; Tate, D. P. J. Polym. Sci., 1972, Part A1, 10, 1319.
9. Worsfold, D. J.; Bywater, S. Can. J. Chem., 1964, 42, 2884.
10. Schue, F.; Worsfold, D. J.; Bywater, S. Macromolecules, 1970, 3, 509.
11. Santee, E. R., Jr.; Chang, R.; Morton, M. J. Polym. Sci., Polym. Lett. Ed., 1973, 11, 449.
12. Santee, E. R., Jr.; Malotky, L. O.; Morton, M. Rubber Chem. Technol., 1973, 46, 1156.
13. Morton, M.; Fetters, L. J. Rubber Chem. Technol., 1975, 48, 359.
14. Rupert, J. Ph.D. Dissertation, University of Akron, 1975.
15. Gebert, W.; Hinz, J.; Sinn, H. Makromol. Chem., 1971, 144, 97.
16. Worsfold, D. J.; Bywater, S. Macromolecules, 1978, 11, 582.

RECEIVED September 28, 1982

22

Viscosity and Aggregation of Alkyllithium Initiated Polymers

H. L. HSIEH and A. G. KITCHEN

Phillips Petroleum Company, Bartlesville, OK 74004

Utilizing a Dynatrol Viscosity System, viscosity response of polymer solutions can be continuously measured. A series of experiments were carried out to measure the change of viscosity response when we introduced a change in chain ends for an anionically polymerized polymer in hydrocarbon solution. It is concluded that the order of viscosity for either polybutadiene or polystyrene molecules prepared with butyllithium is butadienyl end group > isoprenyl end group > styrenyl end group > terminated. Based on coupling experiments with chlorosilanes, it is concluded that higher aggregates exist for polydienes, while polystyryllithium molecules are mostly dimeric. The cross-association between active polymer molecules and alkyllithium molecules forming mixed aggregates is demonstrated and their possible roles in affecting propagation rates and cross-propagation rates are suggested.

It is fairly well understood that alkyllithiums form rather stable aggregates in which carbon-lithium bond order is maximized by the utilization of all valence orbitals of lithium.(1) Polystyryllithium molecules are mostly dimeric in solution.(2,3,4) This has been generally accepted by all the investigators.(5) However, association numbers of two(6) and four(7,8) has been reported for polydienes. Two methods were used in determining these values, viz., light scattering measurements (in vacuo) and viscosity measurements.

In the late fifties and early sixties, we, at the Phillips R&D laboratories, had an extensive research program on the copolymerization of dienes and styrene by direct reaction of the initiator with the monomer mixtures as well as by the incremental addition of monomers.(9) The decrease in solution viscosity was qualitatively apparent in many cases when the active dienyl chain ends were converted to styryl chain ends. Until recently, how-

0097-6156/83/0212-0291$06.00/0

ever, we did not have quantitative measurements to substantiate this observation.

Dynatrol Viscosity Systems are designed for continuous measurement of viscosity in process streams. Dynatrol Viscosity Detectors are installed directly in process vessels without the need for sampling or analysis. Response is immediate and continuous. Utilizing this system, we carried out a series of experiments to study the change in viscosity of polymer solutions when we introduced a change in polymer chain ends.

Experimental

Reactor. The polymerization tests were carried out in a stainless steel reactor with approximately 7.6-liter (2-gallon) capacity. The reactor was built by Bench Scale Equipment, Dayton, Ohio, specifically for Phillips R&D to be used in anionic polymerization studies. The monomer and solvent tanks are connected directly to the reactor and form a closed system. The weights of monomers and solvent tanks can be read directly and the errors are less than ±1% (±40g.) for the solvent and ±0.1g. for the monomers. For this study, a nitrogen pressure of about 350 k.Pa (50 psig) in the reactor was maintained and the impeller mixing speed was 300 r.p.m.. The temperature was controlled by the automatically regulated steam pressure in the reactor jacket.

Materials. Phillips polymerization grade cyclohexane and butadiene were used. Styrene was a commercial polymerization grade. Solvent was dried over activated Alcoa H151 alumina, and monomers were dried over activated Kaiser 201 alumina before they were transferred to the charge tanks. n-Butyllithium and sec-butyllithium were purchased from Lithium Corporation of America. Chlorosilanes were vacuum distilled before use.

Viscosity System. A Dynatrol viscosity system was purchased from Automation Products, Inc., Houston, Texas. The detector (Type CL-10DVT-4) is inserted in the vessel with the probe fully immersed in the process medium. The drive coil is excited at a frequency of 120 cps. This produces a pulsating magnetic field which causes the drive armature to vibrate at a frequency of 120 cps.

Mechanical vibration of the drive armature is transferred along the attached spring rod, through a welded node point to the probe. The probe is driven into mechanical vibration at the same 120 cps frequency. The amplitude of the vibration of the probe depends upon the viscosity of the process medium.

The mechanical vibration of the probe is transferred through a second welded node point along the upper spring rod to the pick-up end of the detector. The pick-up end consists of an armature and coil arrangement which is similar to that of the driver end; one exception being that the stator of the pick-up coil is a permanent magnet. The vibration of the pick-up armature in the field of this permanent magnet induces a 120 cps A-C voltage in the pick-up coil which is proportional in magnitude to the amplitude of the pick-up armature vibration. Since the pick-up armature is being driven by the probe, the magnitude of the voltage generated in the pick-up coil is a measure of the viscosity of the process media. The detector we used had a range of 10 to 1000 centipoises.

The 120 cps output signal from the detector is fed to the converter, where it is converted into a 0-10 MVDC signal compatible with 0-10 MVDC recorder. Span and zero controls are located at the convertor. The laboratory set-up is shown in Figure 1.

Procedure. Cyclohexane solvent (3.8 kg.) was introduced to the reactor first, heated to 50°C and 338 grams of monomer then added. Initiator was added at 50°C and the polymerization allowed to proceed adiabatically until a peak temperature (65°C to 70°C) had been observed. This generally took about 30 to 40 minutes. A second increment of the monomer (338 grams) was added and allowed to polymerize to quantitative conversion. Peak temperature was reached generally in 10-15 minutes. Cooling was applied slowly with constant mixing to 78°C. This generally took 20 additional minutes. This temperature was then held constant (±0.2°C) by controlling the steam pressure in the reactor jacket. The response of the viscometer probe immersed in the polymer solution was recorded continuously. A typical response from the viscometer during the polymerization and the termination of the active ends is shown in Figure 2. Extreme care was taken to maintain a constant temperature, pressure, and solids concentration while recording the viscometer response. For example, when monomers were used for capping, appropriate amounts of solvent were introduced at the same time to maintain the solids concentration.

Because the entire set-up is a closed system, the control and reproducibility of the polymerization are remarkably good. Since the polymer solution can be drained and rinsed out by the dried solvent from the bottom of the reactor, the closed system rarely needs to be opened and exposed to the atmosphere. For this study, the overall scavenger levels (the difference between levels of added RLi and effective RLi) were in the range of less than 10^{-4}, but more than 10^{-5} mole/liter, representing 2 to 6% of the total initiator added. Typically, the alkyllithium initiator concentration is around 1×10^{-3} mole/liter.

The response of the viscometer represents a relative change. An increase (+) in response (a higher number) indicates that the vibration of the probe has been dampened by a more viscous solution. A decrease (-) in response (a lower number) indicates a less viscous solution. The Dynatrol system has shown itself to be reliable, sensitive, and provides reproducible information. This is true when the conditions are carefully controlled.

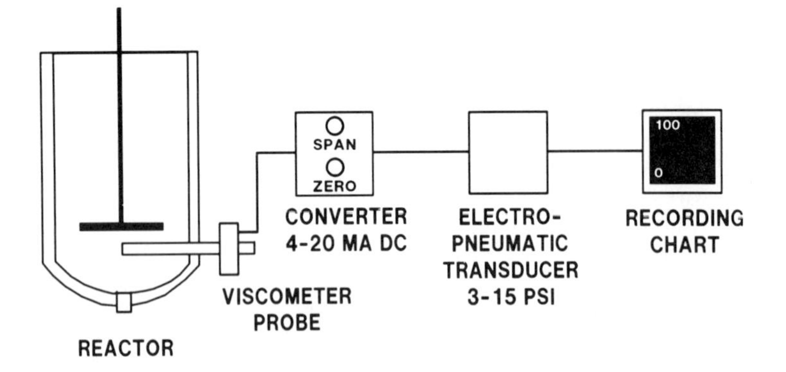

Figure 1. Dynatrol viscometer installation.

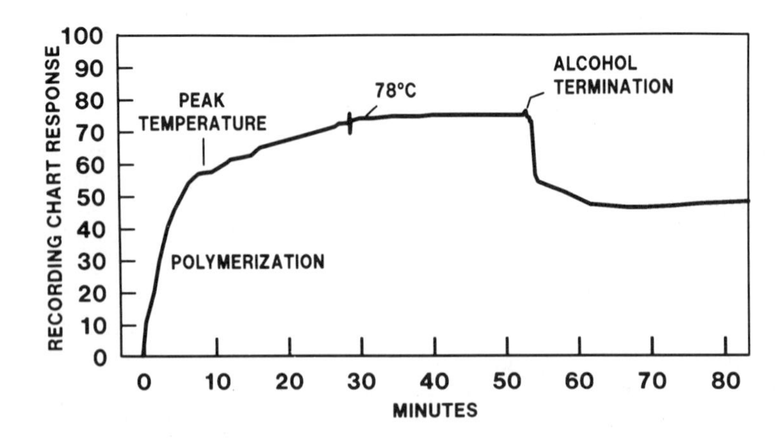

Figure 2. A typical viscosity response curve.

Results and Discussion

In the first series of experiments, polybutadiene samples of about 110,000 molecular weight were prepared in cyclohexane solution at 15% concentration. The viscosity response of the solutions was continuously measured and changes were recorded when terminated with isopropyl alcohol or when capped with isoprene or styrene (Table I). In this table and all subsequent tables, B* and S* mean active ("living") polybutadiene and polystyrene molecules, respectively. Terminated polymers are denoted simply as B and S. BS* and BI* mean polybutadiene molecules with active styrene and isoprene end groups (capped), respectively. SB* and SI* mean polystyrene molecules with active butadiene and isoprene end groups (capped), and BSB* means polybutadiene initially capped with styrene was capped again with butadiene, etc. To convert B* completely to BS* is difficult due to an unfavorable cross-propagation rate.(5) We found that by using 5% of the styrene monomer and adding it in 8 increments resulted in the maximum observed changes. In the same series of experiments, a polybutadiene sample of half of the molecular weight was also prepared for comparison purposes (Table I). Figures 3, 4, and 5 show typical viscosity response charts. They correspond to the data in Table I.

After the completion of polybutadiene experiments, similar experiments were carried out on polystyrene. The results are shown in Table II and Figures 6 and 7. They are consistent with the findings of similar experiments based on polybutadiene. The viscosity is very different depending upon the nature of the end group. This is true whether the polymer is polybutadiene or polystyrene.

Coupling reactions of polybutadienyllithium (B*) with chlorosilanes are well known and well established(10) and are used for the production of controlled long-chain branched polymers.(11) In Table III, the peak molecular weight and viscosity response change, before and after termination and coupling reactions, are reported (also see Figures 8-12). We used mono-, di-, tri-, and tetrachlorosilanes to produce linear, trichain and tetrachain polybutadiene molecules. Generally, about 2-8% of the precursor, for reasons such as premature terminations, incomplete linking, etc., remained uncoupled. Therefore, the peak molecular weight from the GPC curves was used here to illustrate the degree of coupling. Figure 13 shows the GPC curves of the precursor and $SiCl_4$-coupled product which includes the minor amount (~6%) of the uncoupled (precursor) material. The presence of such small amounts of the uncoupled material should have very little effect on the solution viscosity. It is significant that the viscosity of the $SiCl_4$-coupled polymer did not change within experimental error, in spite of the four-fold increase in actual molecular weight. The R_2 $SiCl_2$-coupled polymer with doubling the actual molecular weight of the precursor dropped the viscosity response by 19 units. This decrease in viscosity response is nearly the

TABLE I

VISCOSITY RESPONSE CHANGE[a] OF POLYBUTADIENE[b] BY TERMINATION WITH ISOPROPYL ALCOHOL OR CAPPING WITH ISOPRENE[c] OR STYRENE[d]

Action		Viscosity Response Change
$B_1^* + ROH$	$\dashrightarrow B_1$	-30
$B_1^* + I$	$\dashrightarrow B_1I^*$	-10
$B_1^* + S$	$\dashrightarrow B_1S^*$	-20
{ $B_1^* + I$	$\dashrightarrow B_1I^*$	-12
$B_1I^* + B$	$\dashrightarrow B_1IB^*$	+9
$B_1IB^* + ROH$ }	$\dashrightarrow B_1IB$	-30
{ $B_1^* + S$	$\dashrightarrow B_1S^*$	-21
$B_1S^* + ROH$ }	$\dashrightarrow B_1S$	-10
{ $B_1^* + S$	$\dashrightarrow B_1S^*$	-19
$B_1S^* + B$ }	$\dashrightarrow B_1SB^*$	+14
$B_2^* + ROH$	$\dashrightarrow B_2$	-30
B_1^* versus B_2^*		-21

a. Span 550, zero 420

b. $(\bar{M}w/\bar{M}n) \times 10^{-3}$ = 111/103 for B_1, a typical value
61/57 for B_2

c. About 5% in 2 increments

d. About 5% in 8 increments

{ Denotes a single experiment

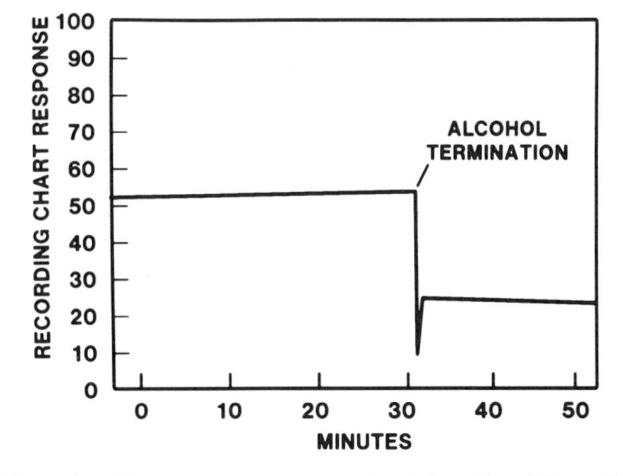

Figure 3. Viscosity response curve of polybutadiene ($B^* \rightarrow B$).

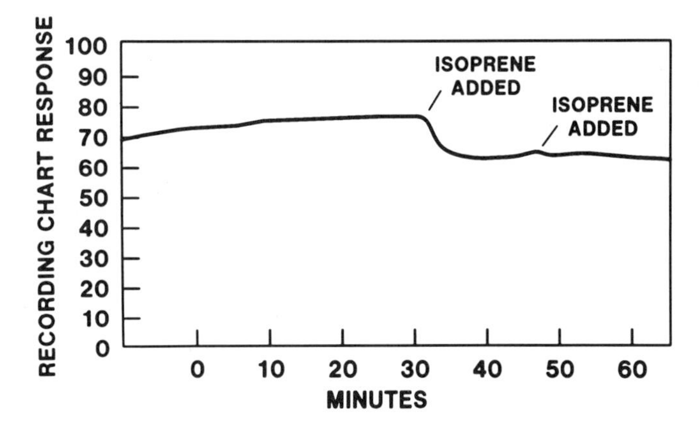

Figure 4. Viscosity response curve of isoprene-capped polybutadiene ($B^* \rightarrow BI^*$).

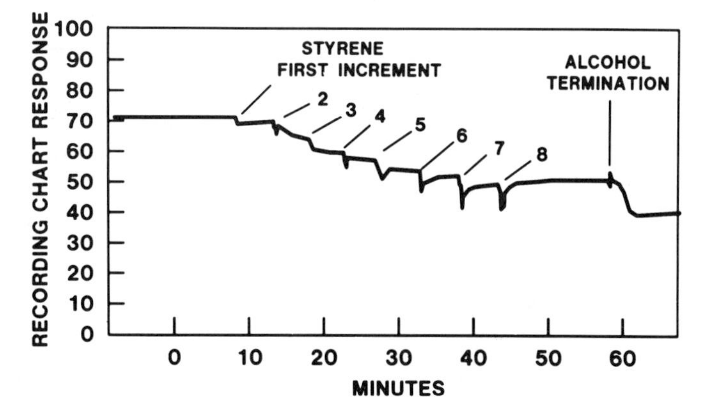

Figure 5. Viscosity response curve of styrene-capped (in eight increments) polybutadiene ($B^* \rightarrow BS^* \rightarrow BS$).

TABLE II

VISCOSITY RESPONSE CHANGE OF POLYSTYRENE

BY CAPPING WITH ABOUT 2% BUTADIENE OR ISOPRENE

Experiment[a]	Action		Viscosity Response Change
A	S* + B	---→ SB*	+30
A	S* + I	---→ SI*	+17
B	{ S* + B	---→ SB*	+39
B	SB* + ROH	---→ SB	-68
B	{ S* + I	---→ SI*	+22
B	SI* + ROH	---→ SI	-46
C	{ S* + B	---→ SB*	+17
C	SB* + ROH	---→ SB	-35
C	{ S* + I	---→ SI*	+9
C	SI* + ROH	---→ SI	-25

a.

Experiment	Span	Zero	Initiation	Average $\overline{M}w/\overline{M}n \times 10^{-3}$
A	550	490	n-BuLi	157/140
B	550	830	sec-BuLi	116/109
C	550	965	sec-BuLi	149/129

{ Denotes a single experiment

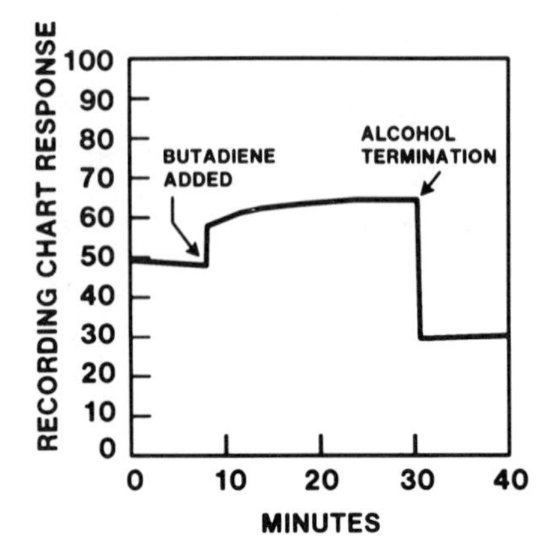

Figure 6. Viscosity response curve of butadiene-capped polystyrene (S → SB* → SB).*

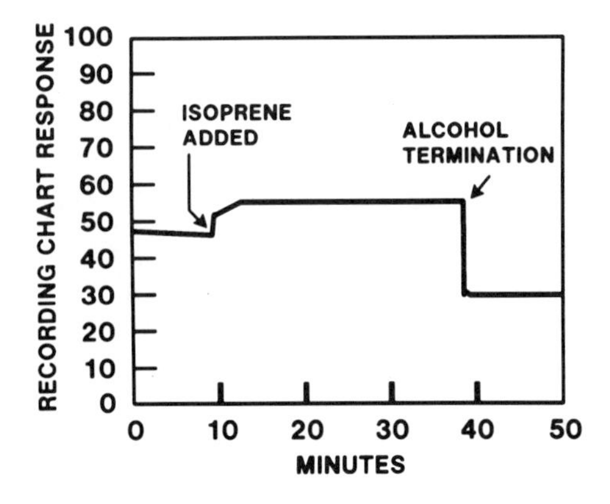

Figure 7. Viscosity response curve of isoprene-capped polystyrene (S → SI* → SI).*

TABLE III

VISCOSITY RESPONSE CHANGE[a] OF POLYBUTADIENE BY TERMINATION WITH H_2O OR COUPLING WITH CHLOROSILANES[b]

Action	(Peak Mol. Wt.)[c] x 10^{-3} Before	After	Viscosity Response Change
B* + H_2O	-	118	-68
B* + $SiCl_4$	110	431	-2
B* + $MeSiCl_3$	113	283	-5
B* + Me_2 $SiCl_2$	134	257	-19
B* + Me_3SiCl	113	117	-64

a. Span 550, zero 420

b. Added in 4 increments - 60%, 20%, 20% and 10% of the calculated stoichiometric amount to allow maximum coupling

c. By GPC

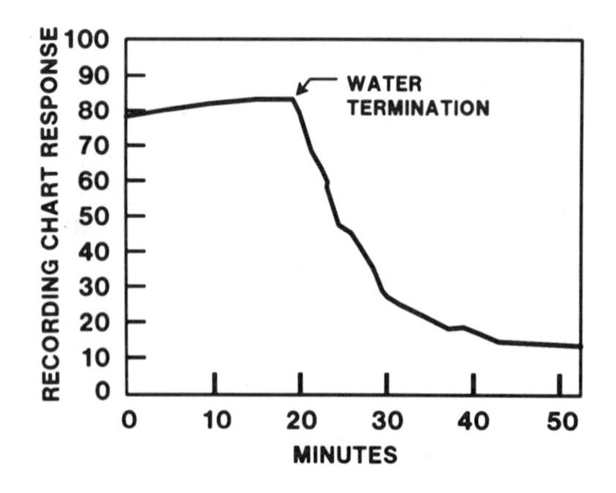

Figure 8. Viscosity response curve of water-terminated polybutadiene ($B^* \rightarrow B$).

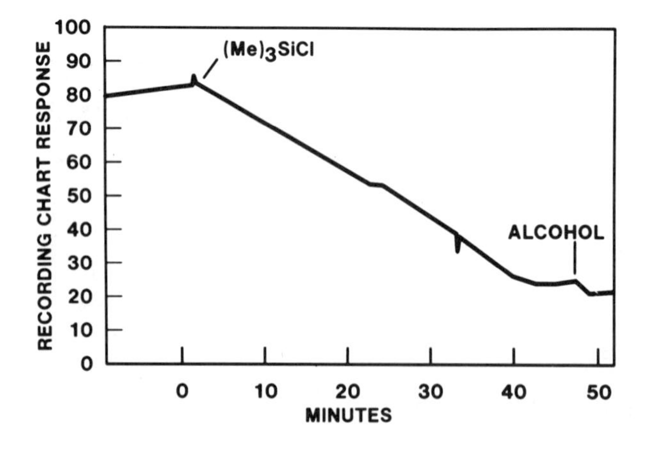

Figure 9. Viscosity response curve of polybutadiene coupled with chlorotrimethylsilane.

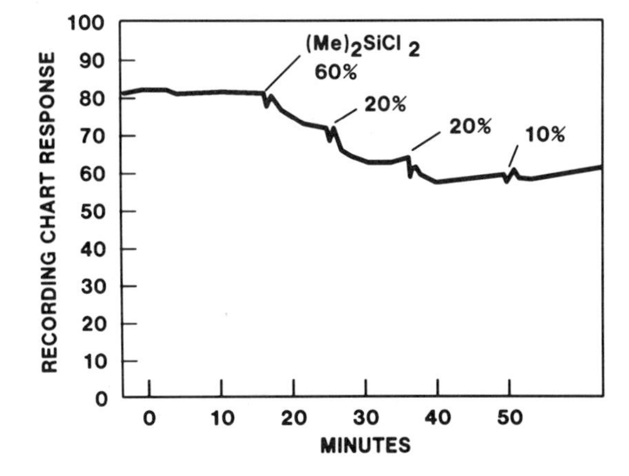

Figure 10. Viscosity response curve of polybutadiene coupled with dimethyldichlorosilane.

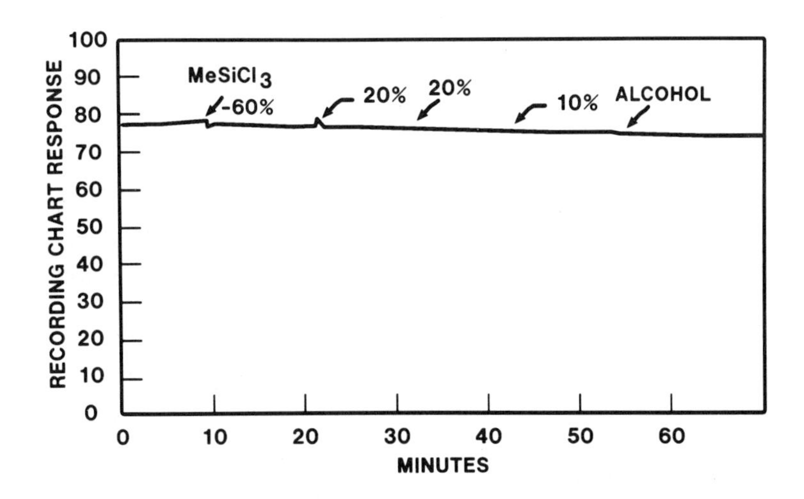

Figure 11. Viscosity response curve of polybutadiene coupled with methyltrichlorosilane.

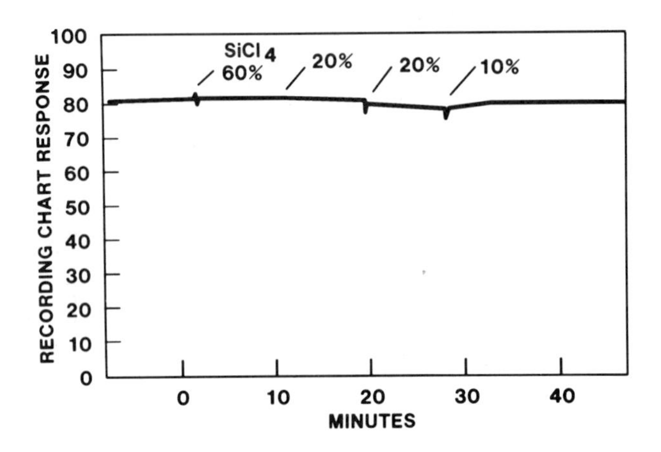

Figure 12. Viscosity response curve of polybutadiene coupled with silicone tetrachloride.

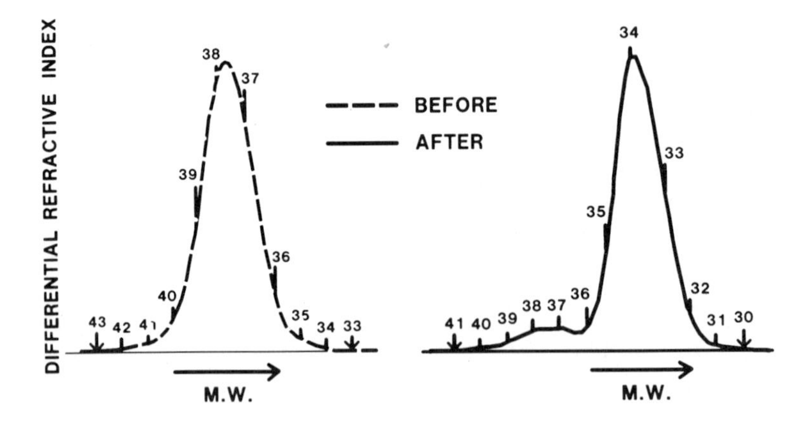

Figure 13. GPC curves of polybutadiene before and after coupling with $SiCl_4$.

TABLE IV

VISCOSITY RESPONSE CHANGE OF THE FORMATION OF POLYMERLITHIUM AND BUTYLLITHIUM CROSS AGGREGATES

Action	Viscosity Response Change[a]	
	B*	BI*
+ 1/6 $(BuLi)_6$[b]	-20	-5
+ 1/6 $(BuLi)_6$[b]	-9	-11
+ 1/6 $(BuLi)_6$[b]	-4	-6

a. Span 550, zero 420

b. Equivalent to the initial initiator level

same as capping a similar polybutadiene with styrene or a "living" polybutadiene with one-half of the molecular weight of the precursor (B_1S^* and B_2S^* in Table I). This is not in agreement with the earlier published results.(12)

All the experiments listed in this report were repeated a minimum of three times. In most of the cases, the individual experiments were repeated five to six times. The results were very consistent and reproducible. There is one phenomenon we observed for which we have no explanation. "Living" polymer terminated with water (or chlorotrimethylsilane) gave consistently lower viscosity (greater viscosity drop) than the equivalent "living" polymer terminated with isopropyl alcohol. Why the alcohol-terminated solution gave higher viscosity, we do not have a clear understanding of it.

To illustrate the difficulties of meaningfully measuring and interpreting the "initial rate of initiation" of alkyllithium with diene and vinyl aromatic monomers, Hsieh and Glaze(5) showed the possible reactions which may occur in the system. The reactions involve the formation and dissociation/association of the mixed aggregates, $R_{n-y}\,P_y\,Li_n$. Thus, with the addition of $(BuLi)_6$ to the aggregated "living" polymer, one should observe viscosity reduction. This is indeed true as shown in Table IV. The B* and BI* in Table IV have a similar polybutadiene main chain as B_1^* in Table I and, therefore, can be directly compared with each other and with data in Table I. The authors wish to propose that the main products are $R_2D_2\,Li_4$ and R_3DLi_4 where R is butyl and D is polydiene. When an equivalent amount of $(n\text{-}BuLi)_6$ was introduced to polystyryllithium solution, viscosity response was also reduced. No further reduction was observed upon the addition of two additional increments of an equivalent amount of $(n\text{-}BuLi)_6$ solution. Since it is generally agreed that polystyryllithium exists as a dimer, it is not surprising to find $RSLi_2$ as the crossed aggregate, where S is polystyrene chain.

Conclusions

1. The solution viscosity of a linear polybutadiene made with butyllithium in the 50 to 150 x 10^3 molecular weight range in 15% cyclohexane solution decreases in the sequence of butadienyl active end > isoprenyl active end > styrenyl active end > terminated.

2. The solution viscosity of a linear polystyrene made with butyllithium in the 100 to 150 x 10^3 molecular weight range in 15% cyclohexane solution decreases in the sequence of butadiene active end >isoprenyl active end > styrenyl active end > terminated.

3. The changes in viscosity by changing the chain ends (capping with another monomer) are reversible and reproducible.

4. Polybutadiene molecules with capped styrene active ends had solution viscosity about the same as unterminated polybutadiene molecules with half of the molecular weight.

5. The viscosity of $SiCl_4$-coupled polybutadiene solution remained essentially unchanged from the unterminated precursor solution.

6. The viscosity of $RSiCl_3$-coupled polybutadiene solution was slightly lower than that of the unterminated precursor solution.

7. When a similar active polybutadiene was coupled with R_2SiCl_2, the solution viscosity dropped significantly. The amount of reduction was about the same as capping the end with styrene.

8. Termination with either H_2O or R_3SiCl gave the same amount of large reduction of viscosity.

9. Cross-associations of polymer-lithium and n-butyllithium, $(RLi)_6$, resulted in reductions of solution viscosity. It is believed that $R_2D_2Li_4$, R_3DLi_4 and $RSLi_2$ are the major products where D is the polydienyl and S the polystyryl moiety.

The inevitable question is what is the association number (aggregation) for polybutadienyl-, polyisoprenyl- and polystyryllithium molecules. As stated earlier, this question had been addressed before by several workers(6-8,13) with contradictory results. After many years of critical examination and discussion (14,15), the discrepancy was not resolved. Our results have clearly shown the differences in solution viscosity caused by these three chain ends. Our technique, unfortunately, does not permit the calculation of the apparent molecular weight. Therefore, it is not possible to report the association number unequivocally. Nonetheless, our findings suggest that the average states of association for these three active polymers are not identical. Furthermore, aggregates of higher than a dimer (polystyryllithium) exist for polydienyllithium under the above polymerization conditions. It is known(16,17) that the rate of initiation in hydrocarbon solvent is in the order: $(\text{menthyllithium})_2 > (\text{sec-BuLi})_4 > (\text{n-BuLi})_6$. It is also known that the rate of propagation is in the order: styrene > isoprene > butadiene. The possible relationship between the rate and degree of aggregation cannot be ignored. The differences in aggregation may also be the key to the mechanism of copolymerization. The so-called "inversion" behavior in copolymerizations of diene and styrene may well be caused by the differences in degrees of aggregation, which in turn, control the cross-propagation rates.

Literature Cited

1. Brown, T.L. in "Advances in Organometallic Chemistry", Volume 3, F. G. A. Stone; R. West Ed. Academic Press, New York, N.Y., 1965, p. 365.
2. Morton, M.; Bostick, E.E.; Livigni, R. Rubber and Plastics Age 1961, 42, 397.
3. Morton, M.; Fetters, L.J. J. Polym. Sci. 1964, A2, 3311.
4. Johnson, A.F.; Worsfold, D.J. J. Polym. Sci. 1965, A3, 449.

5. Hsieh, H.L.; Glaze, W.H. Rubber Chem. & Technol. 1970, 43, No. 1, 22.
6. Morton, M.; Fetters, L.J.; Bostick, E.E. J. Polym. Sci. 1963, C1, 311.
7. Worsfold, D.J.; Bywater, S. Can. J. Chem. 1964, 42, 2884.
8. Sinn, H.; Lundord, C.; Onsanger, O.T. Macromol. Chem. 1964, 70, 222.
9. Hsieh, H.L. Rubber and Plastics Age 1965, 46, No. 4, 394.
10. Zelinski, R.P.; Wofford, F.F. J. Polym. Sci. 1965, A3, 93.
11. Hsieh, H.L. Rubber Chem. & Technol. 1976, 49, No. 5, 1305.
12. Fetters, L.J.; Morton, M. Macromolecules 1974, 1, 552.
13. Al-Harrah, H.M.F.; Young, R.N. Polymer 1980, 21, 119.
14. Szwarc, M. J. Polym. Sci., Polym. Lett. Ed. 1980, 18, 499.
15. Fetters, L.J.; Morton, M. J. Polym. Sci., Polym. Chem. Ed. 1982, 20, 199.
16. Hsieh, H.L. J. Polym. Sci. 1965, A3, 153.
17. Selman, C.M.; Hsieh, H.L. Polym. Lett. 1971, 9, 219.

RECEIVED September 28, 1982

23

The Nature of Stereochemical Control in Metal-Catalyzed Butadiene Polymerization

L. M. STEPHENSON and C. A. KOVAC

University of Southern California, Hydrocarbon Research Institute, Los Angeles, CA 90089

Polymerization of *cis*, *cis*-1,4-dideuterio-1, 3-butadiene by several transition metal catalysts has been studied. The existence of non-stereospecific bond forming events is postulated to signal the involvement of allyl isomerization in the polymerization mechanism. *Trans*-1,4-polymers are accompanied by complete scrambling of deuterium stereochemistry, contrasting with a more specific process to form *cis* polymers. Allyl isomerization is thus implicated as a key event in the formation of *trans*, but not *cis*, polymer.

A long sought goal in mechanistic polymer chemistry has been the determination of those factors which lead to *cis*, *trans* or *vinyl* structures in diene polymers. Various proposals have been made and are summarized in the comprehensive review edited by Saltman[1]. The simplest proposal, advanced by Cossee[2] and Arlman[3], assigns the dominant role to the nature of the diene coordination. In this mechanism bidentate coordination, e.g. **1**, of necessity involving the *cisoid* conformation of the diene, would lead to *cis* polymer.

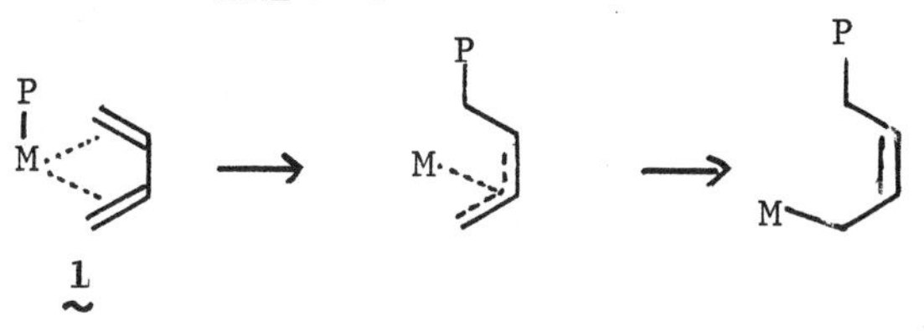

0097-6156/83/0212-0307$06.00/0

Trans polymer backbone in the Cossee-Arlman scheme was then the result of monodentate coordination, e.g. **2**, with the more stable *transoid* conformation adopted.

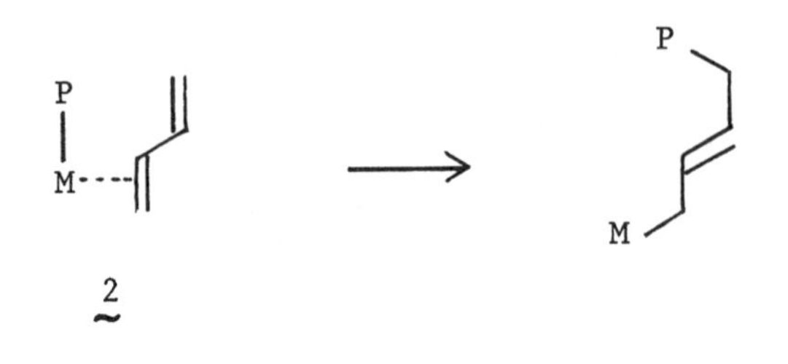

Kormer, Lobach and others[4] were the first to suggest an alternative process which assigned a dominant role to allyl isomerization. In this mechanism the process begins with bidentate coordination, followed by formation of an allyl complex. If this allyl complex has a sufficient lifetime, it will isomerize from the less stable, initially formed, *anti* allyl **3** to the more stable *syn* species, **4**. As shown below, **4** leads to trans polymer. On the other hand the propagation steps, which include coordination of diene and formation of the next bond in the chain, could be the fast steps. Polymer with a *cis* backbone would then be the result.

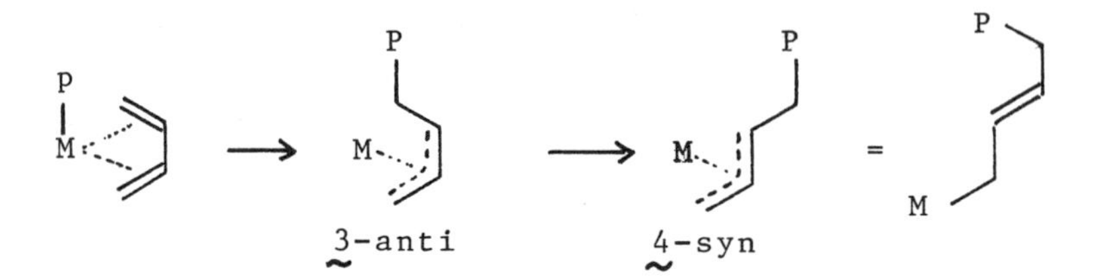

Studies of polymerizing systems and models, particularly using nuclear magnetic resonance techniques, have provided support for both points of view[5].

In this contribution we describe the results of our attempts to devise a chemical probe for the involvement of allyl isomerization during the polymerization reaction. Accordingly, it will be appropriate to briefly review the key aspects of isomerization in transition metal allyls.

The Nature of Allyl Isomerization

Work from the group of Faller[6], in particular, has demonstrated that the most general mechanism for allyl isomerization proceeds by rotation about carbon-carbon bonds in η^1-allyls (also called σ-allyls).

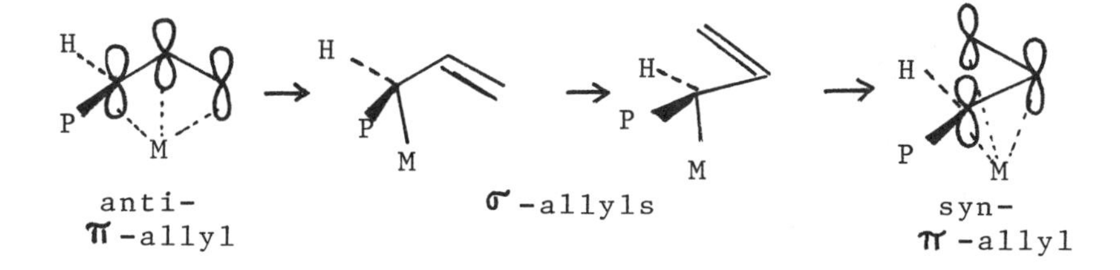

Faller[6] has also demonstrated conclusively that formation of primary σ-allyl species, **5**, from the π (or η^3) species, **6**, is faster (∿ 10x) than formation of secondary σ-allyls **7**. For the case at hand, this would imply that the primary σ-allyl **5**, would be formed much more often than the secondary σ-allyl **7**.

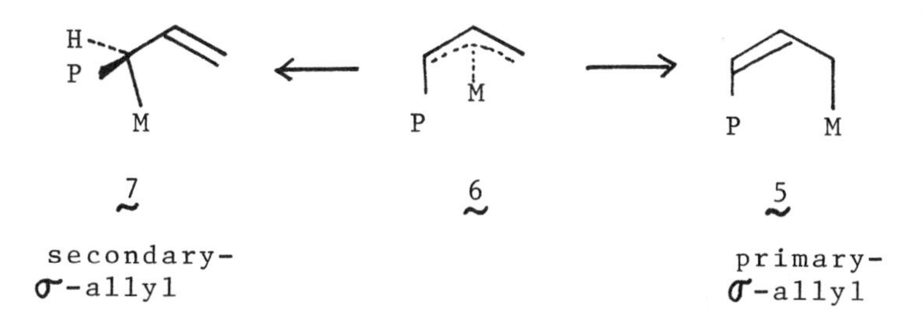

A deuterium label at the primary position could thus be used as a very sensitive probe for allyl isomerization. Formation of the primary σ-allyl would be expected to scramble any initial configuration of the label to a random mixture of syn and anti deuterium.

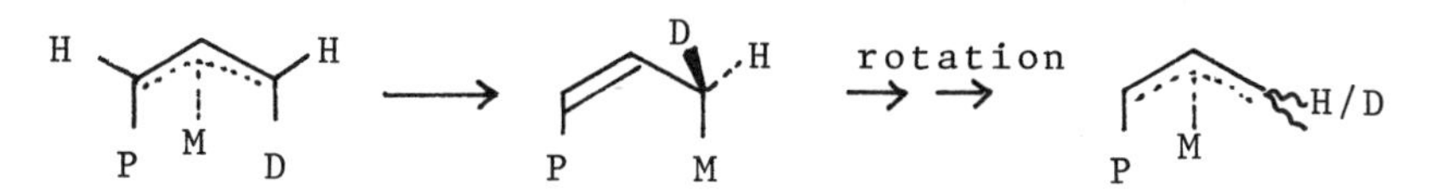

We[7] have previously used this method to establish allyl isomerization events in nickel catalyzed cyclodimerization of butadiene, and describe here our application to the polymerization mechanism.

Results

The use of cis, cis - 1, 4-dideuterio-1,3-butadiene, **8**, in a polymerization reaction will provide direct information about the stereochemistry of the bond forming reaction. We have

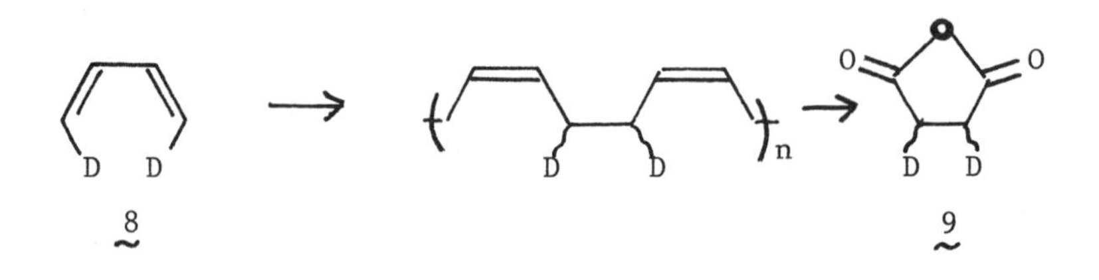

previously developed[8] an analysis for this process, illustrated above. Oxidative conversion of polymer to succinic acid or anhydride, **9**, leads to systems which allow the dl/meso ratio to be assessed easily[8]. Our preliminary analysis of this reaction also revealed that monomer isomerization to cis, trans- and trans, trans-diene also occurred. Thus each reaction also required an analysis of recovered starting material[9]. A previous study by Porri and Anglietto[10], reported while this work was in progress, employed this same method. Close inspection of this work reveals that these authors did not take into account the possibility of monomer isomerization. Thus, their interpretation is less complete than that presented below.

Two catalyst systems, $RhCl_3$[12] and allyl nickel iodide[11] which give high trans polymer were examined and found to give identical, stereorandom results. Examination of recovered monomer shows no isomerization; the stereorandomness of the bond forming event must be inherent to the mechanism.

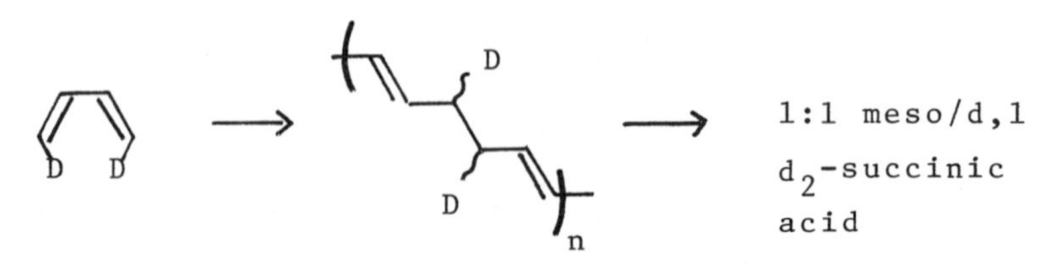

Catalysts which lead to cis polymer show a significantly higher stereoselectivity in the bond forming reaction. When (π-allyl nickel iodide)$_2$ modified with $TiCl_4$[13] is employed, the 1, 2 deuterium stereochemistry is 70% dl, and 30% meso as revealed by analysis of succinic anhydride. In addition, monomer isomerization is extensive, and could account for a large fraction of the meso structures which are formed. The use of (π-allyl nickel trifluoroacetate)$_2$ as catalyst[14] led to a similar result (32% meso), accompanied by little if any monomer isomerization. Thus, it appears that in reactions to form cis polymer, some, but not always all, of the stereochemical information present in the starting diene is preserved in the polymer. In contrast, none of the initial diene stereochemistry can be detected in the trans polymer.

Discussion

While several explanations may be given for these findings, we focus here on our contention that this result supports the importance of allyl isomerization in diene polymerization. We postulate initially that the loss of bond forming stereospecificity in the diene polymerization is the result of allyl isomerization. This is particularly well supported by the completely stereorandom bond formation found for trans polymer. This is satisfactorily accounted for by rapid isomerization of the label, via primary allyls, to a random(1:1) mixture of syn and anti deuterium. Any other mechanism for randomization, for example

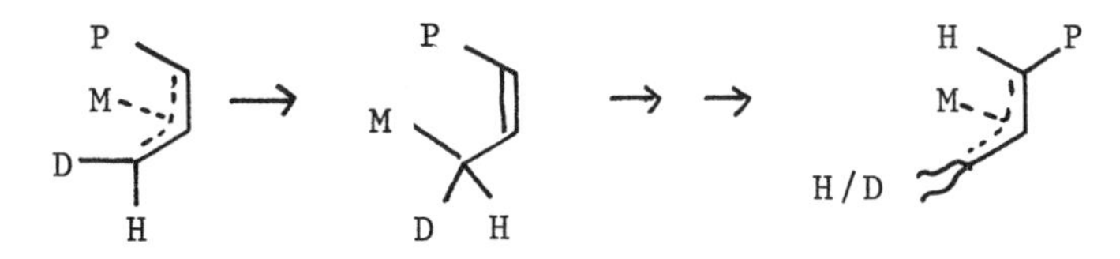

one which allows several organometallic conformers to be involved in the reaction, would be unlikely to result in 1:1 ratios; it would be exceptionally unlikely to find the same ratio from both trans forming catalysts as we do. With this assumption we can then state that allyl isomerization via the primary σ-allyl is slow in those systems which lead to cis polymer. With the added knowledge from Faller's work[6] that isomerization via secondary allyls is some ten times slower, we would anticipate essentially no involvement of such species in cis cases.

A unique mechanism is shown in Figure 1. This scheme starts with an allyl/diene complex with both the growing polymer chain and the deuterium label in an anti configuration initially. Rapid allyl isomerization (left hand branch) leads to trans backbone and stereorandom deuterium; rapid bond formation (right hand

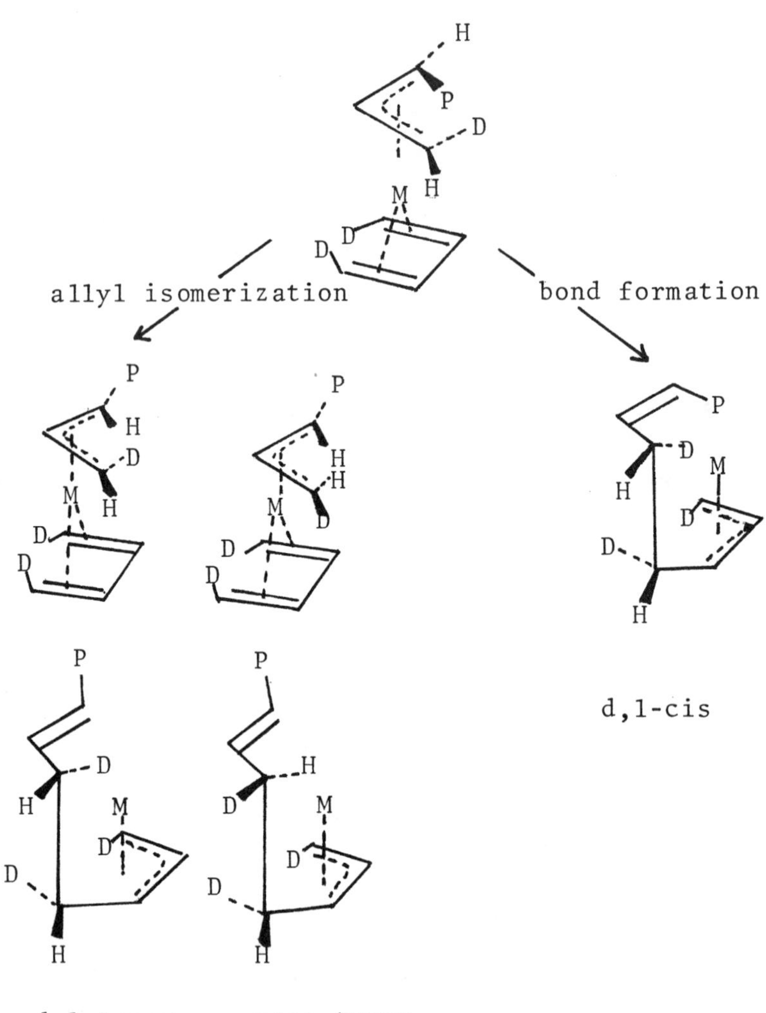

Figure 1. Propagation steps in diene polymerization.

branch) retains the cis backbone and leads to stereospecific placement of deuterium.

Our interpretation most easily lends itself to the idea that the lack of complete stereospecificity in the formation of cis-polymer is due to some isomerization via primary allyls. With the relatively low levels of isomerization seen, we would expect that backbone isomerization occurring via secondary allyls would be essentially unobservable. Indeed little trans polymer is seen, in our two predominantly cis examples.

We cannot exclude the possibility that conformers other than those shown in the right hand branch of fig. 1 are involved in the reaction. Some of these could lead to meso units in the polymer; this feature prevents us from being quantitative in some of these arguments.

We suggest that it is now useful to attempt to rationalize the stereochemical course of diene polymerization reactions in terms of organometallic features which lead to faster or slower allyl isomerization rates. Indications pointing in this direction are available even in the present study. Nickel π-allyl iodide provides trans polymer; however, the simple addition of the Lewis acid species, $TiCl_4$, converts the polymerization reaction into one yielding a high cis polymer. We propose here that the role of the Lewis acid is one which ionizes the iodide away from the nickel center, providing a more positively charged metal site. This would result in tighter allyl binding and slower allyl isomerization rates. This association of cis polymer with electronegative ligands in the catalyst appears to be recognized in commercial application as well.[1] Our current work is oriented toward demonstrating quantitative relationships of this type.

Experimental

General.

Proton NMR spectra were obtained on a Varian XL-200, XL-100, or EM-360 spectrometer, as noted. Chemical shifts are reported in ppm (δ) relative to tetramethylsilane. Experiments for which deuterium decoupling was required were run on the Varian XL-200, using the broadband transmitter and coil as the deuterium decoupler, and observing ^1H through the normal decoupler coil. The instrument was operated unlocked for these experiments.

Carbon-13 NMR spectra were obtained using a Varian XL-200 spectrometer operating in the Fourier transform mode. Polymer samples were run as gels or slurries in benzene-d_6, unless otherwise noted. Operating and processing conditions were standard for all polymer samples. Chemical shifts are reported relative to tetramethylsilane in ppm (δ).

Most deuterated solvents for NMR spectra were purchased from Merck, Sharp, and Dohme, and all were 99+ atom% D.

Analysis of ^1H-NMR spectra was performed using modified versions of the computer programs UEA[15] and DNMR3.[16]

Raman spectra were obtained using a Spex spectrometer with a Spectra-Physics argon ion laser operating at 4880 A. Scattering was measured with a Spex DPC-2 digital photometer in the photon-counting mode. Data are reported in cm^{-1}, relative to the Rayleigh line (20492 cm^{-1}); relative intensities are placed in parentheses following the scattering frequency. Spectra were recorded at 400 milliwatts incident power to give 1 cm^{-1} resolution (slits = 100:200:200:100). Samples were prepared by vacuum distillation of butadiene into glass capillary tubes which were then sealed.

Gas chromatography was performed using a Hewlett-Packard 5700A Gas Chromatograph equipped with a thermal conductivity detector. Output was recorded and measured with a Hewlett-Packard 3380S Integrator.

Synthesis and handling of air- and water-sensitive organometallic complexes was carried out either on the bench using standard Schlenk techniques [17] or in a dry box designed to maintain an atmosphere of less than 1 ppm contaminants. The atmosphere in the dry box was N_2. Solvents for use in polymerizations and synthesis of organometallics--diethyl ether, toluene, heptane, benzene, and dioxane--were carefully dried and degassed prior to use by refluxing under nitrogen with sodium benzophenone[18] then freshly distilled before use. Butadiene was of research purity (99.86%) and was purchased from Matheson. When additional drying was required, anhydrous $CaCl_2$ was used as a drying agent.

IR spectra were obtained using a Perkin-Elmer Infrared photometer, Model 297. Sampling method is reported for individual cases; absorptions are reported in cm^{-1}, and the intensity is designated as s (strong), m (medium), w (weak), or b (broad).

Melting points were obtained on an Electrothermal capillary melting point apparatus. All melting and boiling points are uncorrected.

Cis,cis-1,4-dideuterio-1,3-butadiene.

The dideuteriobutadiene was synthesized by the method of Stephenson, Gemmer, and Current.[9] The dideuteriobutadiene was stored in a pressure vessel in the freezer over anhydrous calcium chloride until use. Raman spectrum: 1171 (0) 1216 (5), 1226 (98) cm^{-1} (93% *cis,cis*-isomer, 7% *cis,trans*-isomer, 0% *trans,trans*-isomer)[9].

General Procedure for Ozonolysis of Polymers to Succinic Anhydride.[19]

Polybutadiene (0.5 g) was placed in a 500 ml 2-necked round bottom flask equipped with a capillary gas bubbler and a magnetic stir bar. *Trans*-polymers were ground to fine powders, while the more elastomeric cis-polymers were cut into small pieces. Anhydrous methanol (25 ml) was added to the flask, and the suspension chilled to -10°C. A Welsbach laboratory ozone generator operating at 90 volts and 8 psig oxygen pressure was used to produce ozone. Oxygen gas was dried by passing through a column of Drierite before entering the ozonizer. A stream of ozone in oxygen was passed through the reaction mixture for 3 to 5 hours, until little or no solid remained in the flask. The methanol was removed on a rotary evaporator without heating, yielding a gummy colorless residue. To this residue was added about 2 ml of formic acid (Mallinckrodt) and 1 ml 30% hydrogen peroxide (Mallinckrodt). The mixture was gently warmed for 1 hour, then refluxed until no peroxides could be detected by starch-iodide paper. Evaporation of the solvent left a gummy solid containing succinic acid, which was not further purified before the next step.

Acetic anhydride (2 ml) was added to the flask and the mixture was refluxed for several hours, then the solvent was removed by rotary evaporation. The last traces of solvent were removed under vacuum at room temperature, then the tar-like brown product was repeatedly sublimed to give 0.2 g (2 mmol) of white, crystalline succinic anhydride. ^{13}C-NMR (dimethyl sulfoxide-d_6, 50 MHz, ^{1}H) resonances at δ 174.1 and 29.0 ppm
Yield = 22% theoretical.

Check for Deuterium Exchange during Polymer Analysis.

A sample of a deuteriated polymer produced by reaction of cis,cis-1,4-d_2-butadiene with π-allyl nickel iodide/$TiCl_4$ catalyst was oxidatively cleaved and cyclized to d_2-succinic anhydride, as described elsewhere in this section. ^{13}C-NMR (chloroform-d, 50 MHz, ^{1}H) resonance at δ 28.02 (t, J=21 Hz) ppm.

A mixture of d_6-succinic anhydride and d_0-succinic anhydride was prepared; this sample exhibited ^{13}C-NMR (chloroform-d,50 MHz, ^{1}H) resonances at 28.04 (t, J=21 Hz) and 28.36 (s) ppm.

The absence of the singlet at 28.36 ppm in the d_2-sample insures that < 5% exchange of the deuterium in the sample could have occurred.

Analysis of Mixtures of d,l and meso-d_2 Succinic Anhydride.

The method of Field, Kovac and Stephenson[8] was used for this analysis.

General Procedure for the Polymerization of Butadiene.

Unless otherwise noted, the following general procedure was used to effect polymerization of butadiene with transition metal catalysts.

Polymerizations were carried out in a 3 oz. Fisher-Porter pressure bottle which was modified to allow pressure measurements and syringe transfers to be made. Typically, catalysts and solvents were introduced into the bottle in the dry box, then the bottle was sealed and removed to the bench. Butadiene was dried by passing it through anhydrous calcium chloride. The amount of butadiene for a given polymerization was measured by condensation into a graduated receiver, then this volume was bulb-to-bulb transferred into the Fisher-Porter bottle containing catalyst and solvent. The reaction vessel was sealed, and warmed to the polymerization temperature by heating in an oil bath which enclosed the entire glass portion of the bottle. Reported reaction temperatures are $\pm$ 5°C. Reactions were stirred with a magnetic stir bar, and the progress of the reactions was monitored by observing the pressure change.

Polymerizations were terminated by cooling the reaction vessel to room temperature and releasing any excess butadiene to the atmosphere. Then the contents of the Fisher-Porter bottle were poured into 100 ml of a 1% solution of hydrochloric acid in methanol to precipitate the polymers. The solid polymers which resulted were leached in this acidic methanol solution to fully remove residual catalyst (at least 5 hours). If the polymer still appeared colored, it was isolated, redissolved in chloroform, and re-precipitated as described above. Finally, polymers were washed with anhydrous methanol, isolated, and dried in a vacuum oven at 50°C.

Polymerization with π-Allyl Nickel Iodide.

π-Allyl nickel iodide was prepared by the method of Wilke.[20] Polymerizations using this catalyst were carried out with either butadiene or cis,cis-1,4-d_2-butadiene as monomer, in the manner described previously in this section. Representative experimental details are given below:

Catalyst:	π-allyl nickel iodide	0.04 g (8.8×10^{-5} mol)
Solvent:	toluene	2.0 ml
Monomer:	cis-cis-1-4-d_2-butadiene	2.0 ml (2.6×10^{-2} mol)
Temperature:	50°C	
Time:	20 hours	
Yield:	0.75 g (51% theoretical)	

When deuteriated monomer was used, samples were taken for Raman spectroscopy before polymerization and at periodic intervals during polymerization so that the isomeric distribution of the monomer could be monitored. Examination of Raman spectra for the experiment described above showed: (1) (d_2-monomer at t=0 minutes) 1216 (10), 1226 (199) cm^{-1}. (93% cis,cis-isomer) (2) (d_2-monomer at t=1200 minutes) 1216 (9), 1226 (206) cm^{-1} (94% cis,cis-isomer).

Geometric isomerism was determined on non-deuteriated polymers. Carbon-13 NMR (methylene chloride-d_2, 50 MHz, ^1H) resonances at δ 130.2 (s) and 32.8 (s) ppm. The absence of a peak at 27.7 ppm indicated that the polymers were 100% *trans*.

A sample of the deuteriated polymer (0.531 g) was ozonized as a suspension in CH_2Cl_2, and the product treated with formic acid-hydrogen peroxide as described previously in this section. Cyclization to d_2-succinic anhydrides and derivatization to form the d_2-N-(o-biphenyl) succinimides were also performed by previously described methods. ^1H-NMR (acetone-d_6, 200 MHz, ^2H) resonances at δ 2.62 (s), 2.62 (d, J=4.4 Hz), 2.84 (s), 2.84 (d, J=4.4Hz), 7.48 (m). Measurement of peak areas by triangulation indicated that the d_2-N-(o-biphenyl) succinimide produced in the above experiment was 55% *meso*-isomer and 45% *d,l*-isomer.

Polymerization with π-Allyl Nickel Iodide/Titanium Tetrachloride.

This catalyst was prepared and used for polymerization in the following manner. A 0.1 M solution of $TiCl_4$ in toluene was made up and stored in the dry box. The desired amount of π-allyl nickel iodide was weighed out, dissolved in 1.0 ml toluene, and placed in a Fisher-Porter tube. In all polymerizations, [Ni] = $3x10^{-3}$ M and the Ni/Ti ratio was ≈ 1. To the π-allyl nickel iodide solution was added an appropriate amount of the stock solution of $TiCl_4$, and the mixture was agitated at ambient temperature for 20 minutes. Almost immediately upon addition of the $TiCl_4$ solution, a brown precipitate formed. After 20 minutes, the solvent (toluene) was added, and the tube was sealed and removed to the bench, where butadiene was transferred into the tube to give the desired monomer concentration. The tube was placed in an oil bath at the polymerization temperature, and the rest of the reaction and work-up was carried out as described in the general procedure detailed previously. Typical experimental details are given below:

Catalyst:	π-allyl nickel iodide	$6.5x10^{-3}$g	($1.4x10^{-5}$mol)
	$TiCl_4$ in toluene (0.1 M)	0.28 ml	($2.8x10^{-5}$ mol)
Solvent:	Toluene	6.0 ml	
Monomer:	1,3-butadiene	2.2 ml	(0.030 mol)
Temperature:	50°C		
Time:	4.5 hours		
Yield:	0.98 g (61% theoretical)		

This polymerization was carried out a number of times with unlabelled butadiene, over a monomer concentration range of

0.25 M to 3.0 M. Relative areas of the ^{13}C-NMR peaks at 32.6 and 27.2 ppm indicated that all the polymers so produced contained 80 to 90% cis-double bonds.

Cis,cis-1,4-d_2-butadiene (2) was used as monomer in two cases. Experimental conditions were the same in both runs: [Ni] - 3.8×10^{-3} M, Ni/Ti - 1.3, [d_2-butadiene] = 3.0 M, T=55°C. Results of both runs are shown in Table I.

These deuteriated polymers were oxidatively cleaved and the oxidation products converted to the d_2-N-(o-biphenyl) succinimides for d,l/meso analysis by ^{1}H-NMR. Run 1 yielded 30% meso-isomer and 70% d,l/isomer, while Run 2 yielded 32% meso-isomer and 68% d,l-isomer.

Table I. Results of polymerization of (2) with π-allyl nickel iodide/$TiCl_4$.

	Run 1	Run 2
Polymerization time	330 min	420 min
Yield	17%	21%
% cis in polymer[a]	78%	83%
Monomer Composition (initial)[b]	93: 7:0	93: 7:0
Monomer Composition (final)[b]	80:20:0	74:26:0

[a] Percentages determined by ^{13}C-NMR.
[b] Reported as relative percentages (cis,cis:cis,trans:trans, trans), as determined from Raman lines at 1226 (cis,cis), 1216 (cis,trans), and 1171 (trans,trans) cm^{-1}

Polymerization of Butadiene with Rhodium Trichloride.

The desired amount of $RhCl_3.3H_2O$ (Alfa) was weighed out and placed into a Fisher-Porter tube. In all polymerizations, [Rh] = 1×10^{-2} M. To this was added a weighed amount of 1,3-cyclohexadiene (Aldrich) and the desired volume of a 2.5 wt.% soln. of sodium dodecyl sulfate (Aldrich) in deionized water. All the above operations were carried out on the bench, without a protective atmosphere. The Fisher-Porter tube was then sealed, the monomer was transferred into the tube, and polymerization and work-up were carried out as described in the general procedure elsewhere in this section. Representative experimental details are given below:

Catalyst:	$RhCl_3.3H_2O$	0.015 g($5.7x10^{-5}$ mol)
	1,3-cyclohexadiene	0.017 g($2.1x10^{-4}$ mol)
Solvent:	2.5 wt.% soln. of sodium dodecyl sulfate in H_2O	3 ml
Monomer:	cis,cis-1,4-d_2-butadiene	1.7 ml (0.022 mol)
Temperature:	25°C	
Time:	31 hours	
Yield:	0.58 g (47% theoretical)	

Geometric isomerism was determined on non-deuteriated polymers. ^{13}C-NMR (benzene-d_6, 50 MHz, ^1H) resonances at δ 129.3 (s) and 32.1 (s) ppm. The absence of a peak at ∿27 ppm indicated that the polymers were 100% trans in structure.

When deuteriated monomer was used, samples were taken for Raman spectroscopy before polymerization and at the end of polymerization to ascertain to what extent monomer had isomerized. Examination of Raman spectra for the experiment described above showed: (1) (d_2-monomer at t=0 minutes) 1216 (10), 1226 (23 (238) cm^{-1}. (94% cis,cis-isomer) (2) d_2-monomer at t=;850 minutes) 1216 (12), 1226 (237) cm^{-1}. (93% cis,cis-isomer).

The deuteriated polymer from the above example was oxidatively cleaved and the oxidation products were converted to the d_2-N-(o-biphenyl) succinimides for d,l/meso analysis. ^1H-NMR (acetone-d_6, 220 MHz, ^2H) resonances at δ 7.48 (m), 2.84 (s), 2.84 (d, J=4.4 Hz), 2.62 (s), 2.61 (d, J=4.4 Hz) ppm. Measurement of methylene peak areas by triangulation indicated that the d_2-N-(o-biphenyl) succinimide produced in the above experiment was 48% meso-isomer and 52% d,l-isomer.

Polymerization with π-Allyl Nickel Trifluoroacetate.

Polymerizations with this catalyst were carried out by the general procedure described previously. Representative experimental details are given below:

Catalyst:	π-allyl nickel trifluoroacetate	0.025 g($5.9x10^{-5}$ mol)
Solvent:	heptane	7 ml
Monomer:	cis,cis-1,4-d_2-butadiene	2 ml ($2.6x10^{-2}$ mol)
Temperature:	42°C	
Time:	3.3 hours	
Yield:	0.68 g (47% theoretical)	

^{13}C-NMR of the polymer (benzene-d_6, 50 MHz, ^1H) showed resonances at δ 57.75 (t, J=20 Hz), 52.38 (t, J=20 Hz), 128.37 (S). [7-5-81-1] Integration of the two triplets indicated the polymer was 95% cis.

When deuteriated monomer was used, samples were taken for Raman spectroscopy before polymerization and at the end of the polymerization to ascertain the extent to which monomer had isomerized. Raman spectra showed: (1) (d_2-monomer at t=0 minutes) 1216 (4.5), 1226 (175.5) cm^{-1}. (96% cis,cis-isomer) (2) (d_2-monomer at t=200 minutes) 1216 (15), 1226 (230) cm^{-1} (91% cis,cis-isomer)

A sample of the deuteriated polymer was oxidatively cleaved and derivatized according to previously described procedures to give d_2-N-(o-biphenyl) succinimide isomers. ^{1}H-NMR (acetone-d_6, 200 MHz, ^{2}H) resonances at δ 2.62 (d, J=4.4 Hz), 2.63 (s), 2.85 (d, J=4.5 Hz), 2.85 (s), 7.48 (m). Measurement of methylene proton peak areas by triangulation indicated that the deuteriated derivative produced in the above experiment was 32% meso-isomer and 68% d,l-isomer.

Acknowledgment

This work has received generous support from the National Science Foundation through Grant 80-12233, and from the Petroleum Research Fund, administered by the American Chemical Society, Grant 11748AC.

Literature Cited

1. See in particular Cooper,W. "Polydienes Coordination Catalysts", p. 21-78 and Teyssie, Ph.; Dawans, F. "Theory of Coordination Catalysts", p. 79, 138 "The Stereo Rubbers", Saltman, W. Ed., John Wiley and Sons: N.Y. 1977.
2. a) Cossee, P. "Stereochemistry of Macromolecules"; M. Dekker: New York, 1967, Vol. I, p. 145.
   b) Cossee, P. J. Catal. 1964, 3, 80.
3. Arlman, E. J. J. Catal. 1966, 5, 178.
4. a) Druz, N.N.; Zak, A.V.; Lobach, M.I.; Vasiliev, V.A.; Kormer, V.A. European Polymer J. 1978, 14, 21.
   b) Kormer, V.A.; Lobach, M.I. Macromolecules 1977, 10, 572.
   c) Druz, N.N.; Zak, A.V.; Lobach, M.I.; Shapkov, P.P.; Kormer,V.A. European Polym. J. 1977, 13, 875.
5. a) Warin, R.; Julemont, J; Teyssie, P. J. Organomet. Chem. 1980, 185, 143.
   b) Dolgoplosk, B.A. Kinetika i Kataliz 1977, 18, 1146.
6. Faller, J.W.; Thomsen, M.E.; Mattina, M.V. J. Am. Chem. Soc. 1971, 93, 2842.
7. Graham, C.R.; Stephenson, L.M. J. Am. Chem. Soc. 1977, 99, 7098.

8. Field, L.D.; Kovac, C.A.; Stephenson, L.M. J. Org. Chem. 1982, 7, 1358.
9. a) Stephenson, L.M.; Gemmer, R.V.; Current, S.P. J. Org. Chem. 1977, 42, 212.
   b) Gemmer, R.V. Ph.D., Thesis, Stanford University, Stanford, Ca., 1975.
10. Porri, L.; Aglietto, M. Makromol. Chem. 1976, 177, 1465.
11. a) Kormer, V.A.; Babitskii, B.D.; Lobach, M.I.; Chesnokova, N.N. J. Polym. Sci. 1969, C16, 4351.
    b) Lazutkin, A.M.; Vashkevich, V.A.; Medvedev, S.S.; Vasilieva, V.N. Dokl. Akad. Nauk,SSSR 1967.
12. Rinehart, R.W.; Smith, H.P.; Witt, H.S.; Romeyn, H. J. Am. Chem. Soc. 1961, 83, 4864; ibid., 1962, 84, 4145.
13. Dolgoplosk, B.A.; Tinyakova, E.I. Izv. Akad. Nauk. SSSR Ser. Khim. 1970, 2, 344.
14. Durand, J.P.; Dawans, F.; Tessie, Ph. J. Polym. Sci. 1970, A1, 979.
15. Ferretti, J.A.; Harris, R.K.; Johannesen, R.B. J. Mag. Reson. 1970, 3, 84.
16. Keeier, D.A.; Bensch, G. "DNNMR3-Program 165, Quantum Chemistry Program Exchange", Indiana University, Bloomington, Indiana, 1970.
17. Shriver, D.F. "The Manipulation of Air-Sensitive Compounds", McGraw-Hill: New York, NY, 1969.
18 Gordon, A.J.; Ford, R.A. "The Chemist's Companion", John Wiley & Sons: New York, NY, 1972.
19. Rabjohn, N. Ed., Organic Syntheses, Coll. V. 4, 1963, 484.
20. Wilke, G.; Bogdanovic, B.; Hardt, P.; Heimbach, P; Keim, W.; Kroner, M; Oberkirch, W.; Tanaka, K.; Steinrucke, E.; Walter, D; Zimmerman, H. Angew. Chem. Int. Ed. Eng. 1966, 5, 151.

RECEIVED October 25, 1982

24

Catalytic Control of Architecture and Properties of Butadiene Block Copolymers

PH. TEYSSIE, G. BROZE, R. FAYT, J. HEUSCHEN, R. JEROME, and D. PETIT

University of Liège, Laboratory of Macromolecular Chemistry and Organic Catalysis, Sart-Tilman, 4000 Liège, Belgium

It is shown how a precise control of the initiation and selectivity of butadiene block polymerization reactions allows the molecular engineering of a variety of new heterophase materials. The typical examples presented illustrate 3 main potentialities of these copolymers. First, the stable emulsification of blends of 2 immiscible homopolymers (PE/PS; PVC/PBD) by the corresponding diblock copolymer. Second, the possibility to obtain high performance thermoplastic elastomers from triblock copolymers containing end-blocks of high cohesive energy density (i.e. nylon-6, polypivalolactone). Third, the opportunity to tailor "multiblock" systems in which the "hard" segment is an ionic group (halatotelechelic polymers), imparting to these systems interesting rheological properties.

Catalytic polymerizations of 1,3-butadiene into essentially 1,4 elastomeric polymers are by now pretty well-mastered processes. It is the purpose of this paper to demonstrate the still very vivid interest in that time-honored backbone, within the frame of a much broader trend developing now in the field of polymeric materials.

It is obvious indeed that industry is not willing any-more to produce many of these necessary materials from new monomers; the key approach here is thus diversification, but starting from basic feedstocks and materials already available at low prices in large amounts. On the

0097-6156/83/0212-0323$06.00/0

other hand, in a time of explosive development of "genetic engineering" techniques in biology, it is only too fair to acknowledge a similar type of achievement in polymeric materials science, i.e. the "molecular engineering of their bulk properties". That recent capability to achieve, through precise (although sometimes small) modifications of molecular structures, a "fine-tuning" of the final bulk properties and macroscopic behaviour of these polymeric materials, has arisen from our rapidly increasing knowledge and mastering of initiation and propagation mechanisms, particularly in terms of selectivity. That allows us in turn to control accurately the molecular architecture of the polymers used, that means also the morphology and further the physico-chemical as well as physico-mechanical behaviour of the corresponding final materials.

All of the examples presented here, leading to different types of properties and applications, are based on copolymers of varying architecture, but all involving essentially 1,4 polybutadiene blocks. Clearly, their properties will thus reflect three important potentialities of heterophase block copolymer-based materials : 1) their "organization" into mesomorphic phases (1), giving rise to original, often anisotropic, properties; 2) their ability to stabilize liquid emulsions (2) and more importantly fine dispersions of corresponding homopolymers (3), so bridging their "compatibility gap" and allowing the development of a new "plasturgy"; 3) their use as high-performance engineering products, and particularly as materials similar to Kraton thermoplastic elastomers (4).

Experimental

All of the important experimental details, as well as the general procedures taken from literature, were reported in the references cited in each section. As a general rule, most of the reagents were purified carefully (particularly for moisture), and all of the reactions carried out either using vacuum techniques or under pure argon atmosphere. Polymer properties were investigated by standard methods, unless otherwise mentioned. Polymer blending was usually performed for 5 minutes at 190°C on a C.A.M.I.L. two-roll laboratory mill.

Results

A preliminary problem. The unsaturated structure of the 1,4 polybutadiene (PBD) backbone is unfortunately rather sensitive to a number of easily occurring secondary reactions, and particularly to different types of cross-linking and oxidation processes (thermal, photochemical, and catalytic). This drawback has been elegantly circumvented by a facile catalytic hydrogenation of the chains, using efficient soluble Ziegler-type complexes (5) prepared from cobalt (or nickel) salts and mixed aluminum alkyls in different ratios. This process has a twofold advantage :

1) it is possible to control its selectivity (varying the reaction conditions and the aluminum-transition metal molar ratio), so that polybutadiene only can be hydrogenated in the presence of other unsaturated chains like polyisoprene or polystyrene, which in turn can be hydrogenated later, if necessary by changing these reaction parameters;

2) the hydrogenation is practically quantitative and yields a product not very different from the corresponding polyethylenes. For instance, starting from an anionically initiated polybutadiene (in apolar medium) containing accordingly ca. 10-15 % 1,2 units, one obtains a product rather similar to low density linear polyethylene LLDPE (in fact a copolymer of ethylene and 1-butene), displaying a moderate semi-crystallinity and a melting point ca 90 to 100°C. On the other hand, the use of a linear pure 1,4 PBD (99 %) obtained by Ziegler-type catalysis (nickel or cobalt complexes) yields a high-density polyethylene-like product, with a higher degree of crystallinity and a melting-point close to 130°C.

Obviously, the resistance of these products towards light, oxygen and other chemicals will be much better, and close to that of the corresponding polyolefins. Moreover, the hydrogenation can be stopped at different conversions opening a much broader range of applications conditions. Industrial developments already include successful materials like Kraton G thermoplastic elastomers.

Polystyrene-hydrogenated polybutadiene diblock emulsifiers (PS-b-PBDh). Combination of different polymers into heterophase systems represents a very attractive route towards new and tailor-made materials (3) displaying most of the properties of the starting products. The incompatibility between partners (a rather general rule) is however responsible for the poor properties of many blends, which display large domain size with poor interfacial adhesion. Accordingly, diblock copolymers (6) containing sequences miscible with the homopolymers to be blended, have been used to alleviate that situation as a

result of their interfacial activity; in that respect, they have proven to be superior (7) to corresponding triblock and graft copolymers respectively (unless the triblock can undergo a tridimensional domain organization providing an additional bonus).
Two different types of (PS-b-PBDh) diblock can be presently synthesized. The first one by classical anionic initiation (s-butyl-lithium) and "living" propagation of the (PS-b-PBD) copolymer (8), followed by the hydrogenation procedure described here; as discussed above, the resulting product will be close to a (PS-b-LLDPE) copolymer. The second one came from the discovery (9) of a "living" polymerization of butadiene into a pure (99 %) 1,4 polymer by a bis η^3-allylnickel-trifluoroacetate) coordination catalyst, followed by styrene polymerization; unfortunately, the length of the polystyrene block is limited (to a M.W. of ca. 20,000) by transfer reactions.
In a general study of the usefulness of the first type copolymers, we have developed ways to drastically improve with them the properties of both low (LD) and high (HD) density polyethylene (PE)/PS blends (7,10). The addition (by hot-milling) of moderate amounts of the suitable PS-b-PDBh copolymer greatly reduces the dispersed particles size at every composition (ca. 2- to 5,000 Å); moreover, it stabilizes efficiently that situation, i.e. even throughout standard processing.
A significant enhancement of both the stress at break σ_B and the elongation at break ε_B, resulting in a striking increase of the total energy at break (E_B values), is noted (Fig.1).
The level of performance obtained depends asymptotically on the amount of copolymer added and, while a significant improvement is already noted for 0.5 % by weight of the additive, most of the performances increase is reached within 2-3 %. As expected, the length of the two blocks has also a great influence on the final properties : although smaller ones already have a great impact on the interfacial situation (smaller domains, higher σ_B), the use of higher M.W. diblocks (i.e. in which each block is similar in size to the corresponding homopolymers to be blended) yields PS-rich blends displaying the mechanical characteristics of an excellent toughened plastic, with a high ε_B value (up to 40 %) a typically ductile behaviour, and a striking resistance towards cryofracture. These features are probably characteristic of the importance of entanglements between chains of homo- and copolymer near the interface.
Interesting and significant differences are also promo-

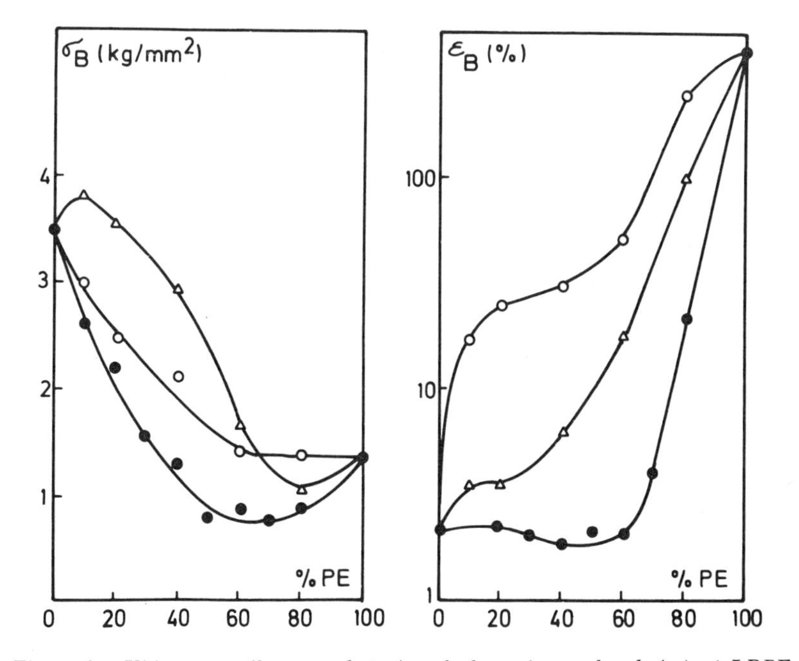

Figure 1. Ultimate tensile strength (σ_B) and elongation at break (ϵ_B) of LDPE (M_n = 40,000)/PS (M_n = 10^5) blends. Key: ●, *without copolymer;* △, *with 9% of a poly(styrene-b-hydrogenated butadiene),* M_n *total = 58,000;* ○, *with 9% of a poly(styrene-b-hydrogenated butadiene),* M_n *total = 155,000. (Reproduced with permission from Ref. 7. Copyright 1981, John Wiley & Sons, Inc.)*

ted by rather subtle modifications in the diblock structure. Compared to "pure" diblocks prepared by consecutive anionic polymerization, "tapered" blocks obtained by anionic polymerization of the comonomers mixture (11) are still more efficient emulsifiers for PS/LDPE blends; owing to their lower melt viscosity and particular miscibility characteristics, they not only act as solubilizing agents of the homopolymers but provide at the interface a "graded" modulus responsible for an improved mechanical response of the overall material (12).

Polycaprolactone block copolymers (PCL-b-PBDh). In order to obtain this new type of macromolecule, an OH-terminated homopolymer, i.e. PBD or PS obtained by anionic initiation combined with oxirane-water termination, was end-capped (an alcohol displacement reaction) with an alkoxide catalyst able to further ensure a perfectly "living" polymerization of CL, thereby yielding the desired block copolymers (13). In this case however the PBD hydrogenation (if needed) can be conducted prior to block copolymerization to avoid any interference from, or secondary reaction with, the polyester block. The final products display a number of attractive features (14): they undergo lamellar mesomorphic organization (periodicity ca. 80 Å) even under the form of hexagonal single crystals (PS-copolymers), and exhibit a ductile behaviour as well as a high resistance to cryofracture. Moreover, they are macroscopically biodegradable, at least when PCL represents the continous phase.

The blending approach described above has been extended to these products, taking advantage of the remarkable miscibility of PCL with other polymers, i.e. PVC, SAN, polycarbonate .... In particular, excellent blends of rigid PVC with PS and PBD have been prepared through hot-milling. Again, they display a very fine morphology (domain size ca 5,000 Å) which is remarkably stable in time, and some improved physicomechanical properties as long as the corresponding molecular parameters have been properly optimized (15,16). It should be stressed here that the need for such an optimization process cannot be over-estimated as rather minute modifications can lead to improvements of one order of magnitude.

Other interesting indications have been obtained on these PVC blends : in the presence of an excess rubbery phase, the presence of the copolymer (i.e. PCL-b-PBD) promotes indeed good impact resistance, indicating a strong anchorage of that rubbery phase in the resin matrix.

In other words, this example emphasizes clearly the very broad applicability of the "emulsifiers" concept depicted above, to practical problems involving large-scale polymers.

The synthesis of high-performance thermoplastic elastomers (TPE) : Nylon 6-PBDh-Nylon-6 triblocks. The interest of TPE has been largely demonstrated by the numerous and active investigations led in the field, and by the commercial success of the Kraton-type products, poly(styrene-co-diene-co-sytrene). It is however a well-known problem that these materials are usually confined to applications under rather mild conditions, due to the relatively low T_g of the glassy phase and/or the mediocre thermal stability of the rubbery phase. In principle at least, answers to that challenge can be offered by the same type of synthesis strategy. As described previously (17), an α,ω-dihydroxylpolybutadiene can be end-capped with isocyanate functions (through reaction with an excess diisocyanate), which are further converted into N-acyllactam groupings (by reaction with caprolactam CLM) active for the polymerization of that CLM monomer into Nylon-6 blocks at both ends of the polybutadiene. Under close control of the reaction conditions, neither the urethane nor the urea linkages formed are broken during the subsequent block copolymerization, and a high yield in block copolymer is obtained. That triblock copolymer displays indeed a set of high performance properties : even at 30 % of PBD phase, the material has a nylon-6 continuous phase, with a crystalline melting point ca. 225°C and a high tensile strength (quite comparable to that of pure nylon-6). On the other hand, the rubbery phase is finely dispersed in very small domains (ca. 250 Å) all over the highly crystalline nylon phase, imparting to the material higher flexibility and hydrophobicity. Although the above product is not a TPE in the strict sense of the term (see ASTM D 1566), it can be prepared with a higher rubber content to meet that type of behaviour.

It has to be stressed however that blocks of so different solubility parameters impose a heterophase situation with strong intermolecular interactions (on the polar side), even in the molten state well above the crystalline melting point : corresponding stained micrographs are very informative in that respect (see ref. (17), fig. 4d). Such a situation raises obviously severe rheological processing problems, which might be difficult to solve. Anyhow, it implies also to work at a temperature which is exceedingly detrimental for the unsaturated 1,4-polybutadiene block; again, preliminary quantitative hydrogenation of the starting polybutadiene is the answer to that problem, even though the difunctional macromolecules form a highly swollen gel in the presence of an excess catalyst.

Halato-telechelic polymers (HTP) : new materials with a dynamic multiblock structure. The general interest of multiblock copolymers is by now well documented particularly in the field of TPE. Based on the efficient cross-linking action of the harder blocks, interesting developments have taken place, leading to some successful industrial materials such as HYTREL polyether-ester thermoplastic elastomers. The very rapid developments of increasingly sophisticated catalysts has also promoted new unexpected achievements; a typical example (18) is the "coding" 1,4 polymerization of butadiene by "tailored" catalysts, namely bis (η^3-allylnickel -X) complexes. Under precisely controlled kinetic conditions, a multistereoblock (poly.cis 1,4-b-poly.trans $1,4)_n$-butadiene can be obtained, that represents the first example of a thermoplastic elastomer (semicrystalline melting point ca. 135°C) obtained in one step from one single monomer.

A completely different approach has been recently developed in that prospect, based on the assumption that "properties similar to those of multiblock copolymers could be reached, in a more versatile manner, by replacing their hard segments by single groupings, provided the molecular characteristics of these groupings promote very strong mutual interactions, at least in the media envisoned for their applications".
That concept had led to the synthesis of so-called "halato-telechelic polymers" (which means a "salt" or "neutralized" telechelic polymer, acidic or basic). Although that is a very general denomination covering all the chains formed by any type of ion-pair coupling in any way, a particularly handy and representative class of such structures can be obtained from the complete neutralization of α,ω-dicarboxylato-polymers (PX), by a di (or multi-) valent metal derivative, (19), according to the general equation :

$$n \ \text{HOOC-PX-COOH} + \frac{2n}{v} \text{MA}_v \leftrightarrows \left[\text{OOC-PX-COO-M}_{\frac{2}{v}} \right]_{\bar{n}} + 2n \ \text{HA}$$

Such a reaction has been performed successfully, starting from anionically prepared telechelic polymers PX, and neutralizing them quantitatively with very reactive metal derivatives such as metal alkyls or alkoxides (20); the latter technique proved to be the most versatile — one, provided a complete elimination of the alcohol evolved in order to displace the reaction but also to avoid solvation by that alcohol : that meets the essential requirement of any stepwise polymerization, i.e., the necessity to ensure a very high (99 %+) conversion to reach the high degree of polymerization necessary for the promotion of the most interesting physical properties, that in turn would be overshadowed if ion solvation by the alcohol takes place (see below). In that way, a broad family of HTP has been synthesized, wherein the nature and size of both the polymer and the ion involved can be systematically modified, as well as that of the solvent and ligands. They represent accordingly a versatile class of materials with a broad potential range of "molecularly engineered" characteristics. The most typical property of hydrocarbon-soluble HTP (i.e., where PX is a polydiene or PS, or polyisobutene) with a high enough MW (> 1,000) to avoid excessive charge density), is the strong dependence of its dilute-solution viscosity upon concentration (20). Although very similar to that of PX at very low concentrations, it increases abruptly and asymptotically between ca. 1 and 2 %, resulting in a gelation phenomenon. This gelation can usually be reduced by increasing the temperature, or upon addition of strong ligands (Fig. 2). The critical gel concentration, C_g, depends essentially on the nature of PX, its end-groups, the solvent, the cation size, and on PX molecular weight following the relationship $C_g = k.\overline{M}_n^{-0.5}$ (where $k^{-1/3}$ is directly proportional to $\left[\frac{(\overline{r^2})}{M}\right]^{0.5}$, i.e., depends on the mean end-to-end distance of the free chain) (21). These are ofcourse clear-cut manifestations of the electrostatic interactions between ion-pairs in a non-polar solvent, leading to aggregates of variable sizes depending on the conditions; for the same reasons, the same type of intermolecular association will obviously occur in the neat material. It is also noteworthy that such a "multiblock" structure is a dynamic one, i.e., carboxylic ligands exchanging around the metal ion resulting in a constant scrambling not only of the ionic aggregates but also of the chain blocks themselves. That situation is in fact responsible for the dynamic mechanical properties of these materials.

With concentrated solutions (over 50 %) at equilibrium as well as with the neat products, it is possible to observe typical SAXS patterns, often exhibiting two diffraction orders with Bragg's spacing in 1:2 ratio sug-

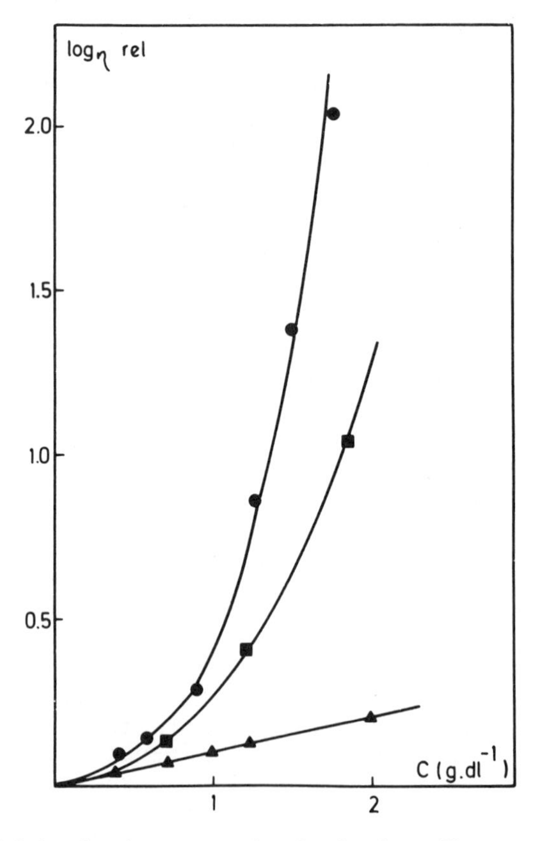

Figure 2. Relative viscosity-concentration plots in toluene. Key: ▲, nonneutralized α,ω-dicarboxylic PBD (M_n = 4,600); ●, Mg salt of α,ω-dicarboxylic PBD, 25 °C; ■, Mg salt of α,ω-dicarboxylic PBD, 80 °C.

gesting the presence of a lamellar organization (22). The calculated lamellar thickness seems independent of the nature of the ion as well as of temperature (below a critical dissociation value), but again strongly reflects the mean dimensions of the free chains.
The rheological behaviour of these products has also been investigated (23). The steady-flow viscosity depends markedly on the shear-rate, and for values higher than ca. 10 sec^{-1} a significant dilatant effect is observed, since preformed ionic cross-links prevent the chains to relax as the deformation time-scale decreases. In terms of dynamic mechanical behaviour, storage (G') and loss (G") moduli have been determined (23) over a range of frequencies for different polymers. At higher frequency (>0.5 $sec.^{-1}$), G' is higher than G", which in turn presents a maximum characteristic of the ionic component relaxation mechanism. These elastic gels have relaxation spectra which indicate again the ions aggregation : a broad distribution of relaxation times appears, the maximum of which depends on the stability of the ionic network. Since these materials display a nice thermorheological simplicity (at least when P_x has a $\bar{M}_n$ lower than 20,000, i.e.,no entanglements interference), master curves have been established. The corresponding shift factors correlate nicely in an Arrhenius-type relationship, allowing the determination of the activation energy of the secondary relaxation mechanism. For different ions, these energies are directly dependent on the ion electrostatic field.

In other words, these properties can be interpreted in terms of a system in which the deformation processes are governed by ionic multiplets'thermal dissociation; however, while electrostatic attractive forces are determinant in the aggregation process, the free conformation of the macromolecule will control the overall morphology of the gels as well as the mean number of ions in the multiplets.

Several interesting applications can be envisioned on the basis of that behaviour, but again saturated-elastomers will be more suitable for most of them, in terms of ageing. Here again, hydrogenated telechelic polybutadiene is a possible answer to that problem, although telechelic polyisobutene might alternatively be used in some cases.

General Conclusion. At this point, it is probably useful and certainly encouraging to stress again that all of these examples confirm the versatility of the molecular engineering techniques presently available, as well as their potentialities in broadening and diversifying the applications of such a time-honored commodity polymer as polybutadiene. That is certainly a worthwhile goal in today's technology and economy.

Literature Cited

1. Sadron, C.; Gallot, B. Makromol. Chem. 1973, 164, 301; Gallot, B. "Liquid Crystalline Order in Polymers," Blumstein, A, Ed.; Academic: New York, 1978; 191.
2. Marti, S.; Nervo, J.; Riess, G. Progr. Colloid. Polym. Sci. 1975 , 58, 114.
3. Paul, D.R. "Polymer Blends," Paul,D.R. and Newman, S., Eds.; Academic: New York, 1978; 168.
4. Holden, G.; Bishop, E.T.; Legge, N.R. J. Polymer Sci. 1969, C36, 37.
5. Falk, J.C. J. Polym. Sci. 1971, A-1,9, 2617; Gillies, G.A. (to Shell Oil Co.), U.S. Patent 3,792,127 (1974).
6. Fayt, R.; Jérôme, R. Actualite Chimique 1980, 21.
7. Fayt, R.; Jérôme, R.; Teyssié, Ph. J. Polym. Sci. Polym. Lett. Ed. 1981, 19, 79.
8. Szwarc, M. "Carbanions, Living Polymers and Electron-Transfer Processes," and references therein, Wiley-Interscience: New York, 1968.
9. Hadjiandreau, P. Ph.D., Thesis, University of Liege, Liege, 1980.
10. Fayt, R.; Jérôme, R.; Teyssié, Ph. J. Polym. Sci. Polym. Phys. Ed. 1981, 19, 1269.
11. Kuntz, I. J. Polym. Sci. 1961, 54, 569.
12. Fayt, R.; Jérôme, R.; Teyssié, Ph. J. Polym. Sci. Polym. Phys. Ed. (in press).
13. Heuschen, J.; Jerome, R.; Teyssie, Ph. Macromolecules 1981, 14, 242.
14. Herman, J.J. Ph.D., Thesis, University of Liege, Liege, 1978.
15. Teyssié, Ph.; Bioul, J.P.; Hamitou, A.; Heuschen,J.; Hocks, L.; Jérôme, R.; Ouhadi, T. ACS Symposium Series No.59, T. Saegusa and E. Goethals Eds., 1977.
16. Heuschen, J.; Jérôme, R.; Teyssié, Ph. French Patent 2,383,208 (1977); U.S.Patent 4,281,087(1981).
17. Petit. D.; Jérôme, R.; Teyssié, Ph. J. Polym. Sci. Polym. Chem. Ed. 1979, 17, 2903.
18. Teyssié, Ph.; Devaux, A.; Hadjiandreou, P.; Julémont, M.; Thomassin, J.M.; Walckiers, E.; Ouhadi, T. "Preparation and Properties of Stereoregular Polymers," Lenz, R.W. and Ciardelli, F., Eds.; M. Reidel: Hingham, Mass., 1980; 144.
19. Moudden, A.; Levelut, A.M.; Pineri, M. J. Polym. Sci. Polym. Lett. Ed. 1977, 15, 1707.

20. Broze, G.; Jérôme, R.; Teyssié, Ph. Macromolecules (in press) and references therein.
21. Broze, G.; Jérôme, R.; Teyssié, Ph. Macromolecules (in press).
22. Gallot, B.; Broze, G.; Jérôme, R.; Teyssié, Ph. J. Polym. Sci. Polym. Lett. Ed. 1981, 19, 415.
23. Broze, G.; Jérôme, R.; Teyssié, Ph. Polym. Bull. 1981, 4, 241; Part VII, J. Polym. Sci. Polym. Phys. Ed. (to be published).

RECEIVED December 23, 1982

25

Synthesis of Block Sequences by Radical Polymerization

W. HEITZ, M. LATTEKAMP, CH. OPPENHEIMER,[1] and P. S. ANAND[2]

Philipps-Universität, Fachbereich Physikalische Chemie, Polymere, D-3550 Marburg, Federal Republic of Germany

Radical polymerization can be used to synthesize block copolymers either by preparing α,ω-bifunctional oligomers (telechelics) and using them in condensation polymerization chemistry or using a precursor containing initiator groups. Problems and requirements in the synthesis of telechelics are discussed in general and specifically for telechelic polybutadiene, polystyrene and polyethylene. Two functional end groups are obtained by use of azo initiators and percarbonates respectively if combination is the only termination reaction. The efficiency of the initiator can be >0.9. Dead-end polymerization is the preferable experimental condition. Block copolymers can be obtained by radical polymerization using polyazoesters as initiators. Fractional decomposition of the polyazoesters in the presence of a first monomer resulted in a precursor containing azo groups. Decomposing the residual azo groups in presence of a second monomer gave block copolymers along with homopolymers. The fraction of block copolymers was 60-80%.

[1] Current address: Bayer AG, Leverkusen, Federal Republic of Germany

[2] Current address: Central Salt & Marine Chemical Research Institute, Bhavnagar, India.

0097-6156/83/0212-0337$06.00/0

Modification of polymers by incorporating block sequences having low glass transition temperatures is a means of changing the mechanical properties and is especially useful for the formation of thermoplastic elastomers if the basic polymer is semi-crystalline. These rubber-like blocks are usually formed by ionic or transition metal catalyzed reactions. Radical polymerization on the other hand is experimentally simpler and applicable to a wide variety of monomers.

In the synthesis of Hytrel, poly(butylene terephthalate) is modified with preformed blocks of poly-(tetrahydrofuran)(1) obtained by cationic polymerization. Prepolymers used in modifying polycondensates must be strictly bifunctional. These products are called telechelics (2). From the analysis of end-groups and number average molecular weight, only average values of functionality are obtained and these values can be the result of a combination of mono-, bi- and tri-functional molecules. The prepolymers can only be considered to be telechelics if it is proven that bifunctionality is caused by one polymer homologous series. That is why chromatographic methods play an important role in this area.

The direct synthesis of block copolymers by radical polymerization is an alternative to using preformed blocks with functional endgroups. Polymers with initiator end-groups (3a,3b), with end groups accessible to radical activation (4,5), and polyinitiators (6,7) can be used as starting materials.

Experimental Section

Materials. All monomers and solvents were purified by standard procedures.

Telechelics. Nitrogen was used as an inert atmosphere. The synthesis is either run in a flow reactor if the reaction time is short (e.g. styrene) or a glass pressure vessel is used as a discontinuous reactor (for butadiene at lower temperatures) or a steel autoclave (with ethylene as the monomer).

The flow reactor consisted of a storage bottle for the initiator/monomer solution through which N_2 is bubbled, a piston pump as used in liquid chromatography, a coiled capillary (i.d. 2 mm., length 43 m., volume 135 ml.) immersed in a thermostat and a back pressure valve at the exit.

Oligobutadiene with two ester-end groups. Eight g. of azobis(methyl isobutyrate) was introduced into a glass pressure vessel ("Laborautokav" Buchi Comp., CH-Uster, Switzerland). Butadiene, 177 g., was condensed into the evacuated vessel. The reaction mixture was heated to 80°C for 24 hrs. Excess of butadiene was distilled off, the final traces at 90°C. in vacuum. The yield was 60.8 g. of oligomer, corresponding to a conversion of 30.4% of butadiene. The product had an $\overline{M}n$ = 1868.

Oligostyrene with two carbonate end-groups. A solution of 40 g. of dicyclohexyl peroxydicarbonate and 136 g. of styrene in 300 ml. of toluene was pumped through the flow reactor at 75°C. and a rate of 1.4 ml./min. The flow reactor was filled with toluene at the beginning and a forerun was discarded. Oligostyrene, 111 g., was obtained from 318 g. of the reaction mixture after evaporation of the volatile fraction at 90°C. in vacuum, $\overline{M}n$ = 1160;Found: C 84.0, H 7.8, O 8.3; calcd. C 84.2, H 7.6, O 7.9. Functionality by CO_2-titration after hydrolysis was 1.9.

Block copolymer of styrene and methyl methacrylate. The polyazoester is made by the reaction of a diol and AIBN. The procedure is described elsewhere (16).

Styrene, 300 ml. (2.61 mol), 408 ml. of methanol and 3.545 g. (13 mmol) of initiator obtained from diethylene glycol and AIBN were mixed and heated to 64°C. in a preheated bath. After a reaction time of 359 min. (30% decomposition of the initiator) the reaction mixture was quenched and the product was removed by filtration, dissolved in benzene, precipitated with methanol and dried at room temp. in vacuum. Prepolymer, 52.2 g., was obtained.

Three g. of prepolymer and 6 ml. of methyl methacrylate (MMA) were dissolved in 100 ml. benzene and heated for 19 hrs. to 75°C. The product was precipitated with methanol, dissolved in benzene and again precipitated with methanol. The product obtained was 5.77 g., corresponding to 49.3% conversion of MMA.

In the precipitation technique, the prepolymer was stirred for 0.5 hr. in cyclohexane. MMA was added to this solution or highly swollen slurry, and immersed in a preheated bath. The product was isolated as given above.

Fractionation. One g. of copolymer dissolved in 25 ml. chloroform was added to 30 g. of glass beads (50 μm diameter). The solvent was removed in a rotary evaporator. The polymer-covered glass was dried in vacuum.

A Soxhlet fitted with a cooling jacket was used for extraction. Homopolystyrene was extracted with cyclohexane, PMMA with acetonitrile. The solvent was removed, and the polymer was dissolved in a small amount of benzene and freeze-dried.

Results and Discussions

Synthesis of Telechelics. The problems of synthesis of telechelics by radical polymerization can be discussed using the scheme of radical polymerization:

$$I \longrightarrow 2\,R\cdot \qquad \text{Initiation}$$
$$R\cdot + M \longrightarrow RM\cdot$$

$$RM\cdot + n\,M \longrightarrow RM_{n+1}\cdot \qquad \text{Propagation}$$
$$RM_n\cdot + TH \not\longrightarrow RM_nH + T\cdot$$

$$2RM_n\cdot \longrightarrow RM_{2n}R \qquad \text{Termination}$$
$$\not\longrightarrow RM_nH + RM_n \text{ (minus H)}$$
$$R\cdot + RM_n\cdot \longrightarrow RM_nR$$
$$\not\longrightarrow RH + RM_n \text{ (minus H)}$$

(I -Initiator; R -fragment containing a functional group; R· -primary radical; RM_n. -macroradical; TH -Transfer agent: solvent, monomer, impurity).

The initiator is decomposed into two radicals. These radicals must add the first monomer unit with high efficiency. This is a basic requirement for an economic synthesis; otherwise a large fraction of the initiator is wasted. In the propagation reaction, transfer reactions must be strictly avoided with the exception of some specific transfer with the initiator. The main point of concern in this synthesis is the termination reaction. Only the combination reaction is allowed. Termination in polymerization is mostly the reaction of two macroradicals. In the formation of telechelics, the reaction with primary radicals cannot be ignored. The ratio of combination to disproportionation reactions may be different in both cases. Even if primary radicals react with each other by combination and macroradicals react by combination, the cross-reaction can be a complete disproportionation (8).

All radical syntheses of telechelics have two things in common: the problem of the efficiency, f, and the overall conditions.

$$P_n = \frac{\Delta [M]}{e/2\ f [I]} \qquad (1)$$

with $\Delta[M]$ - amount of monomer consumed; e - functionality; f - efficiency; $[I]$ - total amount of initiator.
The initiator is completely consumed under the conditions used in the synthesis of telechelics. The quantities of Eq. (1) are easy to determine in the oligomeric range and can be measured very accurately. The definition of the efficiency used in Eq. (1) is different from the usual one. In Eq. (1), f is defined as

$$f = \frac{k_i [R\cdot][M] + k_{tR}[P_n\cdot][R\cdot]}{k_c [R\cdot]^2 + k_i [R\cdot][M] + k_{tR}[P_n\cdot][R\cdot]} \qquad (2)$$

with k_i, k_{tR}, k_c - rate constants of initiation, termination by primary radicals and between primary radicals. At high molecular weights the chain termination by primary radicals can be neglected and Eq. (2) simplifies to the usual definition. The efficiency is highest if bulk polymerization can be used. Polymerizing in bulk results in efficiencies >0.9 with styrene or butadiene as monomers.

The conditions of synthesis should be such that the initiator is completely consumed in the reaction. In other words, the **so-called dead-end conditions are** used (9)

$$v_p = k_p\ (2fk_d/k_t)^{0.5} [I]^{0.5} [M] \qquad (3)$$

$$\ln \frac{[M]_0}{[M]_t} = 2.83\ k_p\ (f/k_t k_d)^{0.5} [I]_0^{\ 0.5} (1-\exp(k_d t/2)) \qquad (4)$$

$$\ln \frac{[M]_0}{[M]_\infty} = K [I]_0^{\ 0.5} \qquad (5)$$

(v_p - rate of polymerization: $[M]_0$, $[M]_t$ and $[M]_\infty$ -monomer concentration at the beginning, at time t, and at the end of the reaction).

Integration of the rate equation of polymerization [(Eq. (3)] - using the first order decomposition of the initiator gives an equation relating conversion with time [Eq.(4)]. At long reaction times, i.e., with respect to initiator decomposition, the conversion of monomer will precisely stop at a predictable conversion [Eq.(5)]. The basic point of the dead end polymerization is that the half-life of the initiator is so short that the monomer is not consumed within 10 half-lives of the initiator. The conversion of the monomer must be limited to values lower than 40% in order to avoid branching. This limitation makes solution polymerization very uneconomical. But the monomer can serve as its own solvent in bulk polymerization with limited conversion.

Eq.(5) by Tobolsky (9) does not take into account the monomer consumption in the initiation step and the termination by primary radicals. This oversimplification may explain the deviations from this equation found for telechelics (10).

These general statements are discussed with reference to three monomers: styrene, ethylene and butadiene.

Styrene. Of course the initiator plays an important role in a successful synthesis of telechelics. Initiators which have no uniform way of decomposition and initiation will fail to give a product with a clear structure of the end-groups. An example is dibenzoyl peroxide. The initiation is caused either by oxybenzoyl radicals or phenyl radicals. So at least three polymer homologous series are to be expected. Fig. 1a gives a GPC of an oligomer obtained with styrene and high amounts of dibenzoyl peroxide. Due to the differences in the end-groups the peaks are not well resolved even at low molecular weight (≙ high elution volume). An aromatic diacyl peroxide will never give a product with a clear structure of the end groups. But the situation is quite different if azo initiators are used (Fig.1b). A clear peak series can be seen in GPC which is in agreement with the analysis giving an ester functionality of two. Nevertheless there are problems with styrene. It has been known for several years that polystyrene radicals terminate about 80% by combination and 20% by disproportionation (11,12,13). This is not in disagreement with a functionality of two. At low molecular weights termination is caused mainly by primary radicals. The trend found in the dependence of functionality on molecular weight may be explained by this fact (14).

Using peroxydicarbonates two carbonate end-groups are fixed at the polystyrene chain (Fig.1c). Even at a molecular weight of 2000 the functionality is

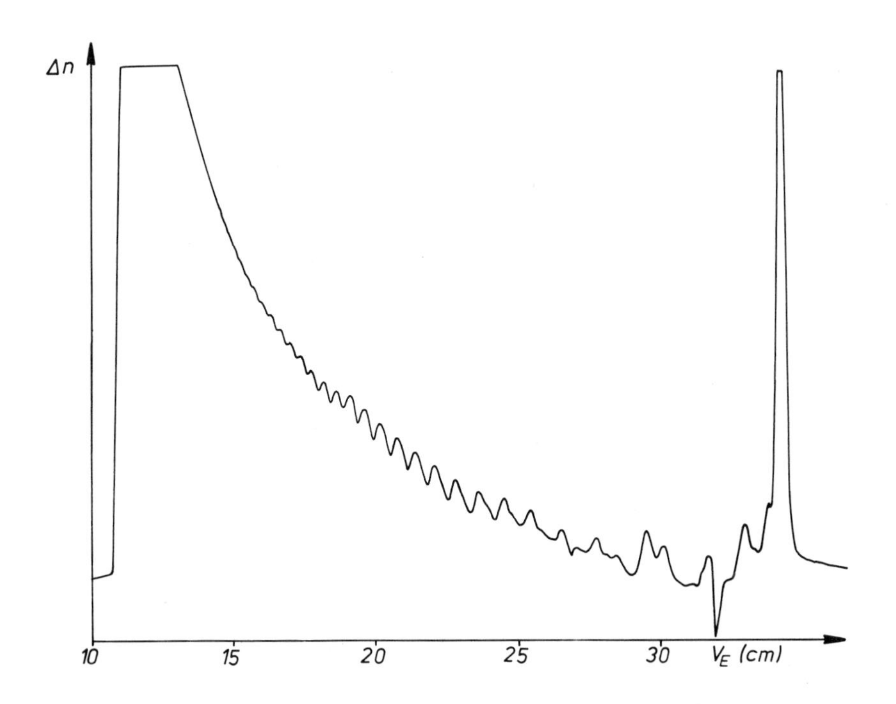

Figure 1a. GPC of dibenzoyl peroxide prepared with large amounts of an initiator. Abscissa shows chart distance in centimeters, ordinate shows differential refractive index.

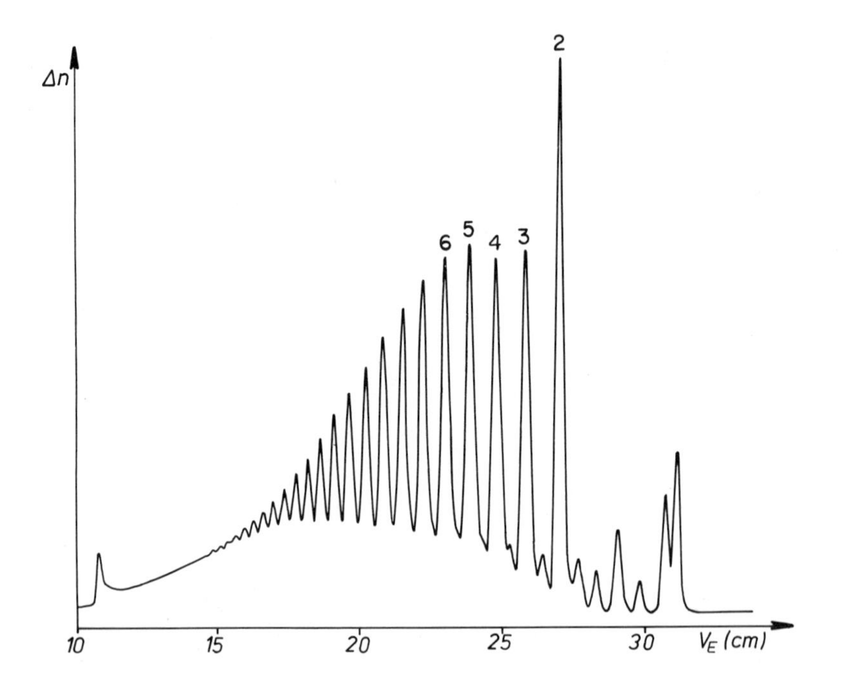

Figure 1b. GPC of azobis(methyl isobutyrate) prepared with large amounts of an initiator. Abscissa shows chart distance in centimeters, ordinate shows differential refractive index.

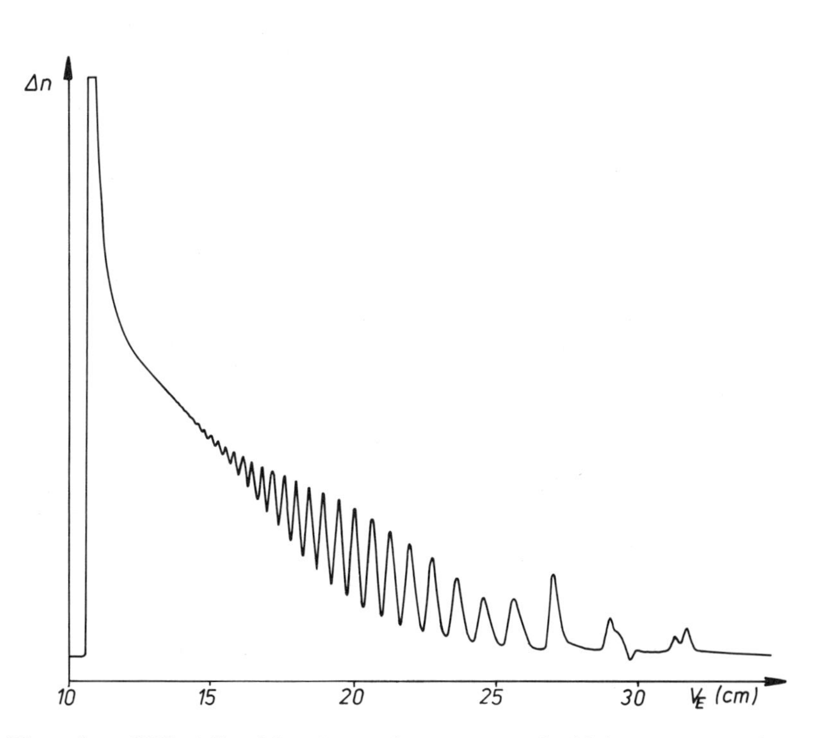

Figure 1c. GPC of dicyclohexyl percarbonate prepared with large amounts of an initiator. Abscissa shows chart distance in centimeters, ordinate shows differential refractive index.

exactly two. The reason for this bifunctionality is a transfer reaction with the initiator attacking the peroxy bond, in contrast to the behavior in presence of ethylene (see below). An appreciable amount of the product is formed by this reaction as can be concluded from the decomposition rate of the initiator.

The efficiency is >0.9 using azo initiators and percarbonates if the monomer concentration is high.

Ethylene. The mechanism of termination in the polymerization of ethylene is mainly by combination. This is known quantitatively for alkyl radicals where $k_d/k_c = 0.12$ (k_c and k_d are rate constants for combination and disproportionation, respectively).

So ethylene should be a good candidate for making telechelics. But all efforts in this direction failed. The reason for this failure is that the macroradicals derived from ethylene are highly reactive alkyl radicals. They will abstract hydrogen from any source in the system. If this cannot be done from the solvent it will be done from the initiator (8,15).

The behavior of ethylene is discussed below with the initiation by percarbonates as an example.

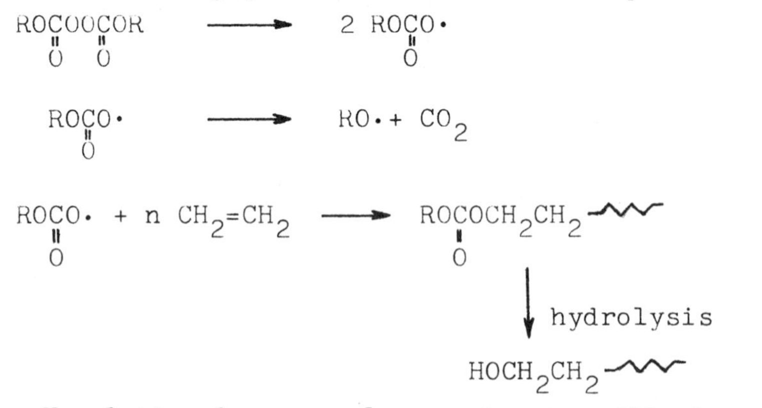

Homolytic cleavage of percarbonate will give primary radicals which in contrast to the behavior of dibenzoyl peroxide will not lose carbon dioxide (exception R=t-butyl (8)). These radicals are fixed as carbonate groups at the end of the chain. Hydrolysis will give a hydroxyl end-group. Chain termination by combination will give diols after hydrolysis. If there is a disproportionation reaction, saturated and unsaturated alcohols with an even number of carbon atoms will result. With dimethyl percarbonate as initiator the diols expected after hydrolysis are present only in small amounts (Fig.2). The surprising result is that

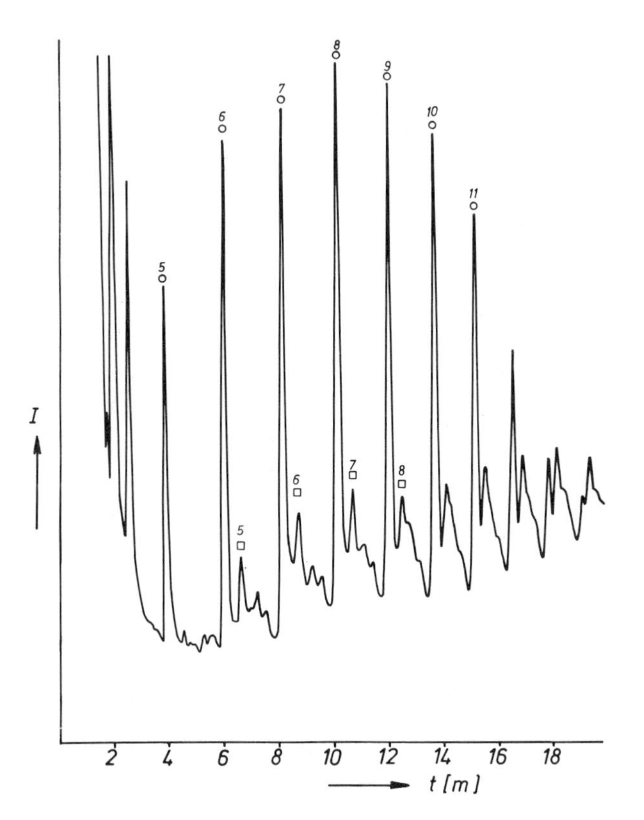

Figure 2. GC of the reaction product of ethylene and dimethyl percarbonate after hydrolysis; numbers correspond to n *(8). Abscissa shows time in minutes. Key:* ○, *$HO(CH_2CH_2)_nH$;* □, *$HO(CH_2CH_2)_nOH$.*

the saturated alcohol is the main product and the unsaturated alcohol is practically not present. Detailed investigation showed that the hydrogen is abstracted from the α-position of the percarbonate which is decomposed with formation of aldehyde, carbon dioxide and carbonate radicals.

$$\sim\!\sim\cdot + \underset{H}{RCH}O\overset{}{\underset{O}{C}}OO\underset{O}{C}OR' \longrightarrow \sim\!\sim H + RCHO + CO_2 + \cdot O\underset{O}{C}OR'$$

In the initiation reaction a carbonate group is fixed at the polyethylene chain-end, and in the termination reaction a saturated end-group is formed. Although these experiments failed to give telechelics they showed that primary radicals from percarbonates will give carbonate end-groups, and telechelics should be obtainable if this side reaction can somehow be avoided.

Butadiene. For synthesizing telechelics which ultimately give elastomeric products, the monomer butadiene is of considerable interest. Fig.3 shows a GPC of a telechelic polybutadiene prepared from azo bis(isobutyronitrile) and butadiene. A clear separation is observed and the analysis gives a nitrile functionality of two. The product is made by bulk polymerization. The most important point with respect to a uniform structure of the end-group and the chain is that the conversion is limited to about 30-40% by proper choice of the dead-end conditions. In fact the monomer is used as a solvent. These conditions cannot be applied with the most commonly used 4,4'-azobis(4-cyanovaleric acid) because of the low solubility in butadiene. In Fig.3, the product with the polymerization degree of 1 is practically not present; this means that cyanoisopropyl radicals immediately react with the first butadiene unit forming allyl radicals. This is the true primary radical in the system. The combination of two such radicals will give a product with polymerization degree of 2. The efficiency is >0.9. The nitrile end-groups can subsequently be converted to carboxyl, hydroxyl or amino end-groups.

AIBN can be quantitatively converted to azo-esters. Bulk polymerization of butadiene with azo-esters results in the formation of telechelic polybutadiene with ester end-groups.

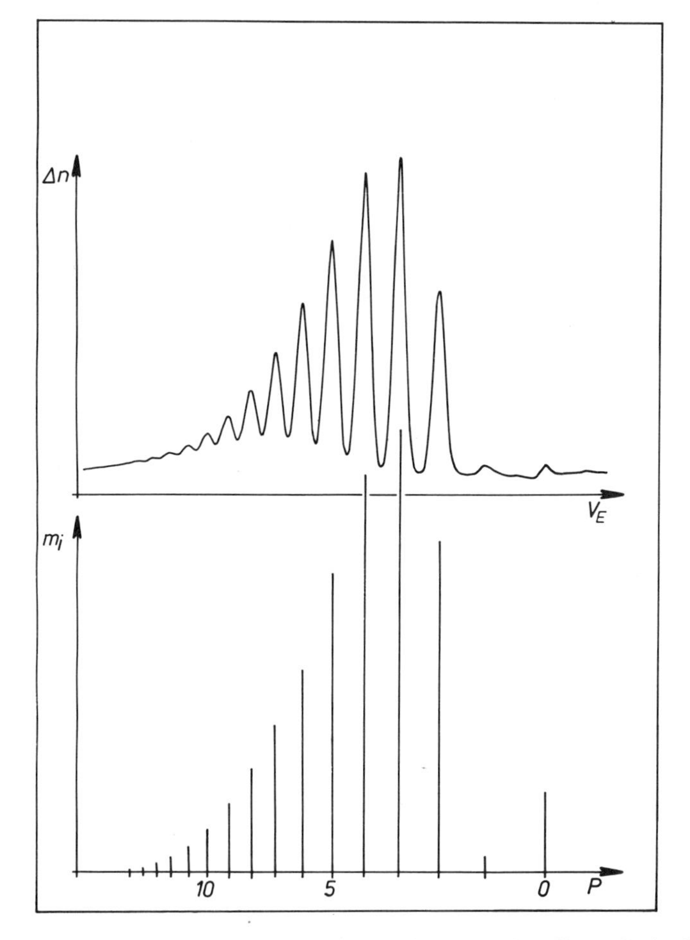

Figure 3. GPC and mass distribution of a telechelic oligobutadiene obtained with high amounts of AIBN (10). P *is degree of polymerization and* m_i *represents mass (relative amount).*

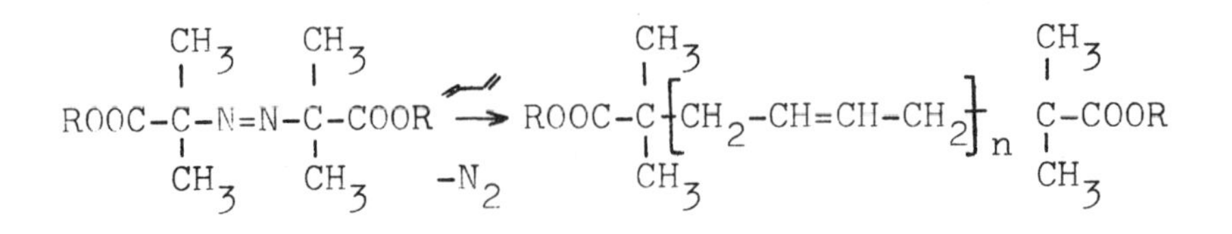

Again the efficiency is >0.9 and the ester functionality is exactly 2. The ester groups can be converted to carboxylic acids. The whole chain can be hydrogenated to obtain telechelic polyethylene with ester or hydroxyl end-groups. These telechelics can replace poly(tetrahydrofuran) in the modification of polyesters. The ester end-groups can be used directly in a transesterification process.

Percarbonates are another class of initiators capable of introducing two functional end-groups into butadiene

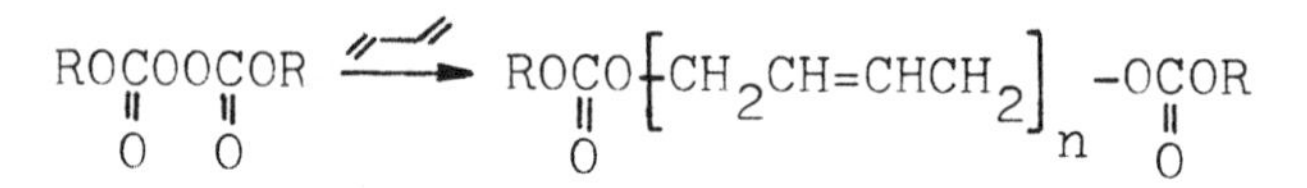

Primary radicals derived from percarbonates will not split off CO_2 and they are fixed as carbonate end-groups with an efficiency of $\cong$0.95. The functionality is exactly two. The end-groups can easily be hydrolyzed, or the chain can be hydrogenated.

Block copolymers. Another possibility to introduce soft segments into polymers by radical polymerization is the direct synthesis of block copolymers using polyinitiators. One class of polyinitiators is obtained from AIBN and a diol under mild conditions. The resulting polyazoesters show decomposition rates similar to low molecular weight azo initiators (16).

$$\text{AIBN} \xrightarrow[\text{HCl/benzene, } 0\text{-}5^{\circ}\text{C}]{\text{HOROH}} \left[-\text{OOC}-\underset{\text{CH}_3}{\overset{\text{CH}_3}{\text{C}}}-\text{N}=\text{N}-\underset{\text{CH}_3}{\overset{\text{CH}_3}{\text{C}}}-\text{COOR}- \right]_n$$

Fig.4 shows a GPC of a polyazoester prepared with a slight excess of the diol. This excess puts a limitation on molecular weight and also produces predominantly the polymer homologous series with two OH-end-groups which is the major peak series of the chromatogram. Small peaks are also present, which means that there are side reactions which will limit the ultimate

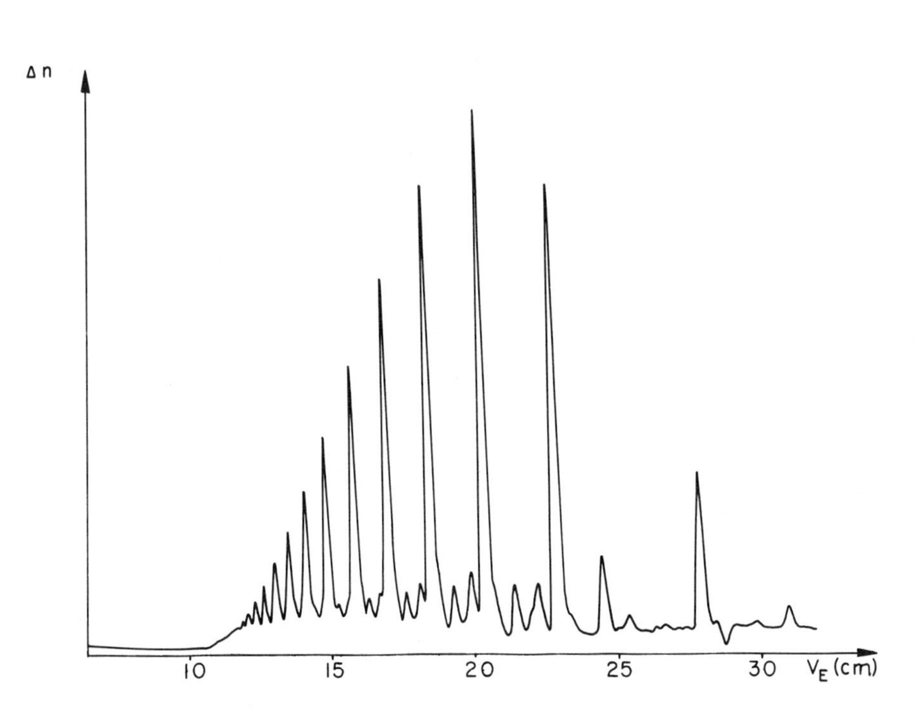

Figure 4. GPC of a poly(azoester) from AIBN and tetraethylene glycol (molar ratio 0.9/1).

molecular weight. Using equimolar amounts of diol and AIBN average degrees of polymerization ∿20 are obtained.

These polyazoesters can be used to produce block copolymers in a two-step procedure. In the first step a fraction of the azo group is decomposed in presence of a first monomer. Thus, an azo group-containing prepolymer is produced. Decomposition of the rest of the azo groups in the presence of a second monomer results in the formation of block copolymers. Bulk, solution, and precipitation polymerization can be used as polymerization techniques. The efficiency in the preparation of the azo group-containing-prepolymer can be as high as 0.8 if bulk polymerization is carried out in such a way that short blocks are obtained. In the second step, typical values for the efficiency are 0.3-0.4.

The type of block copolymer produced in this synthesis is dependent on the termination reactions of both monomers. If both monomers terminate by disproportionation diblock copolymers result. If one reacts by combination and one by disproportionation triblock copolymers are formed and if both monomers terminate by combination multiblock copolymers result. This process is unavoidably accompanied by the formation of homopolymers. So the central question of this synthesis is: How high can the fraction of block copolymers be?

Styrene/methyl methacrylate (MMA) system was studied as a model (17). Styrene mainly terminates by combination. Homopolystyrene can be formed from the cleavage of an azo end-group. But the probability is low that two such macroradicals without an azo group will find each other. The fraction of homopolystyrene in this step is small. In the first step about 50% of the azo groups are decomposed; this means a reaction time corresponding to one half-life of the initiator. In the second step, all the rest of the azo groups are decomposed corresponding to a reaction time of 10 half-lives of initiator decomposition. MMA terminates mainly by disproportionation, therefore ABA block copolymers are principal products.

The results of fractionation of a final product are given in Fig.5. The products of Fig.5a are obtained by solution polymerization. The fraction of homopolystyrene decreases to about 5% with increasing ratio of MMA/prepolymer used in the second step. The fraction of homopoly(methyl methacrylate) increases to about 35% and the block copolymer fraction is 60%, independent of the MMA/prepolymer ratio. The results

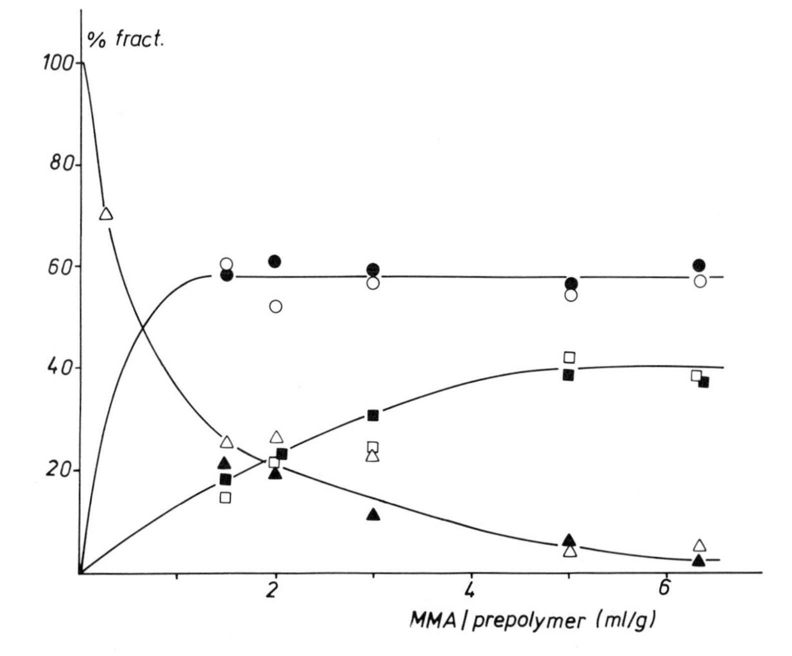

Figure 5a. Fractionation of block copolymers of styrene/MMA obtained by polymerizing MMA (in milliliters) by azogroup-containing-polystyrene (in grams) prepolymer with different MMA/prepolymer ratios (17) using solution polymerization. Key: Open symbols, 85 °C; filled symbols, 75 °C. ▲, △, homopolystyrene; ■, □, homo PMMA; ●, ○, block copolymer.

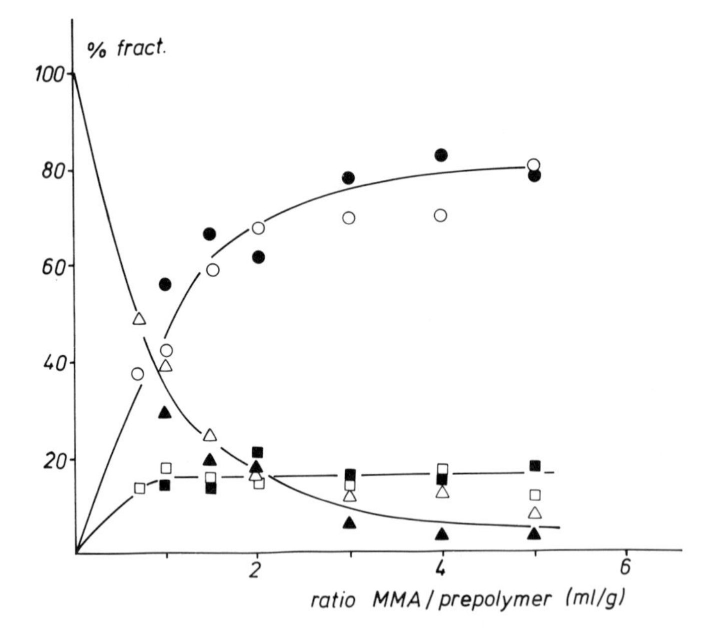

Figure 5b. Fractionation of block copolymers of styrene/MMA obtained by polymerizing MMA (in milliliters) by azogroup-containing-polystyrene (in grams) prepolymer with different MMA/prepolymer ratios (17) *using precipitation polymerization. Key: Open symbols, 85 °C; filled symbols, 75 °C.* ▲, △, *homopolystyrene;* ■, □, *homo PMMA;* ●, ○, *block copolymer.*

are somewhat dependent on the technique used in the preparation.

Fig.5b presents results from precipitation polymerization. Again homopolystyrene decreases, homo PMMA increases but does not exceed ∿20%, and the fraction of block copolymer can be nearly 80% of the product.

Poly(alkyl acrylates) form soft segments. As acrylate polymerization terminates by combination, multiblock copolymers are formed if styrene is the second monomer. These block copolymers show two glass transition temperatures (∿30°C. and 90°C.). Phase separation occurs with domain structures depending on the styrene/methyl acrylate ratio (18).

The advantage of this method of synthesis is that it covers a vast field of monomers which are polymerizable by free radicals. It is possible to produce in the first step an azo group-containing-polybutadiene which is not cross-linked. It is a soluble, highly viscous or solid material depending on the molecular weight produced. This method offers a real possibility that both blocks can be already any statistical/alternating copolymer (5).

Literature Cited

1. Allport, D.C.; Mokajer, A.A. in "Block Copolymers"; Allport, D.C.; James, W.H. eds.; Appl. Sci. Publ. London, 1973.
2. Uranek, C.A.; Hsieh, H.L.; Buck, O.G. J. Polym. Sci. 1960, 46, 535.

3a. Piirma, I.; Chou, L.H. J. Appl. Polym. Sci. 1979, 24, 2051.

3b. Gunesin, B.Z.; Piirma, I. J. Appl. Polym. Sci. 1981, 26, 3103.

4. Bamford, C.H. Europ. Polym. J. Suppl. 1969, 1.
5. Bamford, C.H.; Han, X. Polymer 1981, 22, 1299.
6. Smets, G.; Woodward, A.E. J. Polym. Sci. 1954, 14, 126.
7. Woodward, A.E.; Smets, G. J. Polym. Sci. 1955, 17, 51.
8. Guth, W.; Heitz, W. Makromol. Chem. 1976, 177, 1835.
9. Tobolsky, A.V. J. Am. Chem. Soc. 1958, 80, 5927.
10. Heitz, W.; Ball, P.; Lattekamp, M. Kautschuk Gummi 1981, 34, 459.
11. Olaj. O.F.; Breitenbach, J.W.; Wolf, B. Monatsh. Chem. 1964, 95, 1646.

12. Berger, K.C.; Meyerhoff, G. Makromol. Chem.1975, 176, 1983.
13. Gleixner, G.; Olaj, O.F.; Breitenbach J.W. Makromol. Chem.1979, 180, 2581.
14. Konter, W.; Bömer, B.; Köhler, K.H.; Heitz, W. Makromol. Chem. 1981, 182, 2619 .
15. Guth, W.; Heitz, W. Makromol. Chem. 1976, 177, 3159.
16. Walz, W.; Bömer, B.; Heitz, W. Makromol. Chem. 1977, 178, 2527.
17. Oppenheimer, C.; Heitz, W. Angew. Makromol. Chem.1981, 98, 167.
18. Anand, P.; Stahl, H.G.; Heitz, W.; Weber, G.; Bottenbruch, L. Makromol. Chem.1982, 183, 1685 .

RECEIVED November 22, 1982

BIOLOGICAL INFLUENCE OF INITIATION

26

Initiator–Accelerator Systems for Dental Resins

G. M. BRAUER and H. ARGENTAR

National Bureau of Standards, Washington, DC 20234

Initiators and initiator-accelerator systems for curing acrylic resins employed as dental restorative and prosthetic devices are reviewed. These resins are hardened by a free-radical polymerization that is activated by an initiator and heat or light, or by a redox initiator-accelerator system. Dentures are commonly cured by a heating cycle during which benzoyl peroxide (BP) initiator is decomposed to release sufficient radicals to yield a fully hardened device. Most room-temperature, chemically activated systems employ BP-tertiary aromatic amines. Many other potential redox systems are limited by the instability of the uncured components on prolonged storage or the doubtful biocompatibility of the ingredients. Visible or UV energy-cured systems do not require clinical mixing and allow unrestricted working time.

Acrylic resins are the materials of choice for almost all dental applications wherever synthetic plastics are favored for the restoration of missing teeth or tooth structures. This is not surprising because polymers derived from methacrylate esters fulfill most requisites of a restorative: adequate strength, resilience and abrasion resistance; dimensional stability during processing and subsequent use; translucency or transparency simulating the visual appearance of the oral tissue that it replaces; satisfactory color stability after fabrication; resistance to oral fluids, food or other substances with which it may come into contact; satisfactory tissue tolerance; low toxicity, and ease of fabrication into a dental appliance.

The largest volume of plastics for dental applications is consumed in the construction and repair of dentures. Other uses include artificial teeth, restoratives--especially for anterior teeth, pit-and-fissure sealants, resin cements, orthodontic splints and bonding resins, maxillofacial prostheses, oral implants and mouth protectors.

In dental applications, a fluid acrylic resin formulation is hardened by means of a free-radical initiated polymerization that is effected by one of three means: subjecting a thermally sensitive initiating ingredient to a higher than ambient temperature; subjecting a photoactive ingredient to visible or ultraviolet radiation; or bringing together in the resin solution a binary redox initiating system composed of two chemicals, one a reductant (the "accelerator" or "promotor") and the other an oxidant, i.e., the initiator. Each of these systems has advantages as well as disadvantages and consequently none is strongly preferred over the others.

The ideal polymerization-initiating system does the following: it produces enough free radicals within an acceptable time interval so that a mix with a desirable working and curing time for the specific end use is obtained; yields a polymer with minimum residual monomer and a molecular weight distribution that will result in optimum physical properties of the dental restorative or device; and, finally, produces no undesirable by-products, such as toxic or colored materials. The ingredients of the ideal system meet the following requirements: biocompatible and non-toxic; tasteless and colorless; storage stable for extended periods of time under environmental conditions that may be encountered in transit, in warehouses, in dental depots or in the dental office; compatible chemically and physically with all the other ingredients of the resin which are encountered in storage; and readily available commercially at reasonable prices.

Initiator-accelerator systems for acrylic resins and composites in dental use have been previously reviewed (1). This report updates the information on these systems in dental polymers.

Thermally Initiated Resin Systems Employing Peroxides

Heat-cured denture base materials were introduced into dental use in 1937. These materials are prepared from a powder-liquid slurry. The liquid is methyl methacrylate to which are added a plasticizer, crosslinking agent and inhibitor. The powder is poly(methyl methacrylate) containing approximately one percent initiator, usually benzoyl peroxide. By subjecting this slurry to elevated temperature (about 75°C to 100°C) for one or more hours depending on the temperature employed, sufficient free radicals are produced from the initiator to yield a satisfactory denture. Other initiators have been proposed. These include the thermally less stable diacyl peroxides, e.g., diacetyl-, bis(2,4-dichlorobenzoyl)- or dilauryl, peroxide. The first two of these are available commercially as a 50% paste dispersion in a phthalate ester, the dispersion being safer in handling than the neat material.

Binary Redox Polymerization-Initiating Systems

The thermal decomposition rate of benzoyl peroxide at mouth or ambient temperature is much too slow to cure acrylic monomers. At such temperatures initiator accelerator systems are commonly employed.

Peroxide-Amine System. Although a great number of redox polymerization-initiating systems have been suggested for dental applications, the difficulty in meeting the requirements described above has eliminated all but a few. By far the most popular system is that consisting of a tertiary aromatic amine (nitrogen- and ring-substituted aniline) acting as the accelerator and benzoyl peroxide (BP), as the initiator. This system was originally suggested in the early 1940's by Schnabel (2, 3).

Overview of Experimental Data. Most peroxide-amine systems impart colors to the cured polymers ranging from yellow for N,N-dimethyl-p-toluidine to black for N,N-dimethyl-p-phenylenediamine (4). Highly electron-donating groups in the accelerator molecule usually cause the hardened material to be esthetically unsuitable.

The effect of inhibitor, peroxide, initiator and amine accelerator on the rate of polymerization of poly(methyl methacrylate) slurries has been studied (5, 6). Time required to reach the exotherm indicates aromatic peroxides, especially p-chlorobenzoyl peroxide, to be the most efficient initiators. Although polymerization in the presence of this compound and N,N-dimethyl-p-toluidine (DMPT) is more rapid initially, it is slower after the exotherm than is a system employing BP-DMPT (7). Polymerization using the latter initiator-accelerator gives resins with lower residual monomer.

Most chemically activated denture resins and filling materials employ the BP-DMPT or BP-N,N-bis(2-hydroxyethyl)-p-toluidine (DHEPT) system. Use of DHEPT increases the setting time somewhat. Methacrylate or dimethacrylate monomers using this accelerator have improved storage stability (8) and thus will not gel prematurely even on exposure to elevated temperatures.

Composite restoratives containing acrylic monomers and inorganic reinforcing agents are now being used as filling materials. In this application, large concentrations of crosslinking dimethacrylates are incorporated. These liquids polymerize much more rapidly than less viscous monomethacrylates because of the Tromsdorff autoacceleration or gel effect attributed to lessened translational mobility of growing polymer radicals with increasing viscosity of the medium (9, 10).

The aim of a number of recent studies has been to develop more reactive amines that yield nearly colorless polymers with good color stability and improved biocompatibility.

Amines with more ring substituents, particularly in the 3 and 5 positions, e.g. N,N-dimethyl-*sym*-xylidine, decrease the curing time and improve the color stability of the polymer (11). Accelerating ability is a function of both ring and nitrogen substitution, although ring substitution has the greater influence (12); color stability is influenced more by substituents on the ring. In general, increased substitution and steric hindrance of aromatic amines produces lighter colored materials. Composites containing aromatic amines having a 3,5-dimethylphenyl group discolor less than those with a 4-methylphenyl, and much less than those with an unsubstituted phenyl group. N,N-dimethyl-*p*-*tert*-butylaniline (12, 13) or N,N-*bis*(hydroxyalkyl)-3,5-di-*tert* butylanilines (14) yield hardened resins that have a very light shade and excellent color stability. Tertiary aromatic amines with large substituents on the nitrogen atom and molecular weight above 400 can be effective accelerators (15) yielding composites with rapid curing times. Because of their low volatility and reduced diffusion rate, such accelerators should not readily penetrate body tissues. Thus, they would be anticipated to cause less pulpal irritation or toxic reactions than lower molecular weight amines.

Volatility or diffusion of the tertiary amine accelerator may also be reduced by using polymerizable amines such as those with N-methacryloxyethyl groups that are incorporated into the cured resin (16) or by substituting for the low-molecular weight amine a polymeric tertiary aromatic amine (17). However, substitution of the N-methyl group in the amine by a bulkier methacryloyloxyethyl group slows the polymerization process. Formation of insoluble polymer indicates that the amines copolymerize, yielding a crosslinked polymer. Resins cured with an aminoethyl methacrylate accelerator containing a *p*-tolyl or 3,5-xylyl substituent on the nitrogen atom have color stability similar to those of their low molecular weight counterparts.

The storage stability of the components of composite formulations is mainly limited by the poor shelf-life of the BP ingredient (8). Paste formulations, containing BP, but free of accelerator prematurely harden when stored at 60°C. Formulations using powder-liquid constituents are more stable at these elevated temperatures. After extended storage at room temperature for two years, composite mixes showed delayed setting and decreased mechanical properties in the cured material. The purity of the BP and the amine selected greatly influences parameters such as reaction rate, color stability and biocompatibility (18).

With multifunctional thiols such as pentaerythritol tetra(3-mercaptopropionate), added in concentrations below 10 percent to typical peroxide-amine cured resins, composites with significantly reduced setting time and excellent color stability have been obtained (9). Dodecyl mercaptan yields comparable results (11). The free radical addition of the thiol to the double bond of the monomer probably controls the molecular weight of the polymer.

The polymeric methacrylate materials, even 24 hours after hardening, contain a large number of unreacted groups (20). Their concentration in the cured resin is considerably higher for dimethacrylate monomers than for those with a single methacrylate group in the molecule. Infrared reflectance measurements indicate that the residual methacrylate group in commercial dental composites with dimethacrylate ingredients ranges from 30 to 48 percent.

Peroxide-Amine Mechanism and Supporting Results. Two chemical mechanisms have been proposed to explain how substituents, solvents and external factors affect the polymerization initiating rate. However, the only mechanism that appears to agree with all the experimental data so far reported is that involving an electon transfer as the key step (21, 22). According to this mechanism, the amine and peroxide molecules interact to form a charge-transfer complex (this term is used to indicate a non-polar transition state) consisting of an electron deficient amine and a peroxide with an excess electron. The complex subsequently breaks down to yield an aminium cation, a peroxide free radical sufficiently reactive to combine with a monomer molecule to initiate the polymerization and an inert benzoate anion.

This mechanism is supported by the following facts, which do not seem to be explainable by other mechanisms:

The reactivity of the amine as a polymerization accelerator depends upon the $\sigma+$ value of the meta- or para-substituent of the amine, where $\sigma+$ is the electrophilic substituent parameter previously described and tabulated (23). When the kinetic rate constant (or the reciprocal of the polymerization time with amine and peroxide initial concentrations held constant) is plotted against the $\sigma+$ value on a semi-logarithmic plot (Figures 1 and 2), an inverted "V" shaped curve results. The kinetic data for Figure 1 were taken from a published article describing the polymerization of methyl methacrylate in the presence of BP and various tertiary aromatic amines as shown in Figure 2 (5). Similar plots can be obtained from kinetic data of unsaturated polyester resins (24, 25). This behavior is independent of the identity of the nitrogen substituents of the amine or the monomer. Previous publications from this laboratory have pointed out that the $\sigma+$ value of the amine ring substituent yielding maximum reactivity, i.e., minimum polymerization (or cure) time, occurs at approximately -0.2 in the cases examined (26-29). The optimum value is anticipated to be fairly insensitive to the type of monomer and experimental conditions.

In Figure 2, plots are shown for the polymerization and corresponding gel times. The latter times do not reveal a minimum at the $\sigma+$ value where the polymerization time is minimum, but rather show a decided break in the curve.

Since gel time indicates the clinical working time and

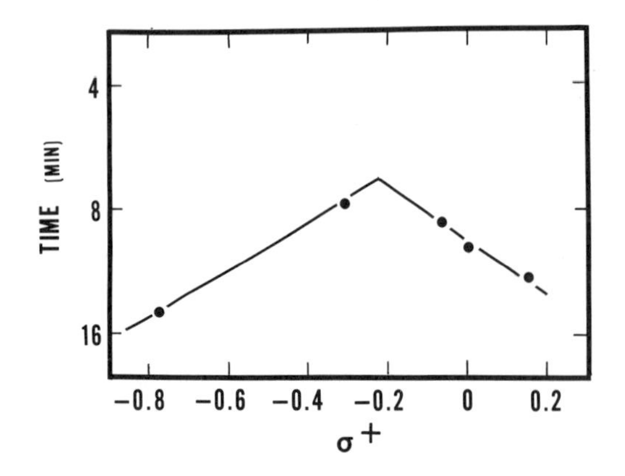

Figure 1. Curing times of methyl methacrylate containing fixed initial concentrations (in mol/L) of benzoyl peroxide and aryl-substituted N,N-*dimethylaniline versus the σ+ value of the aryl substituent of the amine. (Curing times from Ref. 5; σ+ values from Ref. 23.)*

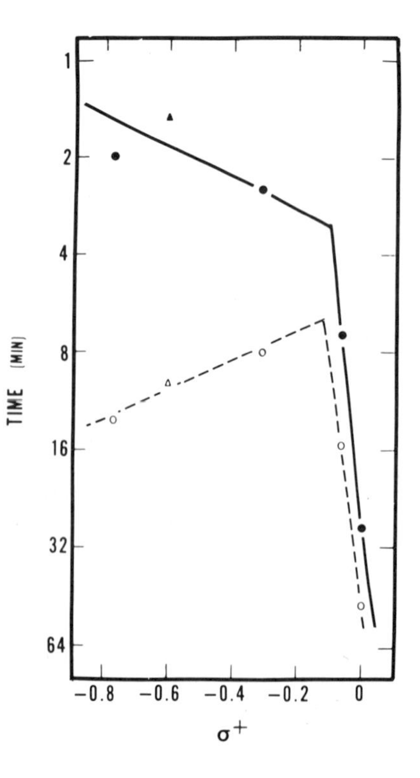

Figure 2. The gel (●) and polymerization (○) times of unsaturated polyester containing a fixed initial concentration (in mol/L) of benzoyl peroxide and aryl-substituted N,N-*bis(3-allyloxy-2-hydroxypropyl)aniline versus the σ+ value of the aryl substituent of the amine. The symbols (▲, △) indicate the corresponding time values for the analogous* N-*substituted 2-naphthylamine which in the usual sense is not a substituted aniline. (Time values from Refs. 24 and 25; σ+ from Ref. 23, except for σ+ value for 2-naphthyl (−0.61), recalculated from Bier (64).)*

polymerization time the period in which a hard polymer is obtained, the ratio of these two values should be near 1.0 so that "snap hardening" results. Ideally, the respective times should be identical but this is never realized. The gel and polymerization times approximate each other most closely when $\sigma+$ is greater than about -0.2; i.e., when the amine does not contain an overly electron-donating ring substituent. If the $\sigma+$ value of the ring substituent is very low, e.g., in such compounds as tertiary aromatic amines derived from p-anisidine or p-phenylenediamine, gelation occurs very rapidly but final cure to the hard state takes a considerable time.

Previous investigations of the spectroscopic behavior of charge-transfer complexes derived from ring-substituted N,N-dimethylaniline have demonstrated the usefulness of the $\sigma+$ parameter in correlating the data (30, 31). The satisfactory fitting of polymerization rates against the $\sigma+$ substituent is indicative of charge-transfer interaction of the aromatic amine.

The effect of solvent upon the rate of reaction of tertiary aromatic amines with BP is not dependent upon the dielectric constant of the solvent or on the monomer formulation (32). The data of previous investigators could be correlated by a simple function of the solvent refractive index as the relevant independent variable (32a). The correlation with the refractive index also indicates charge-transfer complex formation. This is in agreement with the spectroscopic evidence revealing the far greater sensitivity of the electronic energy of charge-transfer complexes between uncharged molecules to the refractive index of the solvent as compared to the dielectric constant (31).

We tested the prediction that tertiary aromatic amine accelerators containing ring substituents with $\sigma+$ values close to -0.2 would be the most effective. A compilation of the $\sigma+$ values of ring substituent (23) listed the p-$CH_2CO_2C_2H_5$ group as having a $\sigma+$ value of about -0.16, suggesting use of a tertiary aromatic amine with this substituent or the corresponding carboxylic acid or its methyl ester as an accelerator.

The overall characteristics of the composites (hardening time, strength and color stability) containing 4-N,N-dimethylaminophenylacetic acid (DMAPAA) or its methyl ester (MDMAPAA) as accelerator ingredients compared favorably to restorative resins cured with commonly used tertiary amines (26). Based on hardening times the approximate order of the accelerating ability of the respective amines was: DMAPAA > N,N-dimethyl-sym-xylidine > DMPT > MDMAPAA >> DHEPT. This order of reactivity is dependent on the components (especially monomers) used in the formulations. Minimal curing time and maximum tensile and compressive strength of the cured material were obtained over a narrow concentration range of accelerator in the liquid. This range, which is dependent on the type of diluent employed, was approximately the same for all amines.

The biocompatibility of these amines as expected from similarity in structures to compounds used medicinally (33, 34,)is good. No mutagenic or cytotoxic effects have been observed using the Ames test for bacterial mutagenicity and the agar overlay test for cytotoxicity with DMAPAA (35).

A second series of amines suggested by theory to be reactive accelerators are the corresponding p-aminophenethanols. The $\sigma+$ value for the p-CH_2CH_2OH group has not been reported but should be somewhat less than that for the p-$CH_2CO_2C_2H_5$ group and be close to -0.2.

The following homologues and derivatives of DMAPAA(I) and N,N-dialkylaminophenethanol (II) have been synthesized and tested as accelerators (37):

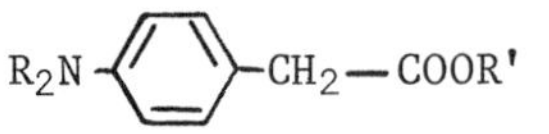

$$R_2N-C_6H_4-CH_2-COOR' \quad (I) \quad \text{and} \quad R_2N-C_6H_4-CH_2CH_2OH \quad (II)$$

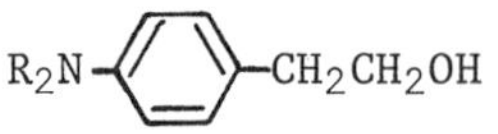

where R=CH_3, C_2H_5 and R'=H,CH_3,C_2H_5.

Composites using a 3 to 1 powder/liquid ratio were prepared containing a silanized barium glass coated with 1% BP powder and 72.4 percent bis(3-methacryloxy-2-hydroxypropyl) bisphenol A (BIS-GMA), 27.6 percent 1,6-hexamethylene glycol dimethacrylate (1,6-HEDMA), 0.2 percent butylated hydroxytoluene (BHT) and various amines in the liquid. Setting times of the formulations varied from 1.5 to 4.0 min. The N,N-diethylaminophenylacetic acid (DEAPAA) was by far the most reactive accelerator with a 3 mmolar amine concentration causing a cure of 4.5 min. Fastest polymerization for the above amines occurred at 17 mmolar concen- accelerator. A composite made from powder coated with 1 percent BP and a liquid with 17 mmolar amine has a molar peroxide to amine ratio of 6.5 compared to a ratio of 1.1 to 1.5 reported as most efficient for curing unfilled resins (4, 36). This much larger excess of peroxide required to obtain minimum setting time should be expected since only a small portion of the peroxide is accessible to the amine. Physical properties (tensile strength 36-55 MPa, compressive strength 245-303 MPa, water sorption 0.5-0.7/cm^2) considerably exceeded the minimum requirement of the specification for dental composite resins (37). If low concentrations are employed in the formulations especially with DEAPAA as accelerator, the cured composites are nearly colorless. No perceptible change occurs in the color of the specimens containing a UV absorber after 24 hours exposure to a UV light source. Because of the excellent overall physical properties, nearly colorless appearance and the potentially better biocompatibility, compositions using these accelerators should yield improved restoratives.

Acrylic monomers are polymerized by carboxylic acids with tertiary aromatic amines (38), e.g., 4-N,N-(dimethylamino)phenylacetic acid (8), presumably via a charge-transfer complex. Neutral amino acid ester accelerators should yield monomer formulations having better shelf-life than the corresponding amino acids. This has been established experimentally (39).

Reactivity of the accelerator employed is dependent on the monomer used in the respective formulation. Whereas compositions containing DEAPAA cure bis-phenol A dimethacrylate monomer formulations fastest, those containing p-(dialkylamino)phenethanol result in the more rapid polymerization of methyl methacrylate monomer-polymer slurries (39). Addition of Bronsted acids (40) to similar systems increases the polymerization rate but, as has been discussed above, reduces storage stability.

Many peroxyesters, hydroperoxides or peroxides are more storage stable than BP. However, their reaction with tertiary amines is generally too slow to give a sufficiently rapid cure for acrylic resins. Composite mixes containing 2,5-dimethyl-2,5-(benzoyl peroxy)hexane, t-butyl perbenzoate, t-butyl hydroperoxides or dicumyl peroxide-tertiary amines do not harden for days. Resins cured with cumene hydroperoxide and DEAPAA harden, but yield highly colored products (39). Faster cures are obtained with t-butyl peroxymaleic acid (TBPM) and primary amines (p-toluidine or p-aminophenethanol), but the shades of the resulting composites are unacceptable. Curing is slower with secondary and tertiary amines. The latter compounds yield materials having higher strength and more desirable shades. Because of the large concentrations of initiator and tertiary amine required for satisfactory cure, the storage stability of components of composite resins containing TBPM and tertiary amines does not equal those initiated by the less thermally-stable BP.

One unusual and surprising characteristic of tertiary aromatic amines is that in addition to acting as accelerators, the same compounds in low concentration in the presence of oxygen and an initiator may act as inhibitors of polymerization (41, 42). This behavior has also been attributed to the ability of the amine to engage in charge-transfer complex formation. Oxygen also inhibits radical polymerization and results in uncured films at the surface of dental sealants (42).

Amine-Free Redox Systems

Many initiator-accelerator systems that contain accelerators other than amine have been suggested for vinyl polymerization, but only a few have been employed in dental resins. Substitution of p-toluenesulfinic acid, alpha-substituted sulfones and low concentrations of halide and cupric ions for tertiary amine accelerators, yields colorless products (43-48). Most of these compounds have poor shelf-life. They readily oxidize in air to sulfonic acids which do not activate polymerization. Lauroyl peroxide, in conjunction with a metal mercaptide (such as zinc hexadecyl mercaptide) and a trace of copper, has been used to cure monomer-polymer slurries containing methacrylic acid (49-50). Addition of Na salts of saccharine to monomer containing an N,N-dialkylarylamine speeds up polymerization (51).

Methacrylate monomers can be polymerized with t-butyl-, or

cumyl hydroperoxide and thiourea derivatives as reducing agent (52, 53, 54). With from 0.5 to 1 percent phenyl-, N-acetyl- or N-allylthiourea composites with excellent color stability have been obtained.

Cumene- or *t*-butyl hydroperoxide or *t*-butyl perbenzoate in conjunction with ascorbic acid or ascorbic palmitate also initiate rapid polymerization, yielding colorless composites with good mechanical properties (55). For commercial application, means must be developed for prevention of oxidation of ascorbic acid or its derivatives on prolonged storage.

Trialkylborane oxides initiate polymerization of acrylic monomers at the moist dentin surface (56, 57). Bonding occurs only between the dentinal collagen and any retention to enamel is mechanical.

Photoinitiated Polymerization. During the last decade, a number of photochemically-cured materials have been introduced to the dental profession. Presently, such resins are available as restoratives, pit-and-fissure sealants, bonding agents and as orthodontic bracket adhesives. The materials are cured by an appropriate light source which, by means of suitable filters, produces radiation in the near ultraviolet region around 360 nm or utilizes visible light of about 470 nm. For UV-initiated material, the acrylic resin is often composed of a dimethacrylate such as BIS-GMA, a polymerizable diluent, and a photoinitiator, usually a benzoin alkyl ether or diketone. BP (58) or phosphite esters (59) may be included to accelerate the cure, which should be substantially completed in 30 to 60 sec. Addition of a photo-crosslinking agent reduces curing time and lowers solubility of UV-cured sealants (60). A single paste comprising a urethane-methacrylate prepolymer or dimethacrylate, a monomeric diluent, an alpha diketone (camphorquinone) initiator, dimethylaminoethyl methacrylate (reducing agent) and silanized glass powder yielded an experimental dental restorative with good physical properties (61).

Light cured materials do not require mixing by the dentist and can be manipulated indefinitely in the mouth until their polymerization is initiated by exposure to radiation. They satisfy the conflicting requirements of long working time and short setting times (snap hardening) that are difficult to achieve with chemical initiator-accelerator systems.

The radiation entering a composite may be scattered at the resin/particle interface as well as be absorbed by the particle and resin. The success of radiation-cured material is a function of the light source, its spectral distribution and intensity, radiation time, the light transmission of the restorative and light absorbancy of the surrounding media (62, 63). Degree of cure decreases slightly below the restoration surface until a depth is reached where it falls off rapidly. Generally, a curing cycle of 30 to 60 sec. ensures polymerization up to a depth of at least 3 mm. Alternatively, in deeper cavities the restoration can be build in consecutive layers since the inter-layer bond is satisfactory. Furthermore, curing continues after the initiating photochemical reaction is cut off as evidenced by the increase in hardness of specimens "aging" from one hour to 24 hours.

Literature Cited

1. Brauer, G. M. "Biomedical and Dental Applications of Polymers"; Gebelin, C. G.; Koblitz, F. F., Ed.; Plenum, New York, 1981; pp 395-409.
2. Schnabel, E. German Patent 736,481, May 13, 1943.
3. Blumenthal, L. "Recent German Developments in the Field of Dental Materials". FIAT field report No. 1185, Field Information Agency, Technical Office of Military Government for Germany (US), 27 May 1947.
4. Brauer, G. M.; Davenport, R. M.; Hansen, W. C. Mod. Plastics 1956, 34, 153-168, 256.
5. Lal, J.; Green, R. J. Polymer Sci 1955, 17, 403-409.
6. Rose, E. E.; Lal, J.; Green, R. J. Am. Dent. Assoc. 1958, 56, 375-381.
7. Cornell, J. A.; Powers, C. M. J. Dent. Res. 1959, 38, 606-610.
8. Brauer, G. M.; Petrianyk, N.; Termini, D. J. J. Dent. Res. 1979, 58, 1791-1800.
9. Pryor, W. A. "Free Radicals", McGraw-Hill: New York, 1966, p. 317-318.
10. Walling, C. "Free Radicals in Solution" John Wiley: New York, NY, 1957, pp 189-193.
11. Bowen, R. L.; Argentar, H. J. Am. Dent. Assoc. 1967, 75, 918-928.
12. Bowen, R. L.; Argentar, H. J. Dent. Res. 1971, 50, 923-928.
13. Argentar, H.; Tesk. J. A.; Parry, E. E. J. Am. Dent. Assoc. 1981, 102, 664-665.
14. Schmitt, W.; Purrmann, R.; Jochum, P. Ger. Offen. 2, 658, 538, June 30, 1977.
15. Bowen, R. L.; Argentar, H. J. Dent. Res. 1972, 51, 473-482.
16. Dnebosky, J.; Hynkova, V.; Hrabak, F. J. Dent. Res. 1975, 54 772-776.
17. Hynkova, V.; Hrabak, F. Makromol. Chem. 1975, 176, 1669-1678.
18. Koblitz, F. F.; O'Shea, T. M.; Glenn, J. F.; DeVries, K.L. J. Dent. Res. 1977, 56B, No. 791.
19. Antonucci, J. M.; Stansbury, J. W.; Dudderar, D. J. J. Dent. Res. Res. 1982, 61, 270.
20. Ruyter, I. E.; Svendsen, S. A. Odontol. Scand. 1978, 36, 75-82.

21. Prior, W. A.; Hendrickson, W. H., Jr. J. Am. Chem. Soc. 1975, 97, 1582-1583.
22. Horner, L.; Brüggemann, H.; Knapp, K. H. Ann. 1959, 626, 1-19.
23. Brown, H. C.; Okamoto, Y. J Am. Chem. Soc. 1958, 80, 4979-4987.
24. Mleziva, J. Plasticke Hmoty. Kaucuk, 1964, 1, 225-231.
25. Mleziva, J. Chem. Prumysl. 1965, 15, 80-85.
26. Brauer, G. M.; Dulik, D.; Antonucci, J. M.; Termini, D. J.; Argentar, H. J. Dent. Res. 1979, 58, 1994-2000.
27. Brauer, G. M.; Stansbury, J. W.; Antonucci, J. M. J. Dent. Res. 1981, 1343-1348.
28. Argentar, H. U. S. Patent 4,243,763, Jan. 6, 1981.
29. Argentar, H. U. S. Patent 4,284,551, Aug. 18, 1981.
30. Kravtsov, D. N.; Faingor, D. N. Izo. Akad. Nauk. SSR, Ser. Khim 1968, 289-296; Chem. Abstr. 1968, 69, 66717.
31. Argentar, H. J. Res. Nat. Bur. Stand. (U.S.), 80A, 173-187.
32. Walling, C.; Indictor, N. J. Am. Chem. Soc. 1958, 80, 5814-5818.
32a. Argentar, H. unpublished results.
33. Stecker, P. G., Ed. "The Merck Index," 3rd ed.: Merck: Rahway, NJ 1968, p. 60.
34. Lapkin, V. A. Farmakol. Tokskol (Moscow), 1974, 37, 660-662; Chem. Abstr. 1974, 82 68082.
35. Jacobsen, A. H.; Pettersen, A. H. personal communication.
36. Bowen, R. L.; Argentar, H. J. Appl. Polymer Sci. 1973, 17 2213-2222.
37. American Dental Association Specification No. 27 for Direct Filling Resins; J. Am. Dent. Assoc. 1977, 94, 1191-1194.
38. Hrabak, F.; Hynkova, V. Macromol. Chem. 1981, 182, 1595-1603.
39. Brauer, G. M.; Stansbury, J. W. unpublished data.
40. Okada, Y.; Kogyo Kagaku Zasshi 1963, 66, 1317-1320; Engl. Summ. 1963, 66, A83.
41. Yates, W. R.; Ihrig, J. L. J. Am. Chem. Soc. 1965, 87 710-715.
42. Ruyter, I. E. Acta Odontol. Scand. 1981, 39, 27-32.
43. Hagger, O. Helv. Chim. Acta 1948, 31, 1624-1630.
44. Hagger, O. Helv. Chim. Acta 1951, 34, 1872-1876.
45. Castan, P.; Hagger, O. U. S. Patent 2,567,803, Sept. 11, 1951.
46. Bredereck, H.; Bäder, E. Chem. Ber. 1954, 87, 129-139.
47. Bredereck, H.; Bäder, E. U. S. Patent 2,846,418, Aug. 5, 1958.
48. Brauer, G. M.; Burns, F. R. J. Polymer Sci. 1956, 19, 311-321.
49. Stern, H. J.; Shadbolt, L. E.; Rawitzer, W. L. British Patent 721,641, Jan. 12, 1955.
50. Shadbolt, L. E. Ger. Offen, 1,937,871, Jan. 29, 1970.
51. Lal, J. U. S. Patent 2,833,753, May 6, 1958.

52. Takaaki, S.; Yuji, M. J. Polymer Sci. 1966, A-1, 4, 2735-2746.
53. Sumitomo Chemical Ltd. Brit. Patent 1,177,879, Jan. 14, 1970.
54. Termin, S. C.; Richards, M. C. U. S. Patent 3,991,008, Nov. 9, 1976.
55. Antonucci, J. M.; Grams, C. L.; Termini, D. J. J. Dent. Res. 1979, 58, 1887-1889.
56. Masuhara, E. Deut. Zahnärztl. Z. 1969, 24, 620-628.
57. Nakabayashi, N.; Masuhara, E.; Michida, E.; Ohmori, I. J. Biomed. Mat. Res. 1978, 12. 149-165.
58. Waller, D. E. U. S. Patent 3,709,866, Jan. 9, 1973.
59. Schmitt, W.; Purrmann, R. Ger. Offen. 2,646,416, May 26, 1977.
60. Brauer, G. M. J Dent. Res. 1978, 57, 597-607.
61. Dart, C. E.; Cantwell, J. B.; Traynor, J. R.; Jaworzyn, J.N. U. S. Patent 4,089,763, May 16, 1978.
62. Cook. W. D. J. Dent. Res. 1980, 59, 800-808.
63. Kilian, R. J. "Biomedical and Dental Applications of Polymers" Gebelin, C. G.; Koblitz, F. F., Ed; Plenum: New York, 1981, pp. 411-417.
64. Bier, A. Rec. Trav. Chim. 1956, 75, 866-870.

RECEIVED October 5, 1982

27

Synthesis of Biodegradable Polymers for Biomedical Utilization

J. HELLER

SRI International, Polymer Sciences Department, Menlo Park, CA 94025

Polymer bioerosion is defined as the conversion of an initially water-insoluble material to a water-soluble material and is discussed in terms of three distinct mechanisms denoted as Type I, II, or III. Solubilization by Type I erosion involves hydrolytic bond cleavage occurring in water-soluble polymers that have been insolubilized by covalent crosslinks. Solubilization by Type II erosion involves reactions of pendant groups, and solubilization by Type III erosion involves backbone cleavage. Solubilization by simple dissolution is not considered. Each type of bioerosion is illustrated with specific examples that include gelatin-based surgical aids, enteric-type coatings, erodible sutures, surgical adhesives, oviduct blocking agents, and controlled drug release.

Polymer bioerosion can be defined as the conversion of an initially water-insoluble material to a water-soluble material and does not necessarily signify a major chemical degradation. In this review we do not consider a simple dissolution process, and we classify the various bioerosion mechanisms into the the three distinct types shown in Figure 1 (1).

In general terms, Type I erosion encompasses water-soluble polymers that have been insolubilized by hydrolytically unstable crosslinks. Type II erosion includes polymers that are initially water-insoluble and are solubilized by hydrolysis, ionization, or protonation of a pendant group. Type III erosion includes hydrophobic polymers that are converted to small water-soluble molecules by backbone cleavage.

Clearly, these three types represent extreme cases, and actual erosion can be a combination of these types. Thus, it is

0097-6156/83/0212-0373$06.00/0

possible to develop a combination of Type I and Type III erosion in which initial hydrolysis involves crosslink cleavage with subsequent backbone cleavage of the high molecular weight water-soluble polymer. Similarly, a combination of Type II and Type III erosion can be developed in which initial solubilization is by ionization, protonation, or hydrolysis, followed by backbone cleavage of the soluble polymer.

Type I Erosion

The most widely used polymer system that bioerodes by Type I erosion or by a combination of Type I and Type III erosion is gelatin that has been insolubilized by heat treatment, aldehyde treatment, or chromic acid treatment.

Aside from the photographic industry, insolubilized gelatin also finds application as surgical aids such as dusting powder for surgical gloves or as sponges or films. Dusting powders are prepared by heating gelatin at 142°C for 25 hours, and insolubilization is believed to occur by formation of interchain amide links (2). Surgical sponges or films are useful in arresting surgical hemorrhage and are prepared by treating sterile gelatin with formaldehyde. Although details are poorly understood, insolubilization occurs by reaction of gelatin amino groups with formaldehyde followed by reaction of the hydroxymethylamino groups (3, 4, 5).

Formaldehyde crosslinked gelatin has also been used as a matrix in controlled drug release applications. However, because insolubilization of water-soluble polymers by crosslinking produces hydrogels that are completely permeated by water, they are clearly unable to immobilize small molecules having appreciable water solubility. Consequently, usefulness of these materials is limited to molecules having extremely low water solubility or to macromolecules that can be physically entangled in the hydrogel so that they can not diffuse out of the matrix even though they are freely soluble in water.

An example of the first application is shown in Figure 2, which shows release of the highly water-insoluble hydrocortisone acetate from a gelatin matrix crosslinked with formaldehyde (1). As indicated by the first-order dependence, drug is released by a simple diffusional process and the crosslinked gelatin simply provides cohesiveness for the drug particles. Even though release kinetics are not constant, useful release over many days is achieved.

The first-order drug release also indicates that matrix erosion makes little or no contribution to drug release, and, because erosion of a formaldehyde-crosslinked gelatin is slow, drug depletion can occur before significant matrix erosion is noted. Such a device may be useful in applications where zero-order kinetics are not important and removal of the expended device is not convenient or desirable.

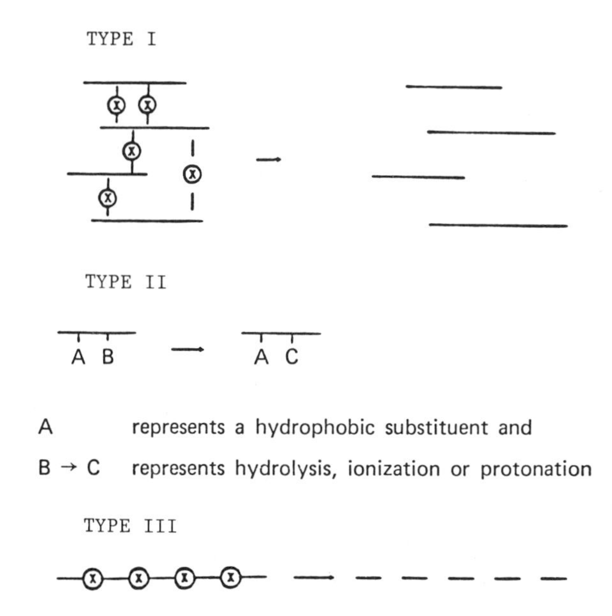

Figure 1. Schematic representation of bioerosion mechanisms. (Reprinted with permission from Ref. 1. Copyright 1980.)

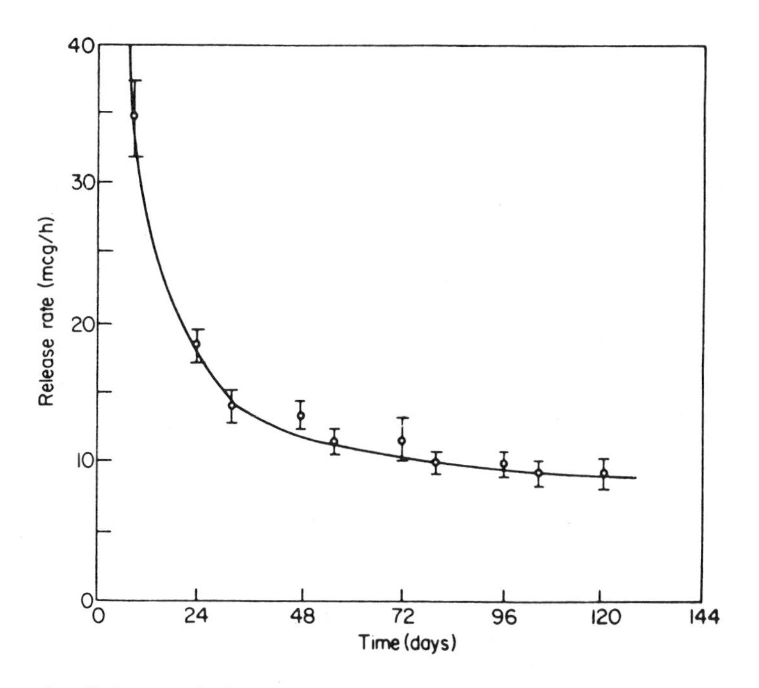

Figure 2. Release of hydrocortisone acetate from a cross-linked gelatin matrix. (Reprinted with permission from Ref. 1. Copyright 1980.)

Release of macromolecules from bioerodible hydrogels has received relatively little attention, and most past efforts have been devoted to the immobilization of macromolecules in hydrogels by physical entanglement (6). However, because of the growing recognition of the therapeutic potential of macromolecules, their entanglement in a hydrogel matrix and controlled release by matrix erosion is becoming recognized as an important methodology (7, 8).

Type II Erosion

Because this type solubilization involves the reaction of a pendant group, bioerosion proceeds without significant changes in molecular weight. Therefore, unless the polymer backbone also undergoes bioerosion, polymers in this category are only useful in topical applications where elimination of high molecular weight, water-soluble macromolecules proceeds with no difficulty.

An important class of polymers that undergo Type II erosion are polymers containing carboxylic acid functions. These solubilize by ionization and consequently are insoluble at low pH and soluble at high pH. A major application for such polymers is as enteric coatings, designed to protect therapeutic agents during passage through the acidic stomach and to abruptly dissolve in the higher pH environment of the intestines.

An interesting example is the partial esters of a methyl vinyl ether and maleic anhydride copolymer (9, 10, 11).

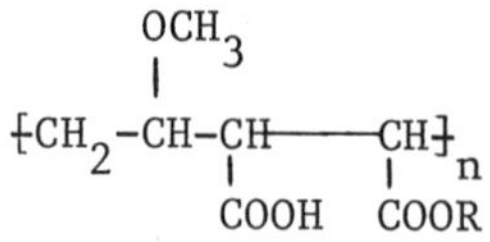

$$\left[CH_2-\underset{\displaystyle OCH_3}{\overset{}{CH}}-\underset{COOH}{CH}-\underset{COOR}{CH} \right]_n$$

These polymers have a characteristic narrow pH range above which they are soluble and below which they are insoluble, and this pH range varies with the size of the R-group in the ester portion of the copolymer. This effect is shown in Figure 3 (12).

This behavior can be readily understood by considering the number of ionized carboxyls that are necessary to drag the polymer chain into solution. With relatively small ester groups, only a low degree of ionization is needed to solubilize the polymer, and hence the dissolution pH is low. As the size of the alkyl group increases, so does the hydrophobicity, and progressively more ionization is necessary to solubilize the polymer resulting in increasingly high dissolution pH. The same argument holds for polymers having the same ester grouping but different degrees of esterification. The higher the degree of esterification, the more hydrophobic the polymer and consequently the higher the dissolution pH.

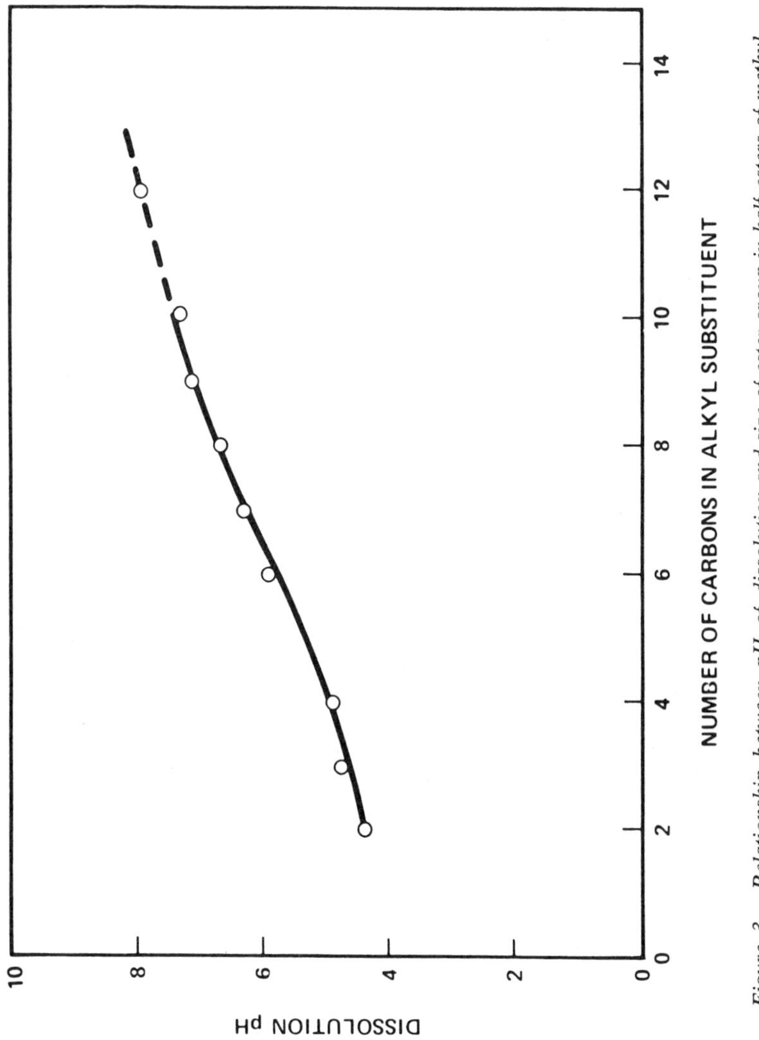

Figure 3. Relationship between pH of dissolution and size of ester group in half-esters of methyl vinyl ether-maleic anhydride copolymers. (Reprinted with permission from Ref. 12. Copyright 1980, John Wiley & Sons, Inc.)

Even though partially esterified copolymers of methyl vinyl ether-maleic anhydride copolymers were orignally designed to dissolve abruptly with an increase in external pH, in a constant pH environment they undergo a controlled dissolution process and are therefore useful materials for the controlled release of therapeutic agents dispersed within them (12).

Figure 4 shows polymer dissolution rate and the rate of hydrocortisone release for an n-butyl half-ester of a methyl vinyl ether-maleic anhydride copolymer film containing the dispersed drug. Each pair of points represents a separate device in which the amount of drug released by the device into the wash solution was determined by uv measurements and the amount of polymer dissolved was calculated from the total weight loss of the device. The excellent linearity of both polymer erosion and drug release over the lifetime of the device provides strong evidence for a surface-erosion mechanism and for negligible diffusional release of the drug. The latter result was independently verified by placing a drug-containing film in water at a pH low enough that no dissolution of the matrix took place and periodically analyzing the aqueous solution for hydrocortisone. None was found over several days.

Type III Erosion

Polymers undergoing Type III erosion have found applications in (a) absorbable surgical sutures, (b) surgical adhesives, (c) contraception, and (d) controlled drug release.

Absorbable Surgical Sutures. The search for an optimum absorbable surgical suture evolved through various forms of collagen to the modern day catgut (13). Although catgut is a strong and effective suture, it suffers from various disadvantages, such as batch-to-batch variation; stiffness, which requires the use of a conditioning storage fluid; and occasional intense tissue reactivity. For these reasons it was desirable to develop a synthetic material that could be tailored to meet ideal suture requirements.

The first synthetic absorbable suture that reached commercial production is Dexon, manufactured by American Cyanamid Co. It is poly(glycolic acid) prepared by the polymerization of glycolide (14). Another absorbable

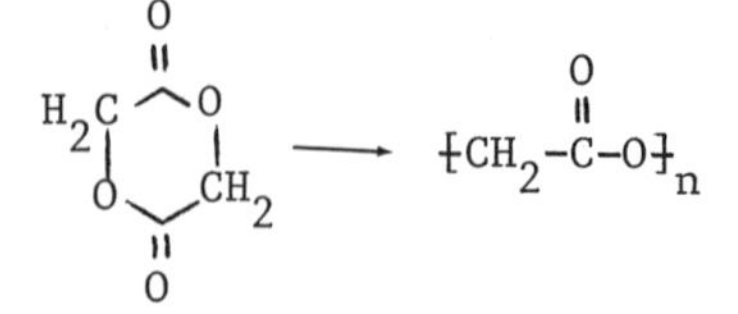

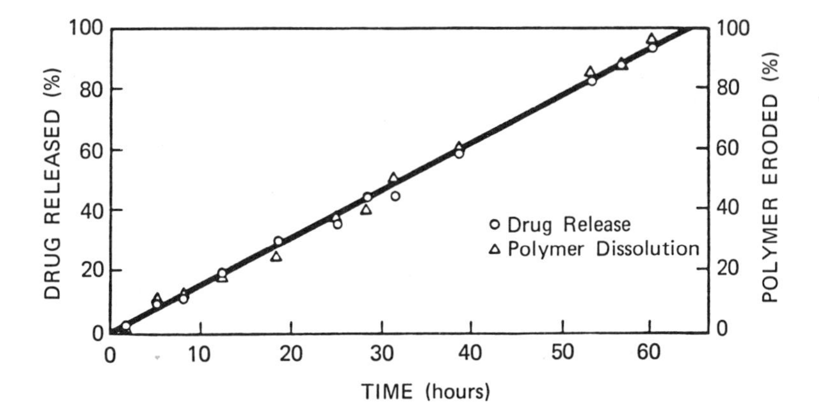

Figure 4. Rate of polymer dissolution and hydrocortisone release from n-butyl half-ester of methyl vinyl ether-maleic anhydride copolymer containing 10 wt% drug. (Reprinted with permission from Ref. 12. Copyright 1980, John Wiley & Sons, Inc.)

suture is Vicryl, manufactured by Ethicon Company. Vicryl is the 10/90 copolymerrpoly(lactide-co-glycolide) and is prepared by the copolymerization of 9 parts L(-) lactide and 1 part glycolide (15).

Both polymers degrade by a simple hydrolysis reaction, and no cellular or enzyme activity is necessary for suture absorption (16, 17).

Surgical Adhesives. In many procedures it is difficult to use conventional suturing techniques. Consequently, procedures whereby incisions or cut ends of arteries can be joined by means of an adhesive would represent a significant surgical advance (18).

Because α-cyanoacrylates contain a double bond substituted by two electron-withdrawing substituents, they are highly susceptible to anionic initiation, and water is basic enough to initiate very rapid polymerization. Therefore, because moisture can initiate polymerization and because the formed polymer is able to firmly adhere to moist surfaces, it has evoked considerable medical interest as a tissue adhesive (19, 20).

However, it has been observed that methyl α-cyanoacrylate in these applications leads to tissue inflamation and cell necrosis, and further research has shown that the polymer undergoes a degradation reaction that occurs both *in vivo* and *in vitro*. Because poly(alkyl α-cyanoacrylates) are structurally similar to poly(vinylidene cyanide), which has been postulated to degrade by a reverse Knoevenagel reaction with evolution of formaldehyde (21), the following degradation mechanism for poly(alkyl cyanoacrylates), which also degrade with evolution of formaldehyde, has been suggested (22, 23):

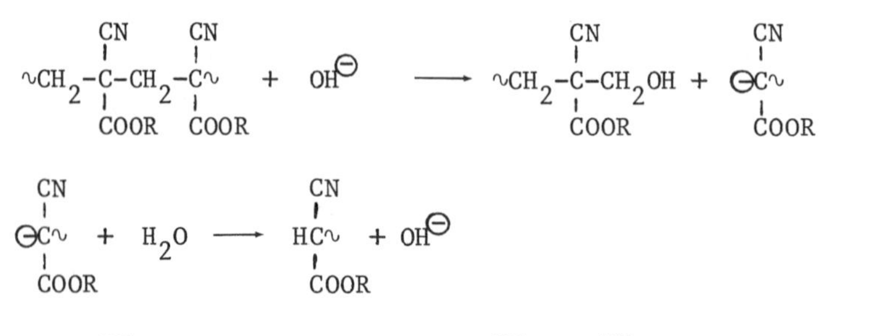

The rate of degradation depends on the size of the ester group, and, as shown in Figure 5, the rate of degradation as measured by formaldehyde evolution of the methyl ester at pH 7.0 is considerably faster than the rate of degradation of the higher esters. A dependence of degradation rate on polymer molecular weight has also been reported (24).

Because poly(methyl α-cyanoacrylate) degrades relatively rapidly to methyl α-cyanoacrylate, which is an intensely necrotizing and pyogenic compound (25, 26), polymerization of this compound is clearly not a viable surgical procedure. However, higher α-cyanoacrylate esters are considerably less toxic, and polymer degradation occurs at a much lower rate so that these materials may have potential as surgically useful materials (27, 28, 29).

Contraception. The rapid polymerization of methyl α-cyanoacrylate and its effect on living tissue has been utilized in female sterilization by oviduct blockage (30). In this procedure methyl α-cyanoacrylate is instilled into the oviduct where it rapidly polymerizes into a solid. Subsequent degradation of the polymer leads to formation of scar tissue, which eventually permanently blocks the oviduct. The overall process is shown in Figure 6 (31).

Major advantage of this method is that the methyl α-cyanoacrylate can be instilled by a specially-designed transcervical delivery device (31) by trained paramedical personnel and does not require a surgical procedure. Thus, this method may be attractive to developing countries with serious population problems.

Controlled Drug Release--Because the degradation products of Type III bioerosion are small, water soluble molecules, the principal application of polymers undergoing such degradation is for the systemic administration of therapeutic agents from subcutaneous, intramuscular or intraperitoneal implantation sites. Application of Type III bioerosion to controlled drug release was first described in 1970 (32) and has since then been extensively investigated. The various types of devices currently under development can be classified into (a) diffusional and (b) monolithic (7).

Diffusional Devices. In these systems a drug-containing core is surrounded by a bioerodible rate-controlling membrane. Thus, these devices combine the attributes of a rate-controlling polymer membrane, which provides a constant rate of drug release from a reservoir-type device, with erodibility, which results in bioerosion and makes surgical removal of the drug-depleted device unnecessary. Because constancy of drug release demands that the bioerodible polymer membrane remain essentially unchanged during the delivery regime, significant bioerosion must not occur until after drug delivery has been completed. Thus polymer capsules will remain in the tissue for varying lengths of time after completion of therapy.

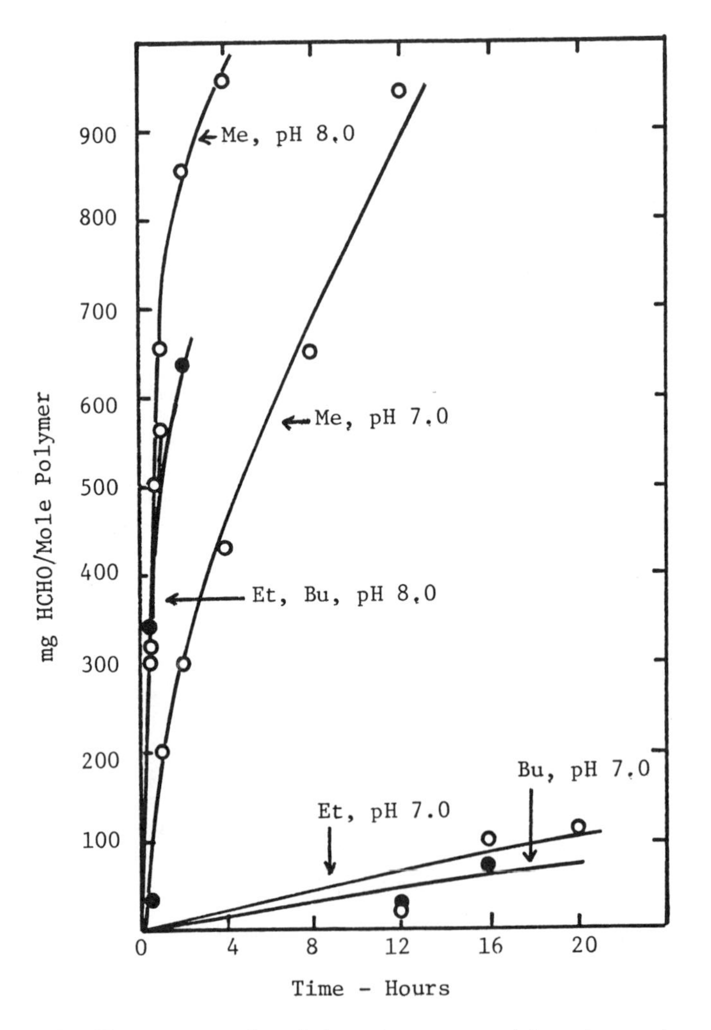

Figure 5. Homogeneous degradation of α-cyanoacrylate polymers in aqueous acetonitrile. (Reprinted with permission from Ref. 22. Copyright 1966, John Wiley & Sons, Inc.)

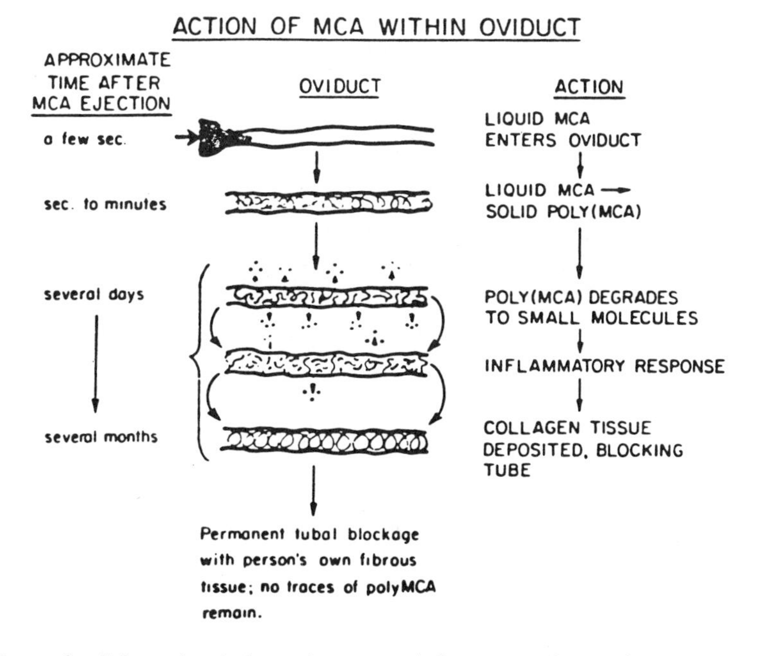

Figure 6. Schematic of the action of methyl cyanoacrylate within an oviduct. (Reprinted from Ref. 31. Copyright 1981, American Chemical Society.)

Two types of erodible diffusional devices are under development: (1) rod-shaped insert and (2) microcapsules. Each of these types of devices has certain advantages and certain disadvantages. Subdermal placement of rod-shaped devices requires a minor surgical procedure, and the device remains visible as a small lump. However, because it can be readily removed should termination of therapy become desirable or necessary, therapy lasting many months is possible. On the other hand, microcapsules can be readily injected through an 18-gauge hypodermic needle, but retrieval of microcapsules is not possible without an extensive surgical intervention. For this reason it has been suggested that the capsules be designed for relatively short-term release, such as three months (33).

Major emphasis for the development of these erodible diffusional systems have been devices that release contraceptive steroids or narcotic antagonists. The polymer systems most extensively investigated for use as a subdermal capsule for the release of levonorgestrel were various aliphatic polyesters and, in particular, poly(ω-caprolactone). The release rate of levonorgestrel from such a device, developed by the Research Triangle Institute and named Capronor, is shown in Figure 7 (34). Clearly, excellent constant daily release over many months has been achieved, and those devices are about to undergo Phase II clinical testing (35).

The polymer most actively investigated for use in a microcapsular delivery system is poly(DL-lactic acid), and release of norethindrone measured as blood plasma level from such microcapsules is shown in Figure 8 (36). These data show reasonably constant blood levels and demonstrate that for a fixed total weight of microcapsules, rate of drug release and duration of therapy can be regulated by capsule size. Furthermore, because each capsule functions as an independent drug delivery system, the rate of drug delivery can also be regulated by variation in the total number of injected microcapsules.

Monolithic Devices --In these systems the drug is homogeneously dispersed within a bioerodible polymer matrix, and release of the drug can be controlled either by diffusion or by polymer erosion. If erosion of the matrix is very much slower than drug diffusion, then release kinetics follow the Higuchi model (37) and drug release rate decreases exponentially with time, following $t^{-1/2}$ dependence over a major portion of the release rate.

If erosion is relatively fast and the drug is well immobilized in the solid matrix so that diffusional release is minimal, matrix erosion determines rate of drug release. However, it is important to distinguish two types of hydrolytic erosion of a solid, hydrophobic polymer. In one, referred to as homogeneous erosion, the hydrolysis occurs at a uniform rate throughout the matrix. In the other, called heterogeneous

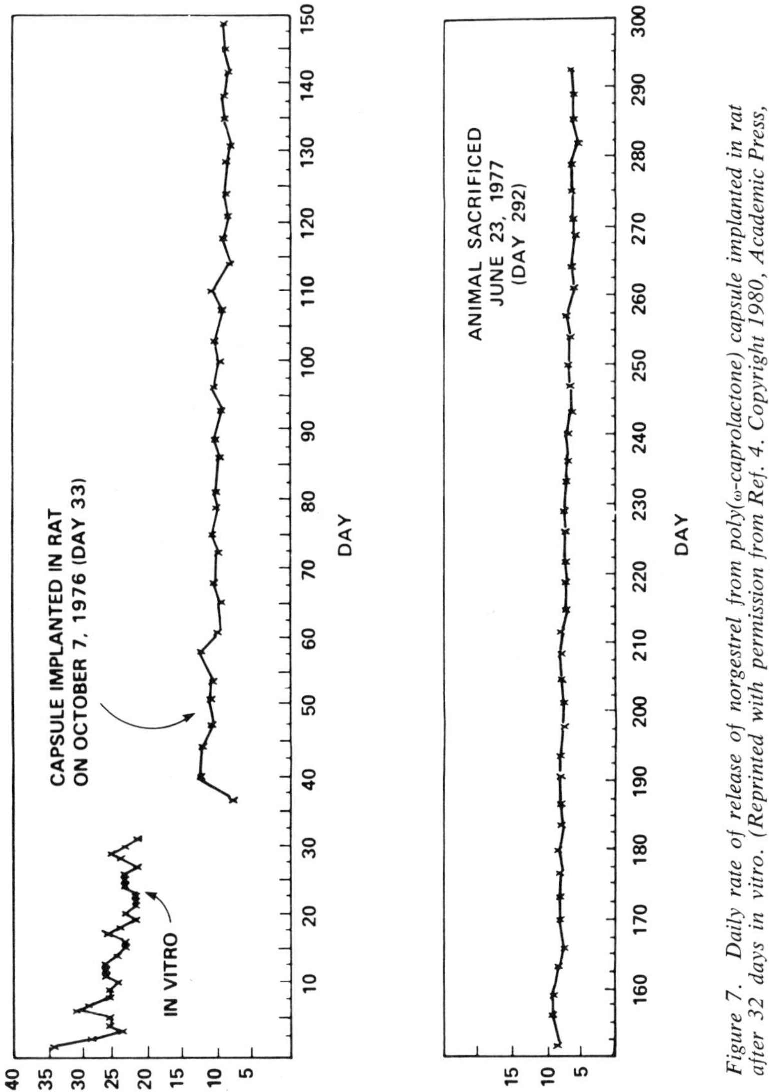

Figure 7. Daily rate of release of norgestrel from poly(ω-caprolactone) capsule implanted in rat after 32 days in vitro. (Reprinted with permission from Ref. 4. Copyright 1980, Academic Press, Inc.)

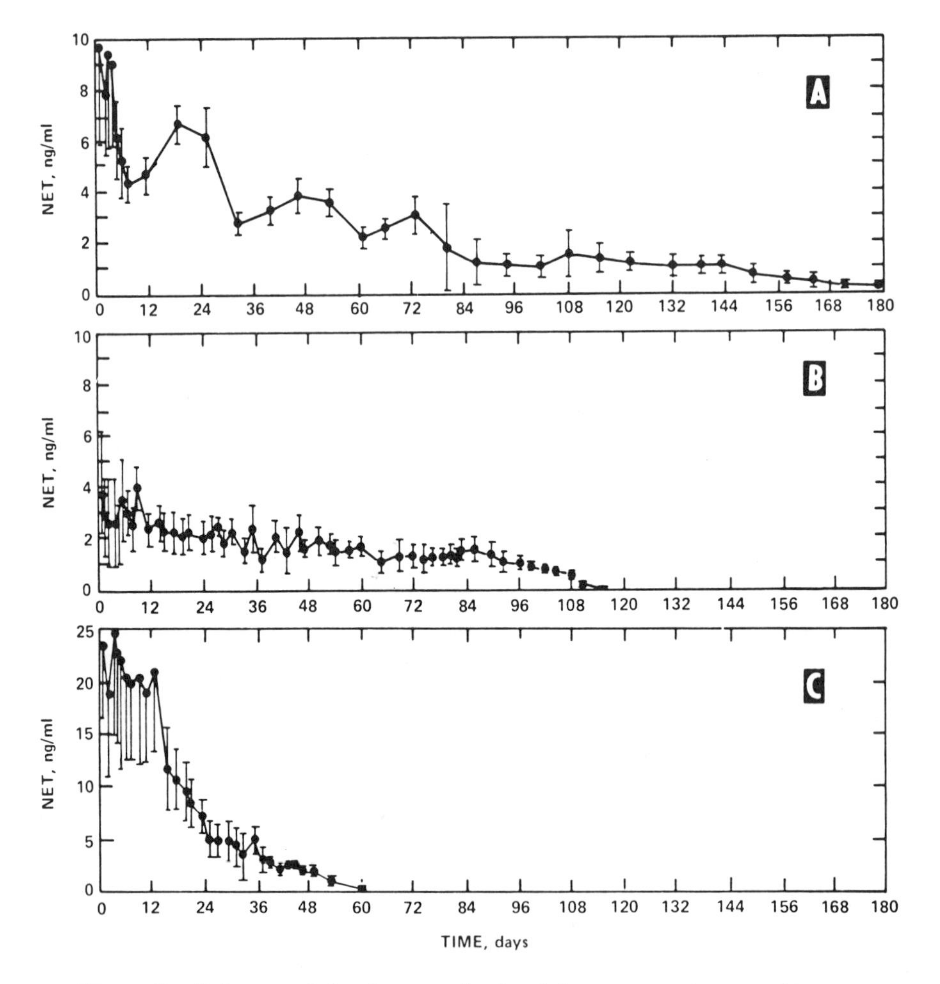

Figure 8. Baboon peripheral serum levels of immune-reactive norethisterone after intramuscular injection of 300 mg of poly(DL-lactic acid) microcapsules containing 75 mg of drug. Size of capsules: A, 10–240 μm; B, 65–124 μm; C, 10–40 μm. (Reprinted with permission from Ref. 36, p. 75. Copyright 1980.)

erosion, the process is confined to the surface of the device and is, for that reason, commonly referred to as surface erosion (38).

Drug release rate for matrices undergoing bulk erosion is nonlinear and difficult to predict because it is determined by a combination of diffusion and erosion. However, drug release from devices undergoing surface erosion is predictable and can lead to zero order-kinetics provided diffusional release of the drug is minimal and the overall surface area of the device remains essentially constant.

Many studies on drug release from bioerodible monolithic devices have been performed and the majority of these studies used poly(lactic acid) or copolymers of lactic and glycolic acids (1, 7). Because these polymers undergo a bulk degradation process, drug release is nonlinear and not amenable to a detailed mechanistic interpretation. Nevertheless, several systems having potential for the prolonged released of various therapeutic agents such as contraceptive steroids, narcotic antagonists, anticancer agents, and antimalarial agents have been demonstrated (39, 40, 41).

Because surface erosion results in constant and predictable rate of drug release, this type of erosion is clearly preferrable to bulk erosion. However, to achieve surface erodibility, a system must be devised in which the rate of polymer degradation at the surface of a device is very much faster than the rate of degradation in the interior.

One approach to surface erodibility is to prepare a polymer that contains linkages that are stable in base but are very labile in acid. Because one such linkage is an ortho ester, poly(ortho esters) are currently under intensive development as monolithic devices for zero order drug release (7).

Poly(ortho esters) were first disclosed in a series of patents assigned to the Alza Corporation (42-45) and were prepared by a transesterification reaction as follows:

EtO OEt (cyclic ortho ester) + HO-R-OH → -[O O-R]-$_n$ (cyclic ortho ester) + EtOH

Although the Alza poly(ortho ester) system has never been structurally identified other than by its tradename Chronomer, and later Alzamer, several publications provide a general description of the use of the polymer for the release of naltrexone (46) and contraceptive steroids (47, 48, 49).

Poly(ortho esters) have also been produced by the addition of diols to diketene acetals (50). Principally because of ease of monomer synthesis, polymers were prepared by the addition of various diols to 3,9-bis(methylene 2,4,8,10-tetraoxaspiro[5,5] undecane), R = H or 3,9-bis(ethylidone 2,4,8,10-tetraoxaspiro [5,5]undecane), R = CH_3.

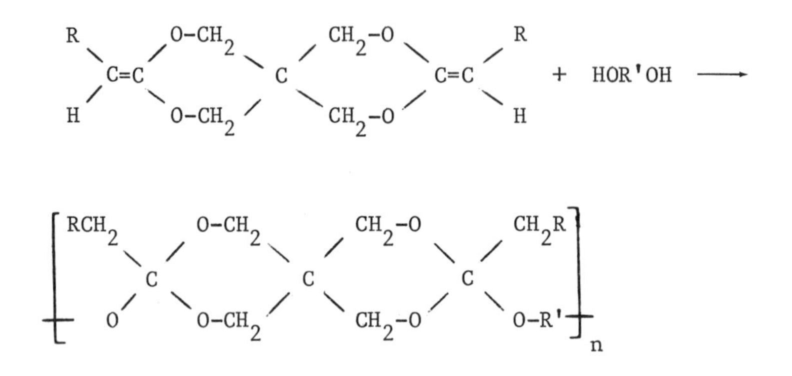

Because poly(ortho esters) are stable in alkaline environments, initial attempts to develop a surface-eroding system were based on the incorporation of sodium carbonate into the bulk material. Polymer erosion was expected to occur only at the outer surface of a solid device in which the incorporated basic salt is neutralized by the external buffer (51, 52). However, it was found that as a consequence of an osmotic imbibing of water caused by the incorporated water-soluble salt, a swelling front develops, and drug is released from the matrix by diffusion from the swollen polymer. Because poly(ortho esters) are stable in base, no polymer erosion takes place. Nevertheless, as shown in Figure 9, constant drug release is achieved.

The use of osmotically active neutral salts such as sodium sulfate, also shown in Figure 9, produces a constant rate of drug release for about 60 days, after which drug release rate accelerates up to drug depletion (53, 54), The number below the arrow indicates weight loss at 160 days. Clearly, polymer erosion significantly lags drug release. Furthermore, the rate of drug release observed with sodium sulfate is not that expected from a simple movement of a swelling front, but instead indicates that the active surface area increases with time. This has been verified by observations of a substantial increase in size and by scanning electron micrograph, which revealed a foam-type interior and heavy surface cratering.

Current work is aimed at producing poly(ortho ester) systems in which drug release and polymer erosion takes place concomitantly by using incorporated agents that are capable of lowering the pH at the polymer-water interface.

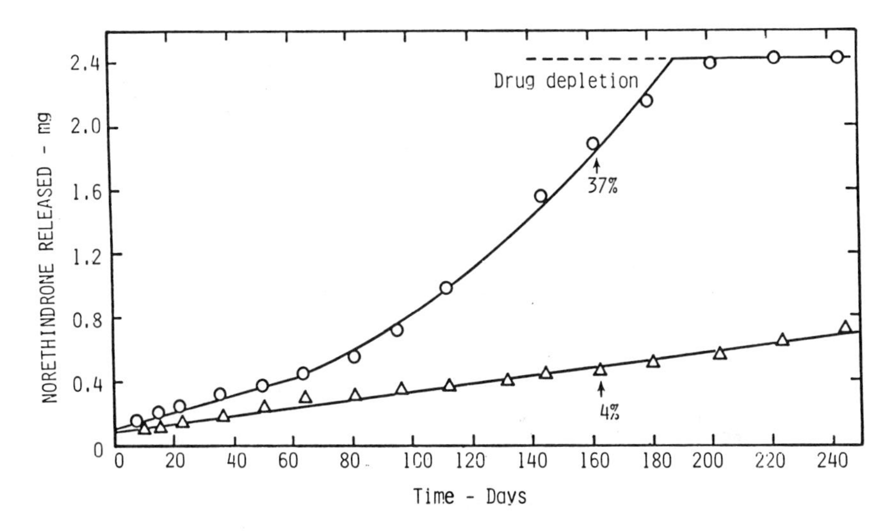

Figure 9. Norethindrone (NE) release from 3,9-bis(methylene 2,4,8,10-tetraoxaspiro[5,5]undecane)/1,6-hexane diol poly(ortho ester), 6.3-mm-diameter discs at pH 7.4 and 37 °C. Key: ○, 10 wt% NE, 10 wt% Na_2SO_4, 0.6-mm-thick disc, total drug content 2.4 mg; △, 10 wt% NE, 10 wt% Na_2CO_3, 1.2-mm-thick disc, total drug content 4.0 mg.

LITERATURE CITED

1. Heller, J. Biomaterials 1980, 1, 51-57.
2. Tobolsky, A. V. Nature 1967, 215, 509-510.
3. Davis, P.; Tabor, B. E., J. Polymer Sci.: Part A 1963, 1, 799-815.
4. Coopes, I. H. J. Polymer Sci. Part A-1 1970, 8, 1793-1811.
5. Robinson, I. D. J. Appl. Polymer Sci. 1964, 8, 1903-1918.
6. Goldman, R.; Goldstein, L.; Katchalsky, E. "Biochemical Aspects of Reactions on Solid Supports"; Stark, G. R., Ed. Academic Press: New York, 1971, pp 1-78.
7. Heller, J. "Pharmaceutical Applications of Controlled Release Drug Delivery Systems"; Langer, R. S.; Wise, D. L., Eds. CRC Press, in press.
8. Heller, J.; Baker, R. W.; Helwing, R. F.; Tuttle, M. E. J. Biomed. Mater. Res. to be published.
9. Lappas, L. C.; McKeehan, W. J. Pharm. Sci. 1962, 51, 808.
10. Lappas, L. C.; McKeehan, W. J. Pharm. Sci. 1965, 54, 176-181.
11. Lappas, L. C.; McKeehan, W. J. Pharm. Sci. 1967, 56, 1257-1261.
12. Heller, J.; Baker, R. W.; Gale, R. M.; Rodin, J. O. J. Appl. Polymer Sci. 1978, 22, 1991-2009.
13. Goldenberg, I. S. Surgery 1959, 46, 908-912.
14. Frazza, E. J.; Schmitt, E. E. J. Biomed. Mater. Res. Symposium 1971, 1, 43-58.
15. Craig, P. H.; Williams, J. A.; Davis, K. W.; Magoun, A. D.; Levy, A. J.; Bogdansky, S.; Jones, J. P. Jr. Surg. Gynecol. Obstet. 1975, 141, 1-10.
16. Salthouse, T. N.; Matlaga, B. F. Surg. Gynecol. Obstet. 1976, 142, 544-550.
17. Chu, C. C. J. Biomed. Mater. Res., 1981, 15, 19-27.
18. Nathan, H. S.; Nachlas, M. M.; Solomon, R. D.; Halpern, B. D.; Seligman, A. M. Ann. Surg. 1960, 152, 648-659.
19. Leonard, F.; Kulkarni, R. K.; Nelson, J.; Brandes, G. J. Biomed. Mater. Res. 1967, 1, 3-9.
20. Leonard, F.; Hodge, J. W. Jr.; Houston, S; Ousterhout, D. K. J. Biomed. Mater. Res. 1968, 2, 173-178.
21. Gilbert, H; Miller, F. F.; Averill, S. J.; Schmidt, R. F.; Stewart, F. D.; Trumbull, H. L. J. Am. Chem. Soc. 1954, 76, 1074-1076.
22. Leonard, F.; Kulkarni, R. K.; Brandes, G; Nelson, J.; Cameron, J. J. J. Appl. Polymer Sci. 1966, 10, 259-272.
23. Wade, C.W.R.; Leonard, F. J. Biomed. Mater. Res. 1972, 6, 215-220.
24. Venzin, W. R.; Florence, A. J. J. Biomed. Mater. Res. 1980, 14, 93-106.

25. Cameron, J. L; Woodward, S. C.; Herrmann, J. B. Arch. Surg. 1064, 89, 546.
26. Woodward, S. C.; Herrmann, J. B.; Leonard, F. Fed. Proc. 1964, 23, 495.
27. Woodward, S. C.; Herrmann, J. B.; Cameron, J. L; Brandes, G.; Pulaski, E. J.; Leonard, F. Ann. Surg. 1965, 162, 113-122.
28. Lehman, R.A.W.; Hayes, G. J.; Leonard, F. Arch. Surg., 93, 441-446.
29. Leonard, F.; Kulkarni, R. K.; Nelson, J.; Brandes, G. J. Biomed. Mater. Res., 1967, 1, 3-9.
30. Stevenson, T. C.; Taylor, D. S. J. Obstet. Gynaecol. Brit. Comm. 1972, 79, 1028-1039.
31. Hoffman, A. S.; Hale, T. J.; Nightingale, J.A.S.; Halbert, S. A.; Buckles, R. G. Presented at Second World Congress of Chemical Engineering and World Chemical Exposition, Montreal, Canada, October 4-9, 1981.
32. Yolles, S; Eldridge, J. E.; Woodland, J.H.R., Polymer News 1970, 1, 9-15.
33. Beck, L. R.; Pope, V. Z.; Cowsar, D. R.; Lewis, D. H.; Tice, T. R., Contracept. Deliv. Syst. 1980, 1, 79-86.
34. Pitt, C. G.; Marks, T. A.; Schindler, A. "Controlled Release of Bioactive Materials"; Baker, R. W., Ed.; Academic Press: New York, 1980, pp. 19-43.
35. Pitt, C. G.; Schindler, A. "Pharmaceutical Applications of Controlled Release Drug Delivery Systems"; Langer, R. S.; Wise, D. L. Eds.; CRC Press, in press.
36. Beck, L. R.; Cowsar, D. R.; Lewis, D. H., "Biodegradables and Delivery Systems for Contraception"; Hafez, E.S.E.; van Os, W.A.A., Eds.; G. K. Hall Medical Publishers: Boston, 1980, pp 63-81.
37. Higuchi, T. J. Pharm. Sci. 1961, 50, 874-875.
38. Heller, J.; Baker, R. W. "Controlled Release of Bioactive Materials"; Baker, R. W., Ed; Academic Press: New York, 1980, pp. 1-17.
39. Wise, D. L.; Schwope, A. D.; Harringan, S. E.; McCarthy, D. A.; Howes, J. F. "Polymeric Delivery Systems"; Kostelnik, R. J., Ed.; Gordon and Breach Science Publishers: New York, 1978, pp. 75-86.
40. Wise, D. L; Gregory, J. B.; Newberne, P. M.; Bartholow, L. C.; Stanbury, J. B. "Polymeric Delivery Systems"; Kostelnik, R. J., Ed.; Gordon and Breach Science Publishers: New York, 1978, pp. 121-136.
41. Yolles, S; Sartori, M. F. "Drug Delivery Systems", Juliano, R. L., Ed.; Oxford University Press, New York, 1980, pp. 84-111.
42. Choi, N. S.; Heller, J. U.S. Patent 4,093,709, June 6, 1978.
43. Choi, N. S.; Heller, J. U.S. Patent 4,131,648, December 26, 1978.

44. Choi, N. S.; Heller, J. U.S Patent 4,138,344, February 6, 1979.
45. Choi, N. S., Heller, J. U.S. Patent 4,180,646, December 25, 1979
46. Capoza, R. C.; Sendelbeck, L.; Balkenhol, W. J. "Polymeric Delivery System"; Kostelnik, R. J. Ed.; Gordon and Breach Science Publishers, New York, 1978, pp. 59-70.
47. Benagiano, G.; Gabelnick, H. L. J. of Steroid Biochem. 1979, 11, 449-455.
48. Pharriss, B. B.; Place, V. A.; Sendelbeck, L; Schmitt, E. E. J. Reproductive Med. 1976, 17, 91-97.
49. Benagiano, G.; Schmitt, E.; Wise, D.; Goodman, M. J. Polymer Sci., Polym. Symp. 1979, 66, 129-148.
50. Heller, J; Penhale, D.W.H.; Helwing, R. F. J. Polymer Sci. Polymer Lett., 1980, 18, 619-624.
51. Heller, J.; Penhale, D.W.H.; Helwing, R. F.; Fritzinger, B. K.; Baker, R. W. Chem. Eng. Progress Symp. Series No. 206, 1981, 77, 28-36.
52. Heller, J; Penhale, D.W.H.; Helwing, R. F.; Fritzinger, B. K. Polymer Eng. Sci. 1981, 21. 727-731.
53. Heller, J.; Penhale, D.W.H.; Helwing, R. F.; Fritzinger, B. K. Controlled Release Delivery Systems"; Roseman, T. J.; Mansdorf, S. Z., Eds.; Marcell Dekker, New York, in press.
54. Heller, J.; Penhale, D.W.H.; Fritzinger, B. K.; Rose, J. E.; Helwing, R. F. Contracept. Deliv. Syst. in press.

RECEIVED October 25, 1982

28

Radiation-Induced Polymerization Reactions for Biomedical Applications

G. R. HATTERY and V. D. McGINNISS

Battelle Laboratories, Columbus, OH 43201

A partial survey of the growing field of radiation induced polymerization methods used in synthesizing both natural and synthetic polymers will be discussed in this paper. To be included are such subjects as:

a) Grafting reaction
 - e.g., e-beam and ^{60}Co to graft substituents amenable to heparization.

b) Crosslinking reaction
 - e.g., ultra high molecular weight polyethylene (UHMWPE) for joint replacement.

c) Plasma polymerization
 - e.g., coating of polymeric surfaces with pinhole free thin layers of material to improve specific mechanical properties.

A review of both past and present work in the field will be included along with several illustrations of various methods utilizing different aspects of radiation chemistry to produce biomaterials.

The interaction of electromagnetic radiation with certain types of organic substrates has found widespread interest in biomedical research related applications. Many such studies involve interaction of electromagnetic radiation with organic substrates to develop crosslinked/insoluble network structures. For example, a preformed thermoplastic polymer upon direct interaction with certain types of ionizing radiation can develop crosslinked or network structures having higher melting points, greater tensile strengths and better chemical resistance than the starting thermoplastic polymer material. It is also possible to impregnate certain thermoplastic polymers with

0097-6156/83/0212-0393$06.00/0

low molecular weight compounds (drugs, chemical agents, etc.) followed by crosslinking via radiation processing techniques to produce a composite structure capable of controlled release of the encapsulated compounds through the polymer network (Figure 1).

Similar types of radiation induced polymerization reactions can be carried out through the use of liquid monomer, oligomer and polymer compositions containing reactive vinyl components (Figure 2).

In general, radiation induced polymerization reactions involve consideration of at least four major variables, namely:

(1) type of radiation source
(2) type of organic substrate to be irradiated
(3) kinetics of the radiation induced polymerization or crosslinking reactions
(4) network formation and final chemical, physical and mechanical properties of the crosslinked structure.

The types of radiaton sources most encountered in biomedical applications are outlined in Table I. Mechanisms associated

Table I. Radiation Sources

^{60}Co	γ-RAYS
UV-Vis	200-700 nm WAVELENGTHS OF LIGHT
PLASMA	RADIO FREQUENCIES

with radiation induced polymerization reactions involve free radical intermediates and can be depicted in the generalized outlines shown in Figures 3 and 4. More complete discussions of radiation processing technologies can be found in references (1-5).

Discussion of radiation induced polymerization reactions in this article will focus on the following biomedical applications:

- Bone Prothesis
- Polymer - Blood Compatibility
- Immobilization of Reaction Centers
- Immobilization of Enzymes
- Controlled Release of Drugs

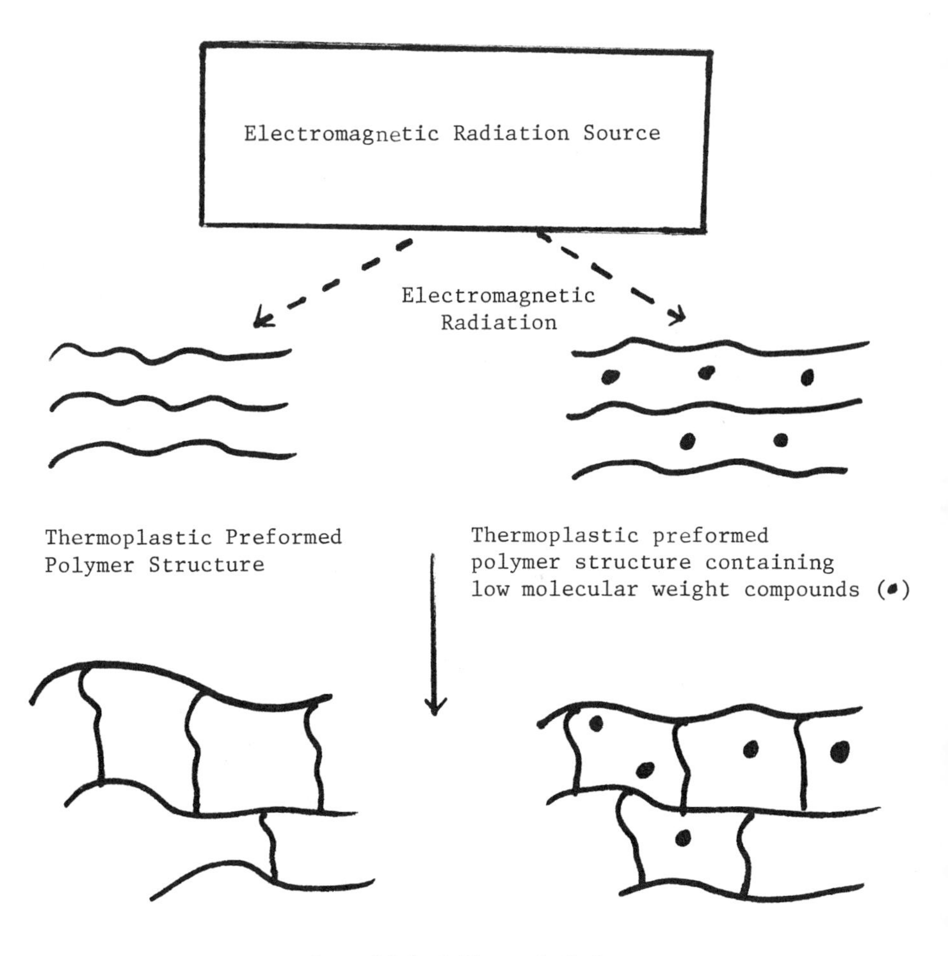

Figure 1. Interaction of electromagnetic radiation with preformed polymer structures.

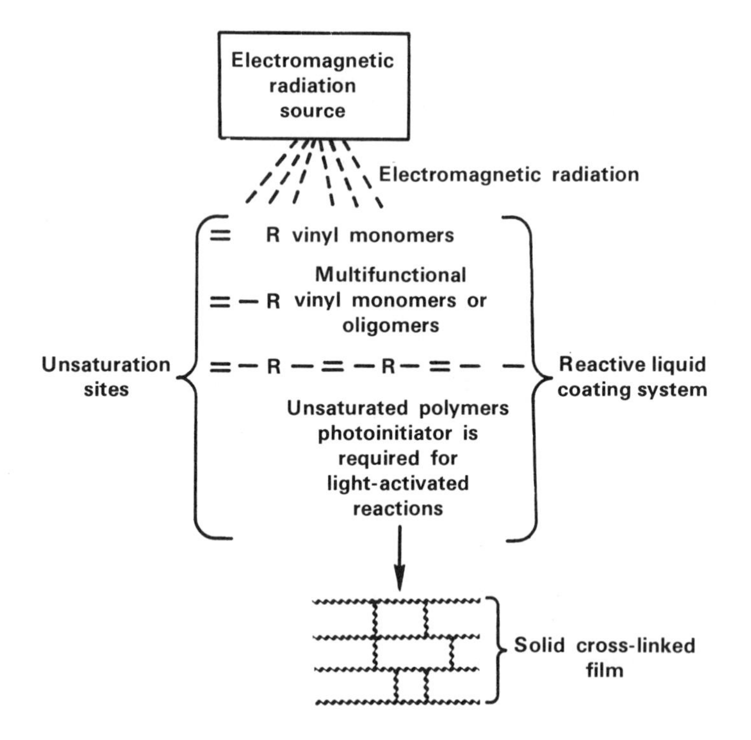

Figure 2. Radiation curing concepts (curing of reactive coatings).

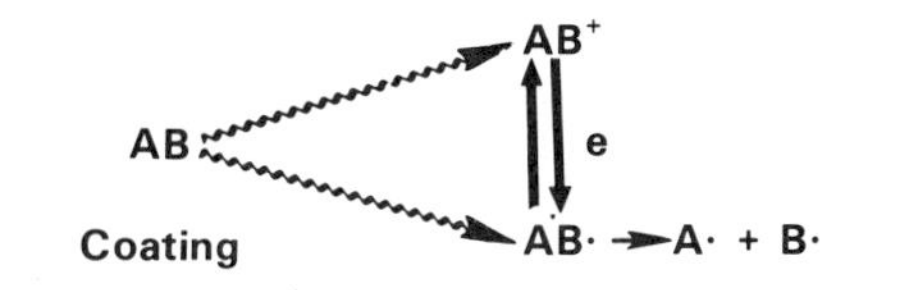

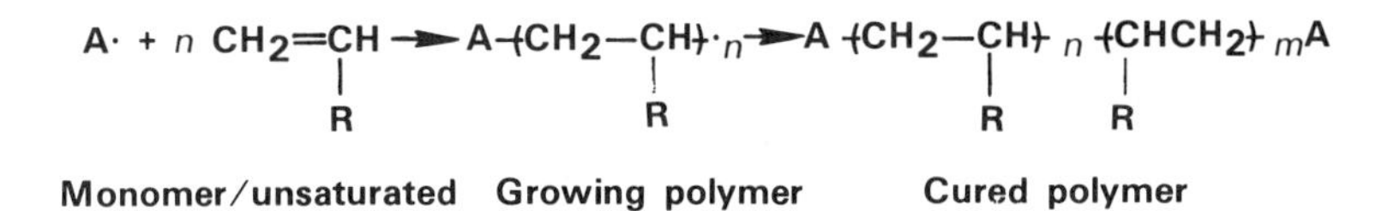

Figure 3. Ionizing or high-energy electron curing mechanisms.

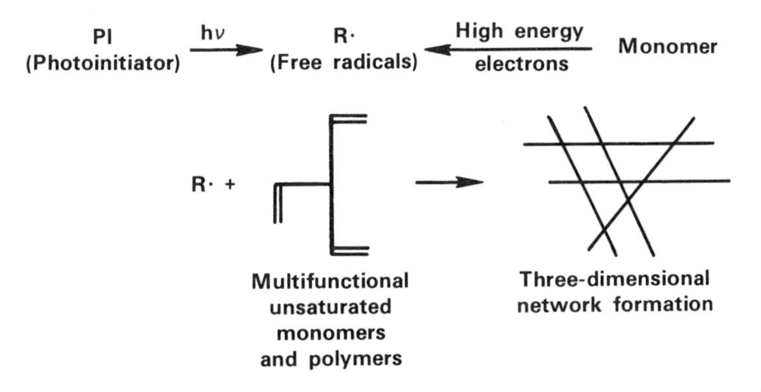

Figure 4. Light-induced or high-energy electron curing mechanisms.

RADIATION TREATMENT OF POLYOLEFINS

Polyolefins, in particular, polyethylenes, have been found to be extremely inert during clinical testing of implant materials. (6,7) These materials are often wholly accepted by the body tissues with little evidence of foreign-body cell reaction in the areas of implantation. For this reason, HDPE (high density polyethylene) has been used in a multitude of systems for biomedical applications. While the clinical properties of these materials are more than acceptable, several problems have arisen with the physical-mechanical properties of the initial material. Several groups have been involved with analyzing the effect of wear on these properties. (7,8,9) Some have attempted to determine whether any increase in hardness or decrease in cold flow could be correlated to irradiation. (7) Since sterilization of most of the joint implants involving polyolefins is done using radiation, most often Co-60, groups have looked at the properties both prior to and following the sterilization dose (10,11) Results in all of these studies have shown slight but inconsequential increases in the mechanical properties. Indeed, studies have shown that significant enhancement of mechanical properties does not occur until the HDPE is subjected to doses greater than 10 times the sterilization dose of 2.5 MRads. (12,13) At these dose rates, coefficient of friction drops noticeably for HMPE.

Very limited work has also been done on the difference between exposure to low flux Co-60 source irradiation and exposure to E-beam radiation of HDPE (10) Results have indicated that the surface of the material is affected differently by E-beam than by similar doses of Co-60. It appears that the E-beam irradiated surface is more susceptible to stress cracking and embrittlement than that of the Co-60 irradiated surface.

Realizing that exposure of HDPE to high levels of ionizing radiation in an inert atmosphere has little effect on the properties of the material, some groups (7,14,15) have begun to study the interaction of HDPE and ionizing irradiation in a reactive atmosphere.

One of the most active groups in this arena has been the University of Pretoria group led by Grobbelaar. (7,16) They have studied the properties of HDPE in the presence and absence of gaseous crosslinking agents (Figure 5). Several significant results can be drawn from their studies on the effect of acetylene and acetylene/chlorotrifluoroethylene on irradiated HDPE.

Their studies indicated that diffusion of the crosslinking agents within the HDPE matrix reached a depth of only 0.3 mm

(Figure 6). Extension of both time allowed for diffusion and total irradiation did nothing to change this value. Thus, they have found that HDPE composite material can be produced with a high crosslink density at the surface to impart increased abrasion resistance and limit cold flow while the limited crosslinking in the bulk of the material leads to continued good shock and embrittlement resistance (Figure 7). It is expected that this work will stimulate interest in studying improvement in HDPE properties through reactive gaseous diffusion. Increased wear life of these materials will be of great assistance to those who must undergo total joint replacement at an early age.

POLYMER-CERAMIC COMPOSITE MATERIALS

The use of ceramic and polymeric constituents in special alloys and blends has become widespread in recent years. These "polymeric cements" have been applied to a variety of uses from pavement repair and maintenance (17) to artificial teeth (18,19,20) and endosseous implants. (21)

Some of these techniques have relied on initiating the polymerization reaction by ionizing radiation. In particular, Kamel (21) at Drexel University has been developing a bone restorative using an alumina- poly(acrylic acid) composite produced by exposing an aqueous mixture of the blend to γ radiation of 1 MRad (Figure 8). The porosity and crosslink density of the system were varied over wide ranges by varying monomer concentration and a heat treatment step to form anhydrides. Other control parameters included both chemical reactions and physical interactions (Figure 9). Variation of these properties caused a concomitant change in mechanical properties and water absorption. Such control allows the composite to be "tailored" to a specific use. This material can be easily fabricated and adapted to a number of different socket geometries while allowing bone growth into the porous material. Initial studies have shown that neither the biocompatibility of the material nor the resistance to body fluid diffusion are affected to a major extent by this crosslinking method. Long term implantation studies are presently underway to determine the ultimate effect of the material on the implant area.

RADIATION INDUCED GRAFTING FOR INCREASED BLOOD COMPATIBILITY

Many groups have studied the grafting of different functionalities onto the backbone structure of different polymers. Some of these have involved covalent or ionic coupling of bio-active compounds to inert substances (22,23) while others have been concerned with exposing the monomer and polymeric substrate to ionizing radiation. (24,25,26,27)

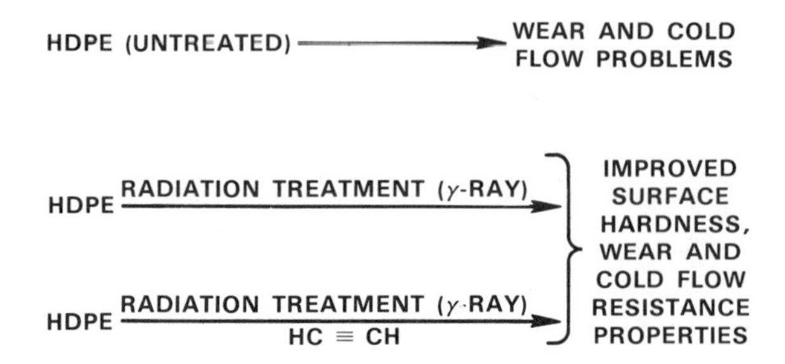

Figure 5. Improved HDPE prostheses through the use of radiation treatment.

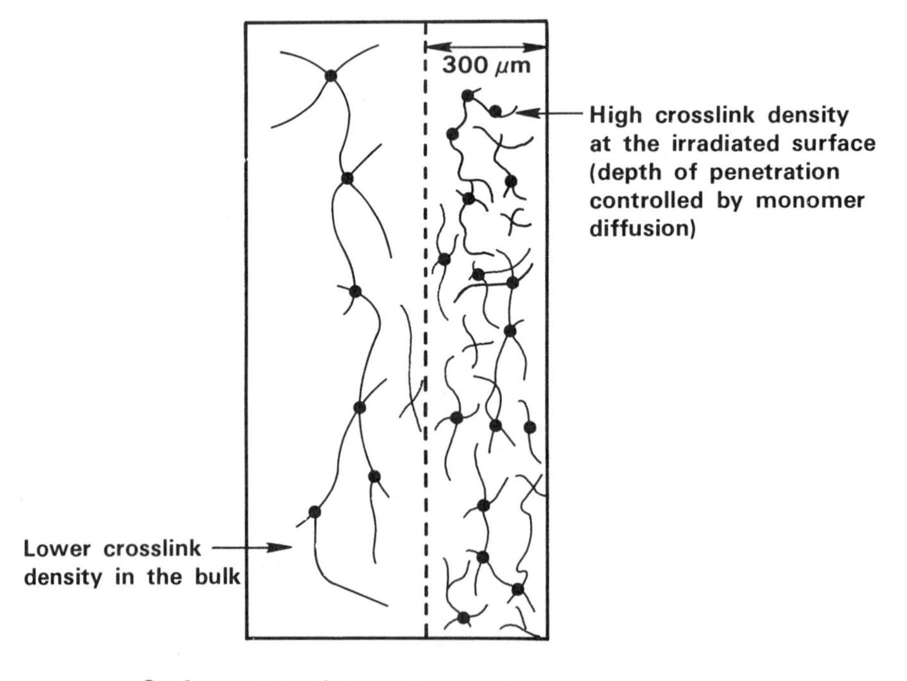

Figure 6. Effect of radiation treatment on bulk and surface of HDPE + CH ≡ CH.

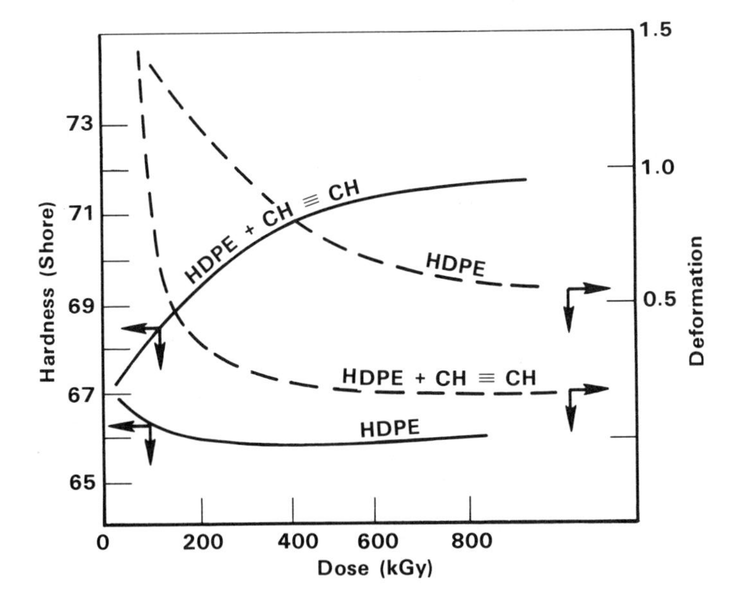

Figure 7. Radiation treatment of HDPE with and without reactive acetylene monomer. (Reprinted from Ref. 7, copyright 1978, and Ref. 16, copyright 1977.)

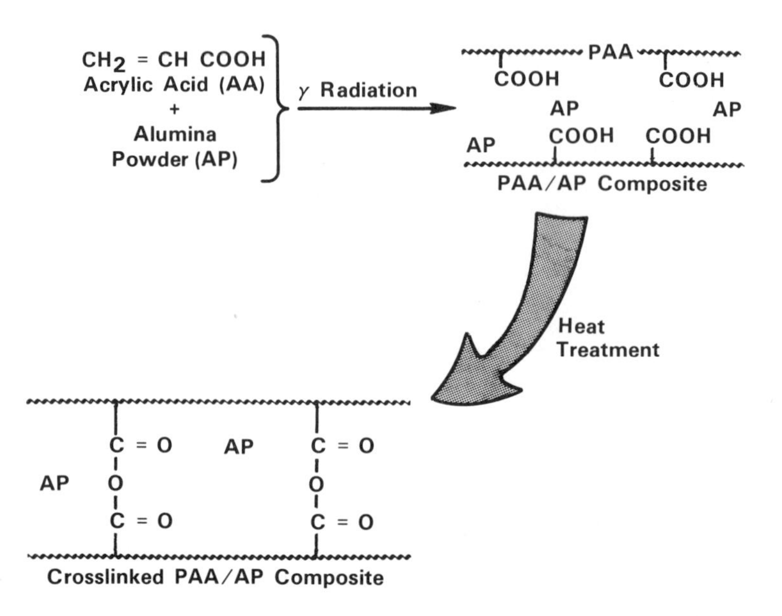

Figure 8. Radiation-processed polyacrylic acid (PAA) composites for bone restoration. (Reprinted from Ref. 21. Copyright 1977.)

OBJECTIVE DEVELOP COMPOSITE MATERIALS HAVING MAXIMUM FLEXIBILITY IN VARIATION OF MECHANICAL PROPERTIES WHILE MAINTAINING BIOCOMPATIBILITY

CONTROL FACTORS

- ANHYDRIDE FORMATION IS RELATED TO CROSSLINK DENSITY
- CROSSLINK DENSITY CONTROLS WATER SWELLING OF COMPOSITE
- STRONG POLYMER—FILLER INTERACTION

Figure 9. Radiation-processed polyacrylic acid (PAA) composites for bone restoration.

A brief survey of the field shows that by far the most prevalent polymer substrate in use is the silicone family.

Wilson, at Bishop College, and Eberhart and Elkowitz at University of Texas (27) have irradiated a silicone substrate in the presence of chloromethylstyrene monomer to produce a reactive graft polymer that can be quarternized with pyridine and reacted with sodium heparin to produce a thromboresistant heparinized product that has a higher blood compatibility than the untreated silicone. The same group has used essentially the same methods to create a heparin grafted polyethylene surface.

The method developed by Wilson, et al, yields an ionically bound heparin moiety that affects the anticoagulant activity of the heparin molecule to a lesser extent than does the traditional method (28) of simple solution coating of material. Tests have shown that the maximum heparin removal rate in normal saline is sufficient to prevent clotting in hollow fiber artificial kidneys as predicted by Schmer (29). Future work will develop around a "purified" heparin prepared by Rosenberg and Lam (30) where one third of the starting mass of heparin contains 85% of the anticoagulant activity.

Several other groups have also studied the blood compatibility of silicones with various radiation grafted copolymer constituents. Chapiro, et al (25) have grafted N-vinyl pyrrolidone using Co-60 onto silicone in both "bulk" and "solution" type reactions. It is interesting to note that similar work using E-beam radiation has not been as successful (26). Chapiro's studies have shown improvement in blood compatibility for samples with a grafting weight increase of greater than 33% (Figure 10). Studies also showed that the grafting percent could be controlled by varying the ratio of n-vinyl pyrrolidone in the solvent. However, it was found that above approximately 30% grafting, the silicone becomes brittle and loses mechanical properties. Attempts are presently underway to limit the depth of grafting of N-vinyl pyrrolidone to just the surface of the tubes so that the original properties of the silicone can be retained.

Other groups have concentrated on the acrylate and methacrylate hydrogel type materials. Dincer (22) has shown that heparin can be attached covalently to crosslinked beads of polymethyl acrylate to yield increased blood compatibility at very low rates of desorption from the surface. These findings are in line with Salzman (32), Merrill (33) and Wong (34) statements that the antithrombogenic effect does not depend on leaching of heparin from a surface into the blood stream. Further work by Hattery (35) showed that radiation grafting of methyl acrylate to silicone followed by heparinization by the method of Flake (23) and Dincer (22) could yield a surface that exhibited good anticlotting properties. Further testing or reporting of these results have been delayed until a satisfactory method for producing a uniform surface coating has been developed.

Another group deeply interested in radiation grafting of acrylates and methacrylates is that of Hoffman, et al at the University of Washington. (26) They have applied Co-60 radiation to the production of hydrogel type biomaterials through grafting of specific monomers onto an inert polymer backbone. This work has looked at immobilization of biologically active molecules such as enzymes, albumins and plasma proteins onto grafted hydrogels of hydroxyethylmethacrylate (HEMA) and other functional grafting agents such as n-vinyl pyrrolidone. They have found that certain metal ions or polar organic solvents can be used to vary the amount and penetration of the graft monomer as well as changing the reaction rate.

These are by no means the only groups working in the field of blood compatibility. However, the ones cited are sufficient to provide insight into the progress of radiation synthesis as related to hemocompatibility research.

IMMOBILIZATION OF REACTION CENTERS

Another area of great interest in biomaterials research has been that of immobilization of reaction centers on an inert substrate to create reaction specific cites. One group (36) has been interested in stabilization of chloroplasts for use in solar energy development (Figure 11). Various hydrophillic and hydrophobic monomers were mixed with isolated chloroplasts in a specific buffer solution (Figure 12). The mixture was cooled to below -24C and irradiated with a Co-60 source to 1 MRad. After irradiation, residual monomer and chloroplast were washed leaving the immobilized product stored in a buffer solution.

The authors found that the hydrophilic monomer did not affect the evolution of oxygen as much as the hydrophobic monomer. In addition, it was reported that concentration of starting material and time after monomer addition decreased the effectiveness of the chloroplast in evolving O_2.

The immobilization technique allowed the chloroplast to remain active more than 7 times longer than the non-immobilized sample at activity levels as high as 40% of initial values. This increase in stable lifetime for photo-activity may be tapped for future use in conversion of solar energy to chemical and elec-

Graft Ratio %	#Tests	Clotted Tubes	Unclotted Tubes
0	16	7	9
16-22	5	4	1
31-39	19	7	12
41-47	9	1	8

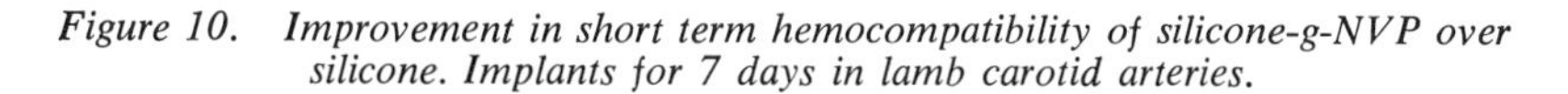

Figure 10. Improvement in short term hemocompatibility of silicone-g-NVP over silicone. Implants for 7 days in lamb carotid arteries.

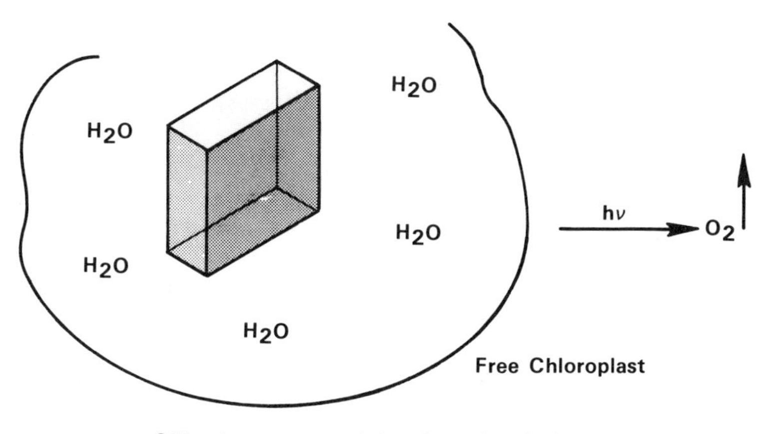

Figure 11. Stabilization of chloroplast. (Reprinted with permission from Ref. 36. Copyright 1981, John Wiley and Sons, Inc.)

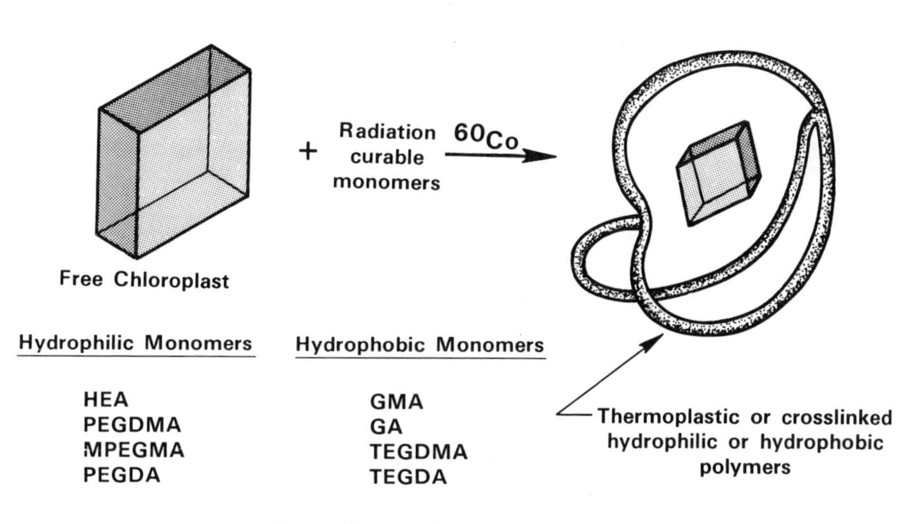

Figure 12. Stabilization of chloroplast.

trical energy. However, experiments presently are only at the lab stage and much work remains to be performed to determine the feasibility of the approach.

RADIATION INDUCED POLYMERIZATION REACTIONS FOR ENZYME IMMOBILIZATION

Several studies have been carried out on the immobilization of enzymes by radiation (ionizing and photochemical) induced polymerization reactions. (37-44) Most of these studies involved the use of combinations of hydrophilic or hydrophobic monomer/polymer substrates for the entrapment of the enzyme catalyst. A listing of typical hydrophilic/hydrophobic polymer materials is contained in Table II. The effects of hydrophilic/hydrophobic polymer properties in enzyme activity are

Table II. Radiation-Induced Immobilization of Enzymes

POLYMER	WATER CONTENT (%)
$-[CH_2-CH]_m-$ N-CH_2-CH_2-CH_2-C(=O)H (ring) PNVP	93.7
$-[CH_2-CH]_m-$ $CONH_2$ PAAm	84.8
$-[CH_2-CH]_m-$ COO $(CH_2)_2$ OH PHEA	45.9
$-[CH_2-C(CH_3)]_m-$ COO $(CH_2)_2$ OH PHEMA	26.0
$-[CH_2-C(CH_3)]_m-$ COO $(CH_2)_6$ OH PHDMMA	13.5
$-[CH_2-C(CH_3)]_m-$ COO $(CH_2CH_2O)_2$ OC $-[(CH_3)C-CH_2]_m-$ PDGDMA	2.5
$-[CH_2-C(CH_3)]-$ $COOCH_3$ PMMA	2.1

shown in Figure 13. In many cases there are very complex relationships among matrix porosity, polymer structure and network architecture. Enzyme initial activity in a hydrophilic polymer structure decreased rapidly with repeated use (enzyme leakage) but the hydrophobic polymer structures retarded loss of enzyme activity through repeated usage. Another advantage of using radiation processing techniques to immobilize enzyme materials on polymer surfaces is the ability to achieve very homogeneous and smooth composite structures which are not attainable by conventional polymerization techniques. (45)

CONTROLLED RELEASE OF MATERIALS FROM RADIATION POLYMERIZED COMPOSITES

The generalized concept for producing composite structures capable of controlled additive release properties involves 1) solution or dispersion of additives in reactive monomer/polymer systems, 2) subjecting the additive/monomer-polymer solution dispersion to radiation, and 3) formation of a crosslinked polymer network which encapsulates the specific agent (Figure 14). Typical monomer and crosslinking oligomers utilized in these types of studies are shown in Table III. The effects of

Table III. Controlled Release of Materials from Radiation-Polymerized Composites

MONOMERS

METHYL ACRYLATE (MA)
METHYL METHACRYLATE (MMA)
2-HYDROXYETHYL METHACRYLATE (HEMA)

CROSSLINKING OLIGOMERS

POLYETHYLENE GLYCOL #600 DIACRYLATE (PEGDA)
POLYETHYLENE GLYCOL #400 DIMETHACRYLATE (PEGDMA)
DIETHYLENE GLYCOL DIMETHACRYLATE (DEGDMA)
TRIMETHYLOLPROPANE TRIACRYLATE (TMPTA)
TRIMETHYLOLPROPANE TRIMETHACRYLATE (TMPTMA)

monomer chemical structure (hydrophilic/hydrophobic) on release of KCl from a cured composite system are shown in Table IV.

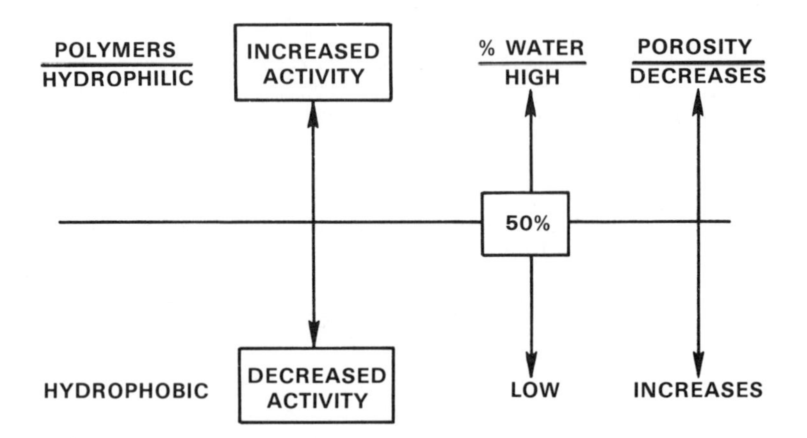

Figure 13. Activity and properties of radiation-induced immobilization of enzymes.

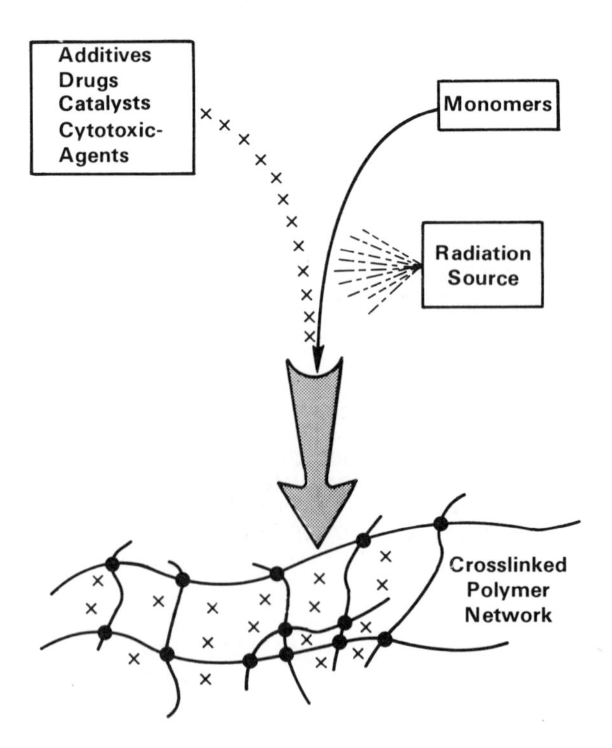

Figure 14. Controlled release of materials from radiation-polymerized composites.

Table IV. Controlled Release of KCl from Radiation-Polymerized Composites

HYDROPHILIC MONOMERS	$W(\%) = \frac{Ww}{Wp + Ww}$	RELEASE RATE ($\mu g/cm^2$ min $^{1/2}$)
HEMA	32.5	4.37
PEGDA	30.7	3.87
PEGDMA	21.6	3.01
HYDROPHOBIC MONOMERS		
MA	7.6	2.41
MMA	6.8	2.06
DEGDMA	3.8	0.69
TMPTA	3.3	0.33
TMPTMA	2.4	0.03

Those polymer composite systems having high W% values (hydrophilic) (Ww is the weight of water absorbed to saturate the polymer and Wp is the weight of the dry polymer) demonstrated higher release rate capabilities than those having low W% values. The functionality of the monomer/crosslinking oligomer also influences the rate of KCl release in that monofunctional monomers release faster than difunctional crosslinking oligomers which in turn have higher release rates than the trifunctional crosslinking oligomers. This may also be due to the fact that trifunctional crosslinking oligomers produce very tight network structures relative to higher molecular weight difunctional crosslinking oligomer structures. (46) The addition of hydrophilic thermoplastic additives to a relatively hydrophobic network can also strongly influence the release of KCl from its structure (Figure 15). (47) An example of another type of drug controlled release composite system is shown in Table V and Figure 16. (48)

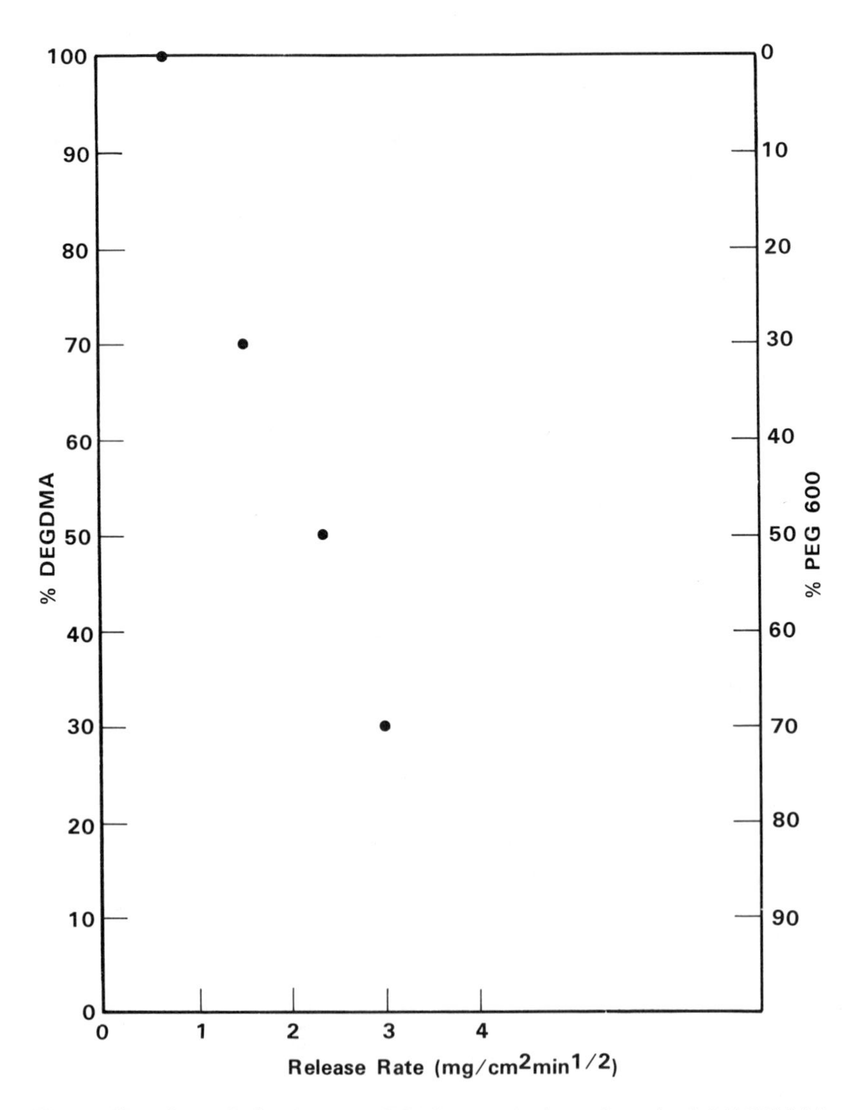

Figure 15. Controlled release of KCl from radiation-polymerized DEGDMA/ polyethylene 600 glycol (PEG 600).

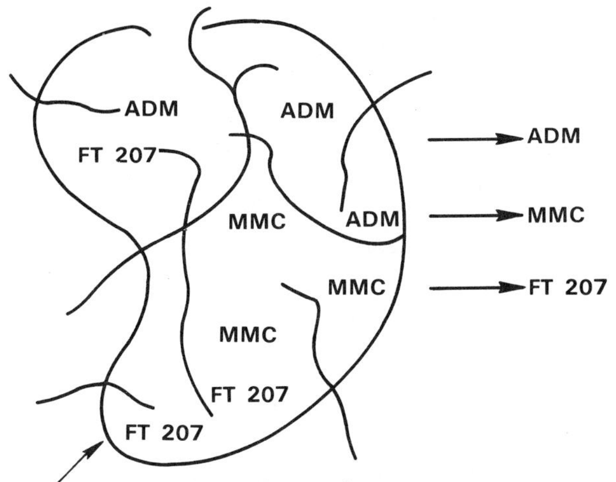

Control Factors Effecting Drug Release Rates

- **Relative concentrations of each drug**
- **Molecular weight of the drug**
- **Content and composition of drugs**
- **Polarity of the composite**
- **Network of the polymer**

Figure 16. Controlled release of multicomponent cytotoxic agents from radiation-polymerized composites.

Table V. Controlled Release of Multicomponent Cytotoxic Agents from Radiation-Polymerized Composites.

CYTOTOXIC AGENTS

1-(2-TETRAHYDROFURYL)-5 FLUOROURALIL (FT 207)
MITOMYCIN C (MMC)
ADRYAMYCIN (ADM)

MONOMERS	POLYMERS
DEGDMA	PMMA
TMPTMA	PEG 1000
	PMAC

Source: Reprinted from Ref. 48. Copyright 1980.

Plasma polymerization reactions have also been utilized to modify hydrogel polymer surfaces for enhanced controlled release capabilities. In one study an ummodified polymer hydrogel exhibited very rapid release rates for an entrapped drug in aqueous medium (Figure 17 and 18). Modification of the hydrogel surface with an Argon ion plasma (crosslinking or casing of the surface) or plasma polymerization of the surface in the presence of tetrafluoroethylene monomer (Figure 19) produced an improved structure for drug controlled release capabilities (Table VI).(49,50)

Table VI. Rates of Drug Release from Unmodified and Modified Hydrogel Structures

HYDROGEL STRUCTURE	DRUG RELEASE RATE
UNMODIFIED	41.8 $\mu g/hr\text{-}cm^2$
CASED	40 $\mu g/hr\text{-}cm^2$
TFE COATED	3-9 $\mu g/hr\text{-}cm^2$

Objective—to modify hydrogel surface so as to control the rate of drug diffusion for release at a specified rate

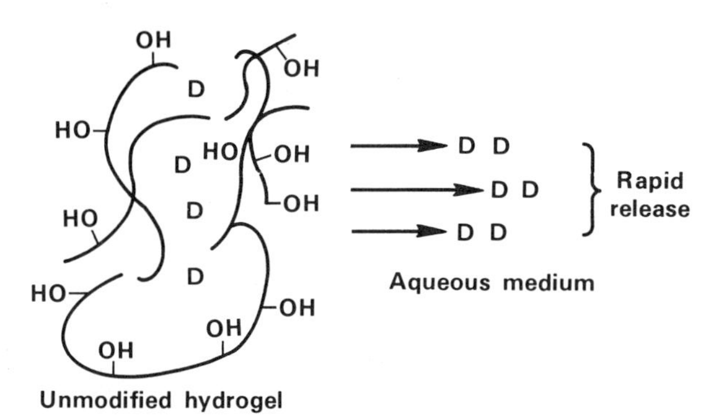

Figure 17. Hydrogel modification by plasma treatment.

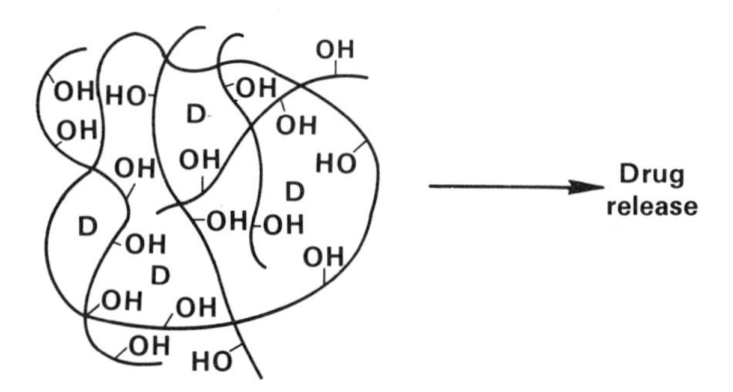

Drug (D) in swollen hydrogel

Drug = pilocarpine, progesterone

Figure 18. Control of drug release rate through hydrogels by plasma treatment. (Reprinted from Ref. 49. Copyright 1977.)

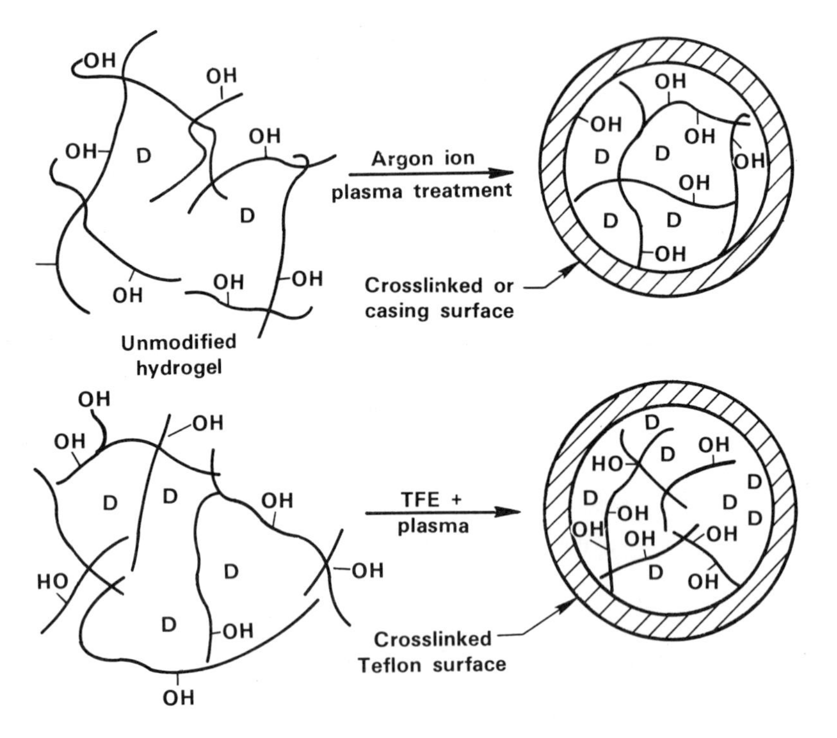

Figure 19. Hydrogel modification by plasma treatment.

LITERATURE CITED

1. Chapiro, A., "Radiation Chemistry of Polymeric Systems", Interscience Publisher, a division of John Wiley and Sons, Inc., New York, 1962.
2 Wilson, J. E., "Radiation Chemistry of Monomers, Polymers, and Plastics", Marcel Dekker, Inc., New York, 1974.
3. McGinniss, V. D., Nowacki, L. J. and Nablo, S. V., ACS Symposium, No. 107, page 51-70, 1979.
4. McGinniss, V. D., National Symposium on Polymers in the Service of Man, ACS, 15th State-of-the-Art Symposium, pages 175-180, 1980.
5. Hollahan, J. R., and Bell, A. T., "Techniques and Applications of Plasma Chemistry", John Wiley and Sons, 1974.
6. Charnley, J., Clin Orth Rel Res., 95 pg. 9 (1978).
7. Grobbelaar, C. J. et al, J. Bone Jt Sug, 60-B, pg. 370 (1978).
8. Dumbleton, J. H., Shen, C., Wear, 37, pg 279 (1976).
9. Nusbaum, H. J., et al., J. Appl Poly. Sc., 23, pg 777 (1979).
10. Hattery, G. R., unpublished report, Battelle Columbus Laboratories, 1980.
11. DuPlessis, T. A., Paper presented at Cong South Africa Assoc. Physicists Med Biology, Bellville, (1977).
12. Dumbleton, J. H., Shem, C., J. Appl. Poly. Sci., 18, pg 3493 (1974).
13. Dumbleton, J. H. et al, Wear, 29, pg 163 (1974).
14. Mitsui, H. et al, Polym. Jour., 13, pg 108 (1972).
15. Hagiwara, M. et al, Poly. Sci, B, Polym. Let., 11, pg 613 (1973).
16. Grobbelaar, C. J. Radiat. Phys. Chem, 9, pg 647 (1977).
17. Nandi, U. S., Personal Communication, 1981.
18. Hodash, M. et al, Oral Surg, 24, pg 831 (1967).
19. Young. F. A., J. Biomed Mater Res. Symp., 2, pg 281 (1972).
20. Reed, O. M. et al, J. Biomed Mater Res. Symp., 2, pg 296 (1972).
21. Kamel, I. L., Radiat. Phys. Chem. 9, pg 711 (1977).
22. Dincer, A. K., Sc. D. Thesis, "Covalent Coupling of Heparin to Synthetic Polymer Surfaces", MIT, 1977.
23. Flake, J., S. M. Thesis, "Aminolysis and Heparinzation of Polymethylacrylate for Biomedical Application", MIT, 1976.
24. Hattery G. R., S. M. Thesis, "Radiation Induced Crosslinking in Polymethyl acrylate", MIT, 1978.
25. Chapiro, A. et al, Radiat. Phys. Chem., 15, pg 423 (1980).
26. Hoffman, A. S. et al, Trans. Amer. Soc. Artif. Internal Organ, 18, pg 10 (1972).

27. Wilson, J. E., et al, J. Macromol Sci. Chem. A16, pg 769 (1981).
28. Grode, G. A., et al, J. Biomed Mater. Res. Symp., 3, pg 77 (1972).
29. Schmer, G., Trans. Am. Soc. Aritif. Intern. Organ, 19, pg 188 (1973).
30. Rosenberg, R. D., Lam. L. H., Ann. N.Y. Acad. Sci., 283, pg 404 (1977).
31. Yosada, H., Reforjo, M. F., J. Polym. Sci, Part A-2, pg 5093 (1964).
32. Salzman, E. W., Blood, 38, pg 509 (1971).
33. Merrill, E. W., et al, J. Appl. Physiol., 28, pg 723 (1970).
34. Wong, P. S. L., et al, Fed, Proc., p. 441 (1969).
35. Hattery, G. R., Unpublished report, MIT, (1977).
36. Yoshii, F., et al, Biotech Bioeng, 23, pg 833 (1981).
37. Kumakura, M. et al. Journal of solid-phase Biochemistry, 2, No 3, 279 (1977).
38. Kaetsu, I., Kumakura, M., and Yoshida, M., Biotech. Bioeng., 21, 867 (1979).
39. Ibid, 863 (1979).
40. Ibid, Polymer, 20, 3 (1979).
41. Ibid, 9 (1979).
42. Ichimura, K., and Watanabe, Poly. Sci. Poly. Chem., 18, 891 (1980).
43. Yoshida, M., Kumakura, M., and Kaetsu, I., J. Macromol. Sci-Chem., A14, No. 4, 541 (1980).
44. Ibid, 555 (1980).
45. Kaetsu, I., et al, Biomed. Mat. Res., 14, 199 (1980).
46. Yoshida, M., Kumakura, M. and Kaetsu, I., Polymer, 19, 1375 (1978).
47. Ibid, 1379 (1978).
48. Kaetsu, I., et al, Biomaterials, 1, 17 (1980).
49. Colter, K. D., Shen, M., and Bell, A. T., Biomat., Med. Dev. Art. Org., 5, No 1, 13 (1977).
50. Ibid, 1 (1977).

RECEIVED November 9, 1982

POLYMER CHEMISTRY

Mechanisms of Electron-Transfer Initiation of Polymerization: A Quantitative Study

MICHAEL SZWARC

University of California, San Diego, Department of Chemistry, La Jolla, CA 92093

Initiation of vinyl, vinylidene or diene polymerization is commonly visualized as an addition of some moiety, X, to a monomer, M, resulting in the formation of a reactive end-group capable of sustaining propagation of polymerization. Depending upon whether X is a radical, a cation, or an anion, the ensuing polymerization is propagated through a radical, cationic, or anionic mechanism, as shown schematically below:

$X^{\bullet} + C{:}C \rightarrow X.C.C^{\bullet}$ radical polymerization

$X^{+} + C{:}C \rightarrow X.C.C^{+}$ cationic polymerization

$X^{-} + C{:}C \rightarrow X.C.C^{-}$ anionic polymerization

In each case one end of a growing polymer is inert while the other is active.

Still another mechanism of initiation was proposed in the 1950's (1), namely, initiation by electron-transfer to monomer. In such an initiation, the electron is transferred from a suitable donor, A or $A^{\overline{\bullet}}$, to a monomer, M, converting it into a monomeric radical-anion

$$A\ (\text{or } A^{\overline{\bullet}}) + M \rightleftarrows A^{+}\ (\text{or } A) + M^{\overline{\bullet}} \qquad K_{tr}$$

The resulting monomeric radical-anions either dimerize into dimeric dianions,

$$2M^{\overline{\bullet}} \rightarrow {}^{-}M.M^{-} \qquad \underline{k}_d$$

e.g.,

$$2Ph.CH{:}CH_2^{\overline{\bullet}} \rightarrow Ph.\overline{C}H.CH_2.CH_2.\overline{C}H.Ph$$

or react with monomer, yielding then dimeric radical-anions,

0097-6156/83/0212-0419$06.00/0

$$M^{\cdot -} + M \rightarrow {}^{-}M.M^{\bullet} \qquad \underline{k}_a$$

e.g.,

$$Ph.CH:CH_2^{\cdot -} + Ph.CH:CH_2 \rightarrow Ph.\bar{C}H.CH_2.CH_2.\dot{C}H.Ph$$

The dimeric dianions initiate anionic polymerization propagated from both ends of the macromolecules,

$${}^{-}M.M^{-} \xrightarrow{M} {}^{-}M.M \;.....\; M.M^{-}$$

whereas the dimeric radical-anions could initiate anionic polymerization from one of their ends and simultaneously a radical polymerization from the other end. However, this is an unlikely event. Most probably, they disproportionate, dimerize, or are reduced to dianions, viz.

$$2\,{}^{-}M.M^{\bullet} \rightarrow {}^{-}M.M^{-} + 2M$$

$$2\,{}^{-}M.M^{\bullet} \rightarrow {}^{-}M.M.M.M^{-}$$

or

$${}^{-}M.M^{\bullet} + A \text{ (or } A^{\cdot -}) \rightarrow {}^{-}M.M^{-} + A^{+} \text{ (or A)}$$

In this paper I will discuss quantitative aspects of the electron-transfer equilibria, both homogeneous and heterogeneous, the fate of the monomeric radical-anions, and the methods leading to the desired thermodynamic and kinetic data. Since flash-photolysis was extensively used in our work, a few words about its basic feature are in place. A detailed description of its usage in our systems is given elsewhere (2).

Principles of Flash-Photolysis

Our experimental set-up is shown schematically in Fig. 1. The investigated solution, usually about 10^{-6} M in the active ingredient, is introduced into a cylindrical, 10 cm. long quartz cell with optically flat end windows. The cell is placed between two parallel flash lamps, separated from them by cuvets containing a light filtering solution that absorbs UV light. A properly collimated beam of monitoring light passes through the cell and is focused on the slit of monochromator. By choosing the desired wavelength, one allows the monochromatic light to reach a photomultiplier, and its output is amplified and fed into an oscilloscope.

On triggering the scope, but not the flash lamps, one gets on the screen of the oscilloscope a horizontal zero-line, showing how much light of a desired wavelength passes through the unphotolyzed solution. Thereafter the flash lamps are

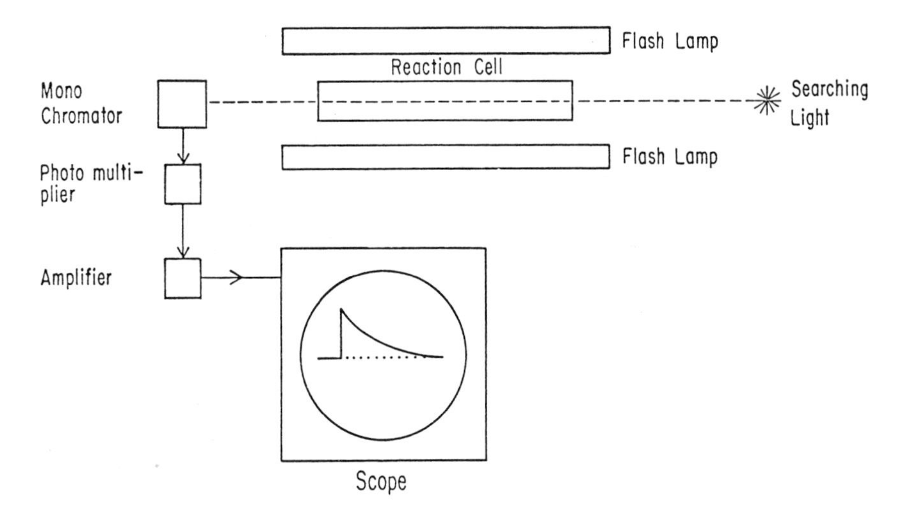

Figure 1. Experimental setup for flash photolysis. Scope registers changes in absorbance resulting from destruction of reagents and formation of products. For systems regenerating reagents, curve displayed on scope returns to its original baseline.

triggered, after the scope has been activated ∿ 100 μsec. earlier. The ensuing photolysis converts some of the original reagents into transient species. The reaction of the latter, taking place in the dark period following a flash, yields the final products. The progress of the dark reaction is revealed by changes in the absorbance of the monitoring light, displayed on the screen of the oscilloscope as a curve giving the intensity of the transmitted light as a function of time.

The systems discussed here are perfectly reversible -- for any wavelength the enhanced absorbance or bleaching decays to a zero-line as shown in Fig. 1. Therefore, flashing could be repeated over and over again, leaving unaltered the ultimate composition of the photolyzed solution. By varying the setting of the monochromator, a family of curves is obtained, each depicting the return of the photolyzed solution to its original state but monitored at a different wavelength. From such curves one calculates the optical density for each wavelength of the photolyzed solution at a chosen time, say 100 μsec. after each flash. This allows one to construct a difference spectrum -- the difference in the absorbance of the transients and the original reagents at that time.

Quantitative Treatment of Homogeneous Electron-Transfer Initiation

A difference spectrum obtained by flashing a THF solution of dimeric dianions of a α-methyl styrene, $K^+,\bar{C}(CH_3)(Ph).CH_2.CH_2.\bar{C}(CH_3)(Ph),K^+$ = $K^+,^-\alpha.\alpha^-,K^+$, is shown in Fig. 2 (3). It fades with time; however, its shape remains unchanged indicating that the transients formed by flash directly regenerate the original dimeric dianions, no other intermediates or products being formed.

The difference spectrum results from the absorbance of the intermediates and bleaching of the dimers. Hence, the spectrum of the intermediates is constructed by adding the known spectrum of the photolyzed dimers to the observed difference spectrum. Such a procedure is illustrated in Fig. 3. The resulting spectrum of the intermediates closely resembles that of α-methyl styrene radical-anions reported by three independent groups (4), who used pulse radiolysis in their studies. It follows that the photolysis leads to direct or indirect photo-dissociation of the dimeric-dianions into radical-anions of α-methyl styrene, α , i.e.,

$$K^+,\ ^-\alpha\alpha^-,K^+ \xrightarrow{h\nu} 2K^+,\alpha^{\overline{\cdot}}$$

and their dimerization takes place in the subsequent dark period.

Analogous studies of flash photolysis of THF solutions of

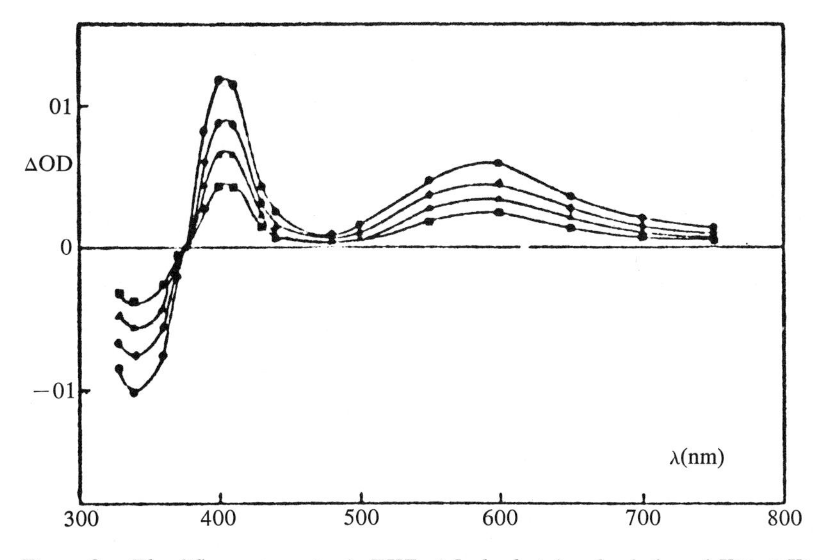

Figure 2. The difference spectra in THF of flash-photolyzed solution of $K^+,{}^-\alpha\alpha^-,K^+$ recorded at 5, 20, 40, and 80 ms after a flash. Key: ●, *5 ms;* •, *20 ms;* ▲, *40 ms;* ■, *80 ms.*

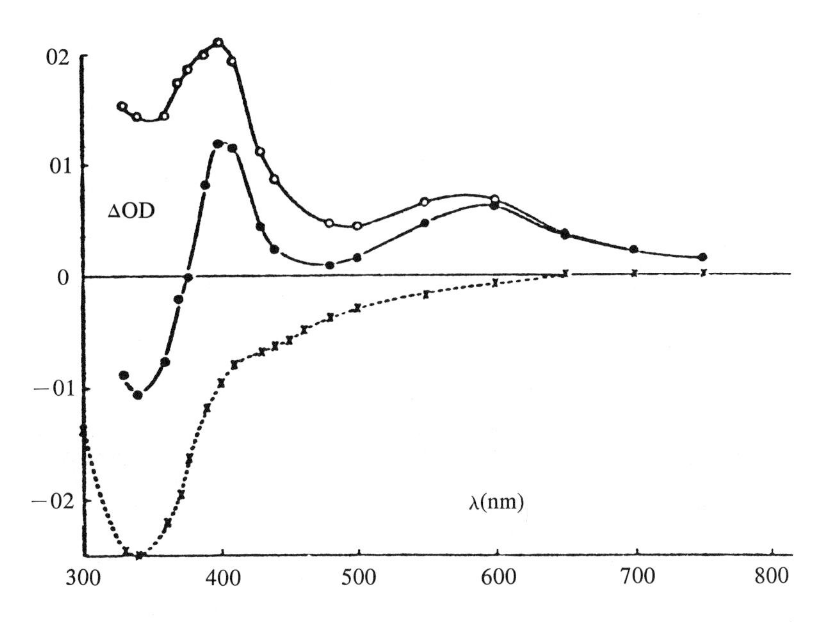

Figure 3. The calculated absorption spectrum in THF of $\alpha^{\overline{\cdot}}$, K^+ deduced from the difference spectrum and the known spectrum of $K^+,{}^-\alpha\alpha^-,K^+$. Key: ○, α^-; ●, *observed difference spectrum (~5 ms after flash);* ×, ${}^-\alpha\alpha^-$.

dimeric dianions of 1,1-diphenyl ethylene,
$Cat^+, \bar{C}(Ph)_2 \cdot CH_2 \cdot CH_2 \cdot \bar{C}(Ph)_2, Cat^+, {}^-D \cdot D^-, Cat^+$
led to the spectrum of the respective transient shown in Fig. 4 (5). Its similarity with the spectrum of 1,1-diphenyl ethylene radical-anions, $D^{\overline{\cdot}}$, reported by Hammill (6) who radiolyzed frozen 2-MeTHF solution of that hydrocarbon, implies that flash-photolysis of those dimeric dianions again leads to their photo-dissociation into the monomeric radical anions, viz.

$$Cat^+, {}^-D.D^-, Cat^+ \xrightarrow{h\nu} 2D^{\overline{\cdot}}, Cat^+$$

For both systems reciprocals of Δ optical density are linear with time. For the α-methyl styrene system, this is shown in Fig. 5 where 1/Δ (opt. density) at 340 nm, 400 nm, and 600 nm is plotted vs. time, and for the 1,1-diphenyl ethylene system in Fig. 6 giving plots of 1/Δ (opt. density) at 390 nm, 470 nm, and 750 nm vs. time. Linearity of those plots proves the bimolecular character of the processes through which the transients regenerate the dimeric dianions. Slopes of those plots provide, therefore, the values of the respective dimerization constants, k_d, divided by $\varepsilon \ell$, where ε is the respective effective molar absorbance and ℓ is the length of the cell. The effective molar absorbances were determined by various methods (3, 5) and thereafter the dimerization constants of $D^{\overline{\cdot}}, Cat^+$ and $\alpha^{\overline{\cdot}}, Cat^+$ were calculated. Their values, listed in Table 1, show their dependence on the nature of cation -- increasing with its radius. Significantly, the dimerization of radical-anions is slower than of small, neutral radicals the latter combination being diffusion controlled. Apparently, the repulsion of the negatively charged particles (for free ions) or of the unfavorably oriented dipoles (for ion-pairs) leads to the inertia. Surprisingly, $\alpha^{\overline{\cdot}}, Cat^+$ dimerize much slower than $D^{\overline{\cdot}}, Cat^+$, an unexplained finding.

The dimerization is therefore rate determining in the initiation of polymerization by aromatic radical-anions. Its rate is

$$k_d K_{tr}^2 \left\{[A^{\overline{\cdot}}, Cat^+].[monomer]/[A]\right\}^2$$

because due to the extremely short relaxation time of the electron-transfer equilibrium (this relaxation time is given by $\tau = 1/\left\{k_f[M] + k_b[A]\right\}$, k_f and k_b denoting the forward and backward rate constants of the electron transfer. Thus, τ is at the most 1 μsec. and the equilibrium concentration of the monomeric radical-anions is unperturbed by their dimerization. A numerical example is illuminating.

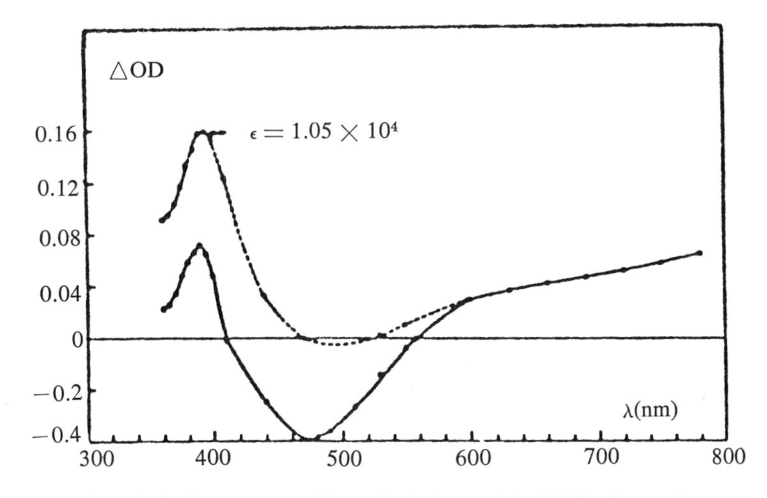

Figure 4. Flashphotolysis of ⁻DD⁻ with D (excess) in THF. Absorption spectrum of D·⁻, Na⁺ in THF.

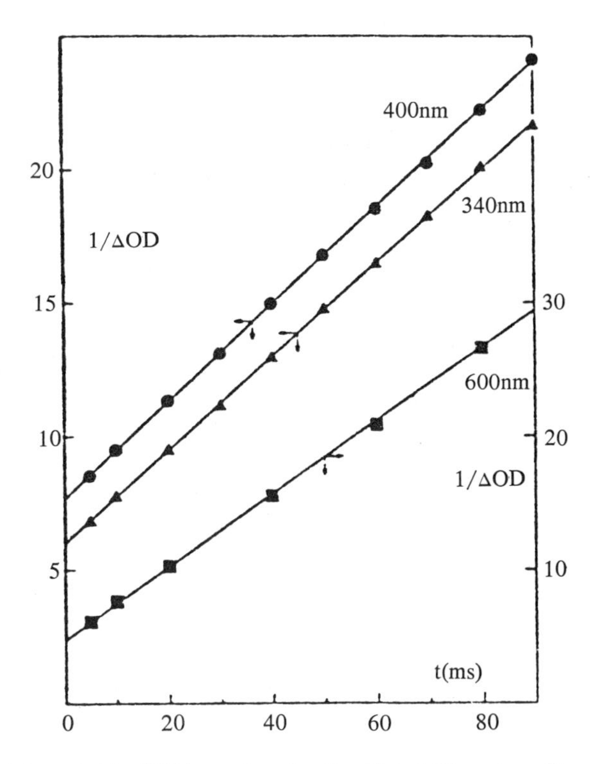

Figure 5. Plots of 1/Δ (OD) vs. time for the K⁺,⁻αα⁻,K⁺ system. Recorded at λ 340, 400, and 600 nm.

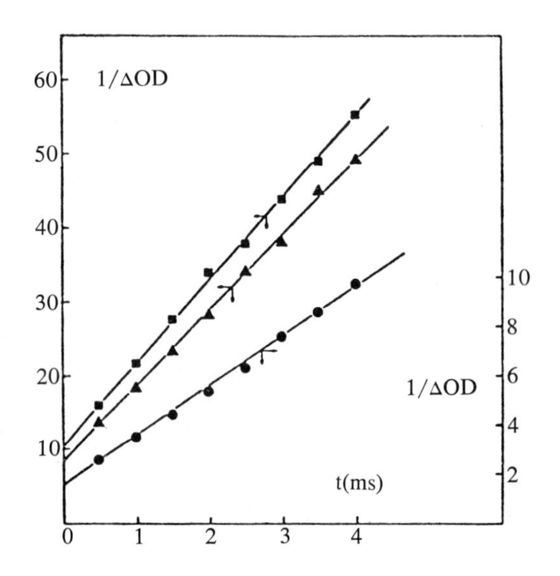

Figure 6. Flashphotolysis of $^{-}DD^{-}$ with D (excess in THF. Typical plots of 1/Δ (OD) vs. time obtained by monitoring the dark reaction proceeding in the presence of a large excess of D at 470, 390, and 750 nm. Key: ■, 750 nm (slope = 1.13×10^4); ●, 470 nm (slope = 1.98×10^3); ▲, 390 nm (slope = 1.02×10^4).

TABLE I

Dimerization Constants of $D^{\overline{\cdot}},Cat^+$ and $\alpha^{\overline{\cdot}},Cat^+$ in THF at 25°C

Cat^+	$10^{-8}\underline{K}_d$ Msec($D^{\overline{\cdot}},Cat^+$)	$10^{-7}\underline{K}_d$ Msec($\alpha^{\overline{\cdot}},Cat^+$)
Li^+	1.2	-
Na^+	3.5	0.2
K^+	10.0	1.0 - 1.2
Cs^+	30.0	-

Consider the initial conditions of a 1 M THF solution of a monomer with 10^{-4}M of an initiator $A^{\cdot-},Cat^{+}$. For as low K_{tr} as 10^{-5}, the electron-transfer equilibrium, established within a few μsec, converts 27% of $A^{\cdot-},Cat^{+}$ into monomer$^{\cdot-}$,Cat^{+}. Even for as low k_d as 10^6 $M^{-1}sec^{-1}$, it takes less than 0.2 sec to produce 95% of all the dimeric dianions expected in the quantitative conversion. In that time less than 50 monomer molecules are added to each of the formed growing centers, provided the propagation constant is not larger than 250 $M^{-1}sec^{-1}$ - a rather high value, while at the completion of polymerization 10^4 molecules are added.

The dimerization competes with monomer addition to monomeric radical-anions,

$$M^{\cdot-} + M \rightarrow \cdot M.M^{-} \qquad k_a$$

Most likely k_a is smaller than the propagation constant, i.e., $k_a < 250$ $M^{-1}sec^{-1}$. However, since the concentration of the monomer is at least 10^4 times larger than that of the monomeric radical-anions, the addition competes efficiently with the dimerization. Nevertheless, the resulting dimeric radical-anions, $\cdot M.M^{-}$, play no role in the polymerization because their diffusion controlled disproportionation (rate constant $\sim 10^{10}$ $M^{-1}sec^{-1}$) destroys them as soon as formed. Hence, radical propagation is imperceptible in such systems.

The dimeric dianions are stable. The rate of dissociation of $K^+,{}^-\alpha.\alpha^-,K^+$ was determined by the following procedure (7). α-methyl styrene perdeuterated in phenyl groups, α_{5D}, was prepared and converted into dimeric dianions, $K^+,{}^-\alpha_{5D}.\alpha_{5D}{}^-,K^+$. Their THF solution was mixed with a solution of ordinary protic dimers, $K^+, \alpha.\alpha^-,K^+$. The reversible dissociation-association

$$m = 238,\quad K^+,{}^-\alpha.\alpha^-,K^+ \underset{}{\overset{k_{diss}}{\rightleftharpoons}} 2\alpha^{\cdot-},K^+$$

$$m = 243,\quad K^+,{}^-\alpha.\alpha^-_{5D},K^+$$

$$m = 248,\quad K^+,{}^-\alpha^-_{5D}.\alpha^-_{5D},K^+ \underset{}{\overset{k_{diss}}{\rightleftharpoons}} 2\alpha^{\cdot-}_{5D},K^+$$

forms mixed dimers, $K^+,{}^-\alpha.\alpha_{5D},K^+$. For a 50:50 mixture the rate of mixed dimers formation is equal to 1/2 that of dissociation, k_{diss}. The mixture was kept at a constant temperature, and at desired time intervals (12h, 24h, etc.) aliquots were removed and protonated by methanol. The resulting hydrocarbons were isolated and analyzed by mass-spectrometer. The analysis gives the fraction of the homo-dimers, masses 238 and 248, converted into mixed dimers, mass 243, in a predetermined time interval, and this allows the calculation of the first order dissociation constant, $\underline{k}_{diss} = 6.10^{-8}$ sec^{-1} at 25°C. In conjunction with the determined association constant, it gives the equilibrium constant of the dissociation, $\underline{K}_{diss} \sim 10^{-14}$ M. Assuming a plausible value of 15 e.u. for ΔS of dissociation, one finds the heat of dissociation $\Delta H \sim 24$ kcal/mole.

Another approach led to the dissociation constant of Na ,$^{-}$DD^{-},Na^{+} (8). Since 1,1-diphenyl ethylene does not add to its dimeric dianions, a solution of Na^{+},$^{-}$DD^{-},Na^{+} was mixed with radioactive 1,1-diphenyl ethylene and the kinetics of exchange was investigated. The results led to the upper limit of the respective dissociation constant, namely $\underline{k}_{diss} < 10^{-7}$ sec^{-1}.

Flash-photolysis led also to the determination of electron-transfer equilibrium constants, e.g.,

biphenylide$^{\cdot}$ (B$^{\bar{\cdot}}$),Na^{+} + 1,1-diphenyl ethylene (D) $\rightleftarrows$

biphenyl (B) + 1,1-diphenyl ethylene radical-anion$^{\bar{\cdot}}$ (D$^{\bar{\cdot}}$),Na^{+}; K_{tr}

This is achieved by flash-photolyzing a THF solution of Na^{+},$^{-}$DD^{-},Na^{+} in the presence of a mixture of biphenyl and 1,1-diphenyl ethylene of known composition. Neither hydrocarbon reacts with the dianions, although both being at much higher concentration than Na^{+},$^{-}$D.D^{-},Na^{+}. Flash photolysis photo-dissociates some of the dimers into D$^{\bar{\cdot}}$,Na^{+} and then the equilibrium

$$D^{\bar{\cdot}},Na^{+} + B \rightleftarrows D + B^{\bar{\cdot}},Na^{+} \qquad 1/K_{tr}$$

is rapidly established due to relatively high concentrations of B and D ($\sim 10^{-3}$ M). In the dark period following a flash, the photolyzed dimers are regenerated by the reaction

$$2D^{\bar{\cdot}},Na^{+} \rightarrow Na^{+},^{-}DD^{-},Na^{+} \qquad \underline{k}_d$$

Since $[D^{\bar{\cdot}},Na^{+}] = X/(1+[B]/[D]K_{tr})$, where 1/2 X is the concentration of the dissociated dimers,

$$-d(1/X)/dt = \underline{k}_d/(1+[B]/[D]K_{tr})^2$$

i.e., plots of the reciprocal of the optical density at any chosen wavelength are again linear with time. This is shown by Fig. 7 and their slopes give $\{k_d/(1+[B]/[D]K_{tr})^2\}\ell\ \varepsilon$. In the absence of the added biphenyl, an analogous plot has a slope $\underline{k}_d/\ell\ \varepsilon$ (Fig. 8), and hence the ratio of both slopes gives $(1+[B]/[D]K_{tr})^2$. Since [B]/[D] is known, K_{tr} is derived from that ratio.

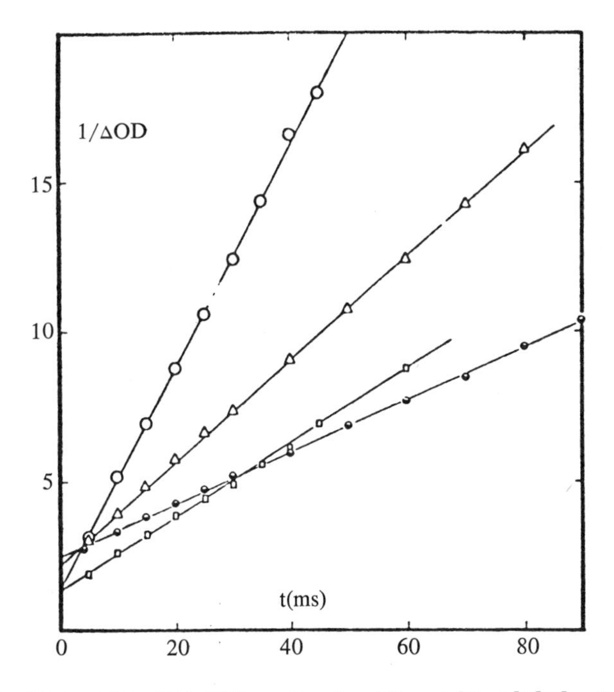

Figure 7. Plots of 1/Δ (OD 470) vs. time for different [Trph]/[D] ratios. Slopes must be corrected when used for plot shown in Figure 8 because of change in effective extinction coefficient at 470 nm due to absorbence of $Trph^{\div}, Na^{+}$. Key to [Trph]/[D] ratios: ○, 204; △, 434; □, 593; and ◒, 847.

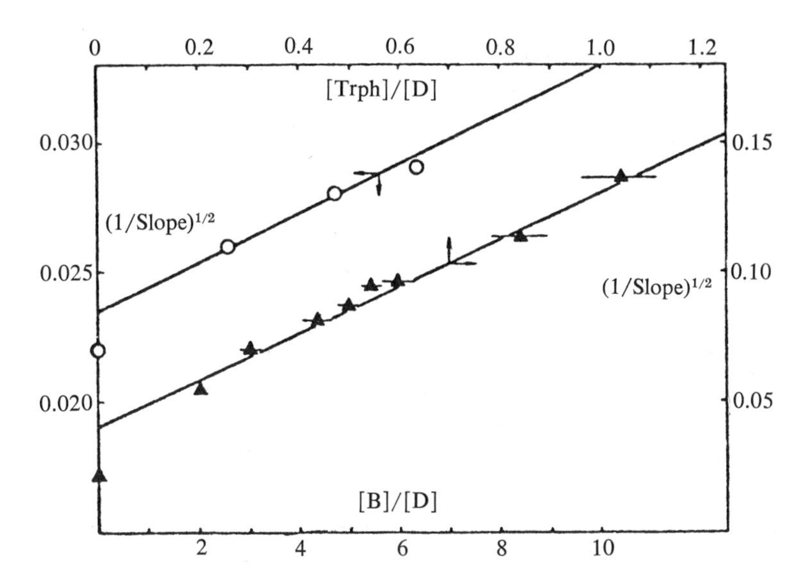

Figure 8. Square root reciprocals of corrected slopes of lines 1/Δ (OD 470) vs. time (see caption to Figure 7) plotted as functions [Trph]/[D] (▲) [B]/[D] (○).

Electron-transfer to monomer is not the only mode of initiation by radical-anions, although it is unique for non-polar monomers. With some polar monomers, especially cyclic ones, the initiation resembles protonation. For example, the reaction of sodium naphthalenide with ethylene oxide follows the route (9)

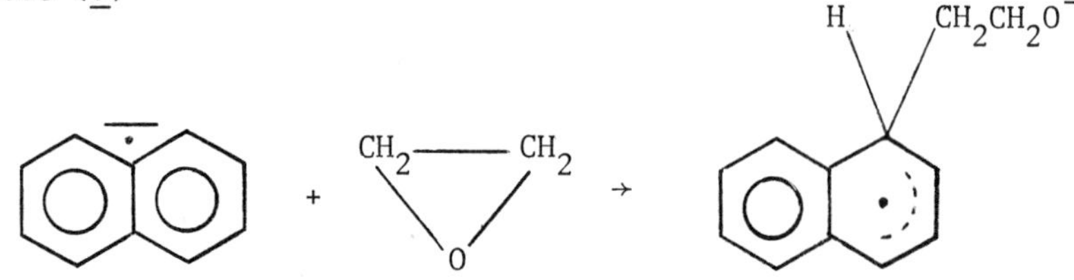

and the reduction of the adduct by naphthalenide followed by the addition of another molecule of epoxide yields the para or ortho diadduct

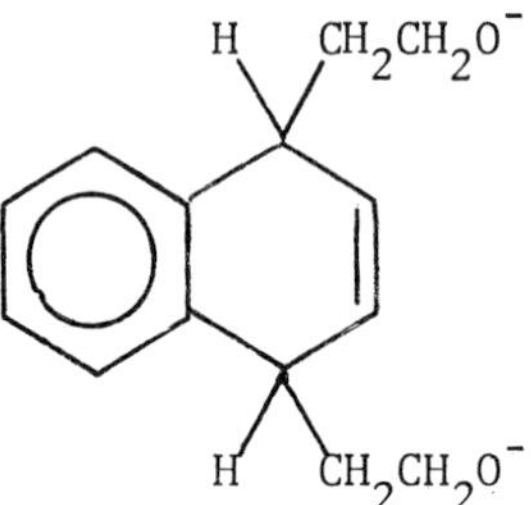

The bimolecular rate constant of the first addition is ∿ 1 M^{-1} sec^{-1} (10). The subsequent propagation is due to alkoxide ions. A similar process was reported for the initiation of cyclic-tetra-dimethylsiloxane by salts of naphthalenide (11).

The evidence for this mechanism is two-fold: the presence of aromatic moiety in the resulting poly-glycol and the quantitative analysis of the solution left after precipitation of the polymer, demonstrating only one-half of the utilized naphthalenide being converted into naphthalene. (It has been claimed that at very low concentrations of the naphthalenide, polymers with only one growing group were formed (10). Hydrogen abstraction from solvent was proposed as an explanation. However, terminating impurities become significant at very low concentrations of initiators and their action may account for the observations.)

Another variant of initiation by naphthalenide was proposed by Sigwalt (12), who studied the anionic polymerization of propylene sulphide. Formation of propylene in the course of this reaction led him to a following mechanism:

sodium naphthalenide + propylene sulphide →
naphthalene + NaS• + propylene

NaS• dimerize and the formed sodium disulphide, NaSSNa, initiates the propagation. It would be advantageous to demonstrate the presence of S-S bond in the resulting polymer.

Heterogeneous Electron-Transfer Initiation

Reactions of solid alkali metal particles with monomer or its solutions lead to heterogeneous electron-transfer initiation. The transfer takes place on the surface of the metal to the adsorbed molecules of the monomer and yields adsorbed monomeric radical-anions with the positively charged particles acting as counter-ions. Detachment requires not only desorption of the adsorbed radical-ions but also removal of metal cations from the metal lattice.

The hindrance of desorption does not affect the mobility of radical-anions on the metal surface. Hence, their dimerization with formation of still adsorbed dimeric dianions is very likely, and these may grow and form living oligomers. Degree of polymerization of the attached oligomers depends on their lifetime on the surface, and the lifetime is shortened by a cationsolvating solvent that facilitates removal of the cation from the metal lattice and therefore the desorption. This is demonstrated by Overberger (13), who studied the co-polymerization of styrene and methyl methacrylate initiated by a fine suspension of particles of metallic lithium.

Styrene and methyl methacrylate compete for the sites on the metal surface, the former being favored by the high polarizability of its π electrons. Hence, only styrene polymerizes on the surface, yielding a block of living polystyrene. The eventually desorbed living polystyrene initiates in solution polymerization of methyl methacrylate because this monomer is greatly preferred to styrene in anionic co-polymerization. The greater hindrance of desorption, the higher the percentage of styrene in the resulting block-polymer. Indeed, as demonstrated by the Overberger study, the size of polystyrene block increases with decreasing solvation power of the medium and the reactions performed in the absence of ethers yields co-polymer with 28% of styrene at 1% conversion. On the other hand, the facile removal of Na^+ cations from the sodium lattice might explain the formation of homo-poly-methyl-methacrylate in similar experiments involving sodium dispersion instead of lithium dispersion.

Further evidence of monomer adsorption on a metal surface is provided by the extensive studies of Richards and his co-workers (14). As is well known, alkyl bromides in tetrahydrofuran vigorously react with alkali metals, say lithium, yielding the Wurtz coupling products. The violent reaction slows down on addition of aromatic monomers like styrene, and the nature of the products is drastically changed (15). For example, when an equimolar mixture of ethylbromide and styrene reacts in tetrahydrofuran with metallic lithium, the alkyl capped dimer, $C_2H_5.CH(Ph).CH_2.CH_2.CH(Ph).C_2H_5$ forms 90% of the products, about 5% appear as $C_2H_5.CH_2.CH(Ph).C_2H_5$, the remainder being a mixture of an alkyl capped trimer and butane. The tail-to-tail structure of the capped dimer was demonstrated by the n.m.r. technique.

It seems that styrene and ethylbromide compete for the sites on the lithium surface. The adsorption of styrene possessing easily polarizable π electrons is more favorable than that of ethylbromide. Hence, Wurtz coupling is hindered, while electron-transfer to the adsorbed styrene yields radical-anions and their mobility on the surface allows for their dimerization. Eventually, the dimeric dianions are desorbed; and since their reaction with ethylbromide is faster than propagation, the ethyl capped dimers are the main products.

Further support for the proposed mechanism is provided by the results of experiments involving phenylbromide instead of ethylbromide (16). The polarizable π electrons of this aryl compound allow it to effectively compete with styrene for the sites on the lithium surface and thus the Wurtz coupling reaction becomes dominant. Similar results were obtained with ethyltosylate. Although the reaction of tosylate with living polystyrene is rapid and quantitative, yielding ethyl capped polymers, its reaction with the monomer and metallic lithium produces only 10% of the ethyl capped polymers, the remainder being evolved as butane. Again, the aromatic nature of tosylate allows it to compete with styrene for the lithium sites.

An interesting extension of this picture is provided by the behavior of p-xylylene dibromide (17). With butadiene as monomer and tetrahydrofuran as solvent, their reaction on metallic lithium leads to an unusual co-polymer,

$$-(BD)_{n}-CH_2.C_6H_4.CH_2.CH_2.C_6H_4.CH_2-(BD)_{m}-$$

(BD-butadiene moiety)

Its composition is determined by the initial ratio of the reagents in the feed. N.m.r. analysis confirmed the above structure and showed that the p-xylylene moieties in the polymeric chains exist exclusively as dimers.

Unexpectedly, the vicinyl dihalides react differently (18). For example, an equimolar mixture of styrene and 1,2-dibromoethane reacts on lithium metal yielding a head-to-head, tail-to-tail polystyrene, ethylene and lithium bromide. Apparently, the adsorbed styrene is reduced and dimerized to the dianions,

$$^{-}CH(Ph).CH_2.CH_2.CH(Ph)^{-}$$

and the latter rapidly react with the dibromide yielding the unconventional head-to-head, tail-to-tail polymer with elimination of ethylene

$$\underline{n}\{^{-}CH(Ph)\cdot CH_2\cdot CH_2\cdot CH(Ph)^{-} + {}^{-}CH(Ph)\cdot CH_2\cdot CH_2\cdot CH(Ph)^{-}\} \xrightarrow{BrCH_2CH_2Br}$$
$$-[CH(Ph)\cdot CH_2\cdot CH_2\cdot CH(Ph) - CH(Ph)\cdot CH_2\cdot CH_2\cdot (Ph)]_{\underline{n}/2}^{-} + \underline{n}C_2H_4 + 2\underline{n}LiBr$$

Other linking agents were investigated, e.g., dibromo-dimethyl-silane and dichlorophenylphosphine (19). Interesting products were obtained with diepoxides, namely

$$^{-}M.M.CH_2.\underset{\displaystyle OLi}{\underset{|}{CH}}.R.\underset{\displaystyle OLi}{\underset{|}{CH}}.CH_2.M.M.^{-} \qquad \text{etc.,}$$

yielding the respective poly-ols on hydrolysis.

Electron-Transfer Step-Wise Reaction

An example of such a reaction is discussed. Reaction of bis(1,1-diphenyl ethylene) linked by a chain of aliphatic hydrocarbon yields on reduction with alkali metals or suitable electron donors, e.g., naphthalenide, a product of poly-dimerization. For example:

$$CH_2{:}\underset{\displaystyle Ph}{\underset{|}{C}}-C_6H_4-CH_2-C_6H_4-\underset{\displaystyle Ph}{\underset{|}{C}}{:}CH_2 \xrightarrow{+2e}$$

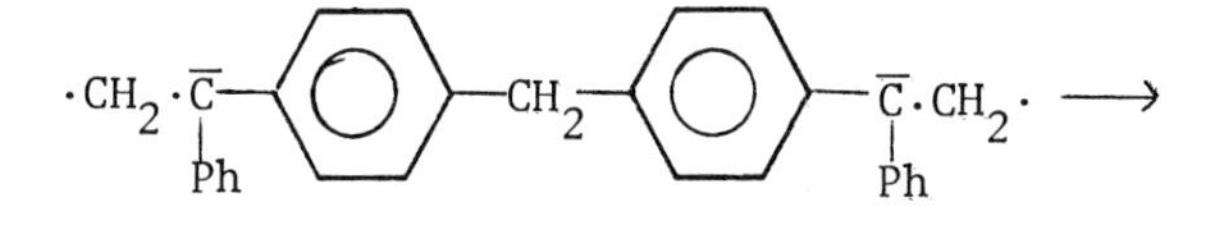

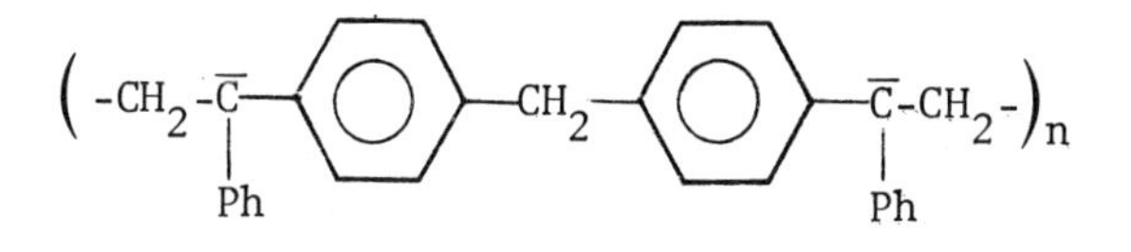

The resulting poly-carbanions are protonated and yield the respective hydrocarbon (20). Other examples are provided by the work of Hocker and his co-workers, and those discussed in the preceding section.

Literature Cited

1. (a) Szwarc, M.; Nature 178, 1168 (1956).
(b) Szwarc, M.; Levy, M. and Milkovich, R.; J. Amer. Chem. Soc. 78, 2656 (1956).
2. Rämme, G.; Fisher, M.; Claesson, S. and Szwarc, M.; Proc. Roy. Soc. A327, 467 (1972).
3. Wang, H. C.; Levin, G. and Szwarc, M. ; J. Phys. Chem. 83, 785 (1979).
4. (a) Katayama, M.; Hatada, M.; Hirota, K.; Yamazaki, H. and Ozawa, Y.; Bull. Chem. Soc. Japan 38, 851, 2208 (1965).
(b) Schneider, C. and Swallow, A. J.; Makromolek. Chem. 114, 155, 172 (1968).
(c) Metz, D. J.; Potter, R. C. and Thomas, J. K.; J. Polymer Sci. A5, 877 (1967).
5. (a) Wang, H. C.; Lillie, E. D.; Slomkowski, S.; Levin, G. and Szwarc, M.; J. Amer. Chem. Soc. 99, 4612 (1977).
(b) Lillie, E. D.; Slomkowski, S.; Levin, G. and Szwarc, M.; ibid., 99 4608 (1977).
6. Hamill, W. H.; "Radical Ions", p. 321, Kaiser, E. T. and Kevan, L., Eds., Interscience (1968).
7. Asami, R. and Szwarc, M.; J. Amer. Chem. Soc. 84, 2269 (1962).
8. Spach, G.; Monteiro, M.; Levy, M. and Szwarc M.; Trans. Faraday Soc. 58, 1809 (1962).
9. Richards, D. H. and Szwarc, M.; Trans. Faraday Soc. 55, 1654 (1959).
10. Kazanskii, S.; Solovyanov, A. A. and Entelis, S. G.; Europ. Polymer J. 7, 1421 (1971).
11. Morton, M.; Rembaum, A. and Bostick, E. E.; J. Polymer Sci. 32, 530 (1958).
12. (a) Boileau, S.; Champetier, G. and Sigwalt, P.; J. Polymer Sci. C16, 3021 (1967).
(b) Favier, J. P.; Boileau, S. and Sigwalt, P.; Europ. Polymer J. 4, 3 (1968).
13. Overberger, C. G. and Yamamoto, N.; J. Polymer Sci. B3, 569 (1965); A4, 3101 (1964).
14. Richards, D. H.; Polymer 19, 109 (1978).
15. Davis, A.; Richards, D. H. and Scilly, N. F.; Makromolek. Chem. 152, 121, 133 (1972).
16. Cunliffe, A. V.; Paul, N. C.; Richards, D. N. and Thompson, D.; Polymer 19, 329 (1978).
17. Richards, D. H. and Scilly, N. F.; Brit. Polymer J. 2, 277 (1970); 3, 101 (1971).
18. Richards, D. H.; Scilly, N. F.; Chem. Comm., p. 1515 (1968).
19. (a) Richards, D. H. et al.; Europ. Polymer J. 6, 1469 (1970).
(b) Cunliffe, A. V.; Hubbert, W. J. and Richards, D. H.; Makromolek. Chem. 157, 23, 39 (1972).
(c) Richards, D. H. et al.; Polymer 16, 654, 659, 665 (1975).
20. Höcker, H. and Lattermann, G.; J. Polymer Sci. Symposium 54, 361 (1976).

RECEIVED January 10, 1983

30

Initiation of Polymerization with High-Energy Radiation

VIVIAN STANNETT
North Carolina State University, Chemical Engineering Department, Raleigh, NC 27650

JOSEPH SILVERMAN
University of Maryland, Institute for Physical Science and Technology, Laboratory for Radiation and Polymer Science, College Park, MD 20742

High energy photons and electrons interact with organic liquids to produce excited species, positive molecule ions and electrons. Most of the electrons rapidly recombine with their geminate cation radicals to form additional excited molecules. Some of the latter lose their excitation energy by collision with other molecules; the remainder break into free radicals. For this reason most of the polymerization reactions of vinyl monomers in bulk, emulsion and in solution are initiated by free radicals. A small proportion of the ejected electrons have sufficient energy to escape the coulombic forces of their corresponding positive molecule ions. These free electrons lose their energy to the surroundings, become thermalized, and can generate anionic species either by simple capture or by dissociative electron capture. Both the positive and negative ionic species can, under suitable conditions, initiate cationic and anionic polymerizations, respectively. An important point of fundamental interest is that the early positive and negative species are paramagnetic, and that the mechanisms by which they are converted to propagating carbonium ions and carbanions are as yet not well understood.

The subsequent chain growth of the radical, cationic and anionic initiating species has led to new fundamental information of considerable value to both radiation and polymer chemists. Apart from the fundamental interest in radiation induced polymerization, there is considerable practical interest including industrial exploitation. The initiation process has essentially zero activation energy in marked contrast to chemical initiation. This has interesting kinetic consequences as will be discussed later. In addition, it leads to an overall lower activation energy for the polymerizaton itself. This, coupled with the ease of control of the rate of initiation including, if necessary, the rapid removal of the source, lessens the chance of run-away

0097-6156/83/0212-0435$06.00/0

polymerizations. There are a number of other practical advantages of radiation: for example, the lack of any residual catalyst or catalyst fragments and the virtually unlimited range of chain initiation rates which can be readily obtained. There are some possible disadvantages, in particular, changes in the polymer itself brought about by the concurrent radiolysis. In any practical radiation polymerization process, the required dose is so low, however, that these effects would normally be negligible.

Radiation Initiated Free Radical Initiation

All organic liquids, including vinyl monomers, produce free radicals on irradiation. The G value (radicals produced per 100 eV of absorbed radiation energy), which is the radiation yield, depends on their structure and can vary from 0.7, for pure styrene, for example, to an apparent value of 19 for carbon tetrachloride in styrene solution.

In the case of bulk monomers, it is easy to calculate the rate of initiation from the G values, the dose rate, and the physical constants of the monomer. If the $k_p/k_t^{\frac{1}{2}}$ values are known for the monomer in question at the temperature used, the rates of initiation can be calculated from the observed steady state rates of polymerization. Thus for the simplest case,

$$R_i = \frac{R_p^2 k_t}{k_p^2 (M)^2} \tag{1}$$

R_i combined with the measured dose rate can be then used to calculate the G value for radicals; this method was developed by Chapiro (1). Styrene is a particularly well behaved monomer and the $k_p/k_t^{\frac{1}{2}}$ values are known over a wide temperature range. It can be used, therefore, to measure the rates of initiation in solution. Other methods, for example, the use of radical scavengers such as DPPH, can also be used to obtain a direct value for the rate of radical production. With bulk monomers the R_i values are a function only of the dose rate and the kinetics are quite comparable to those of the benzoyl peroxide initiated system, for example.

Radiation initiation for solution vinyl polymerization is more complex. A special feature of radiation is that it attacks all components of the system including the solvent. With a monomer-solvent mixture, therefore, the rate of initiation can be represented by a linear equation as follows:

$$R_i = \dot{D} \; [A(M)G_m(i) + B(S)G_s(i)] \tag{2}$$

Where $\dot{D}$ is the dose rate and $G_m(i)$ and $G_s(i)$ are the G values for initiation for the monomer and solvent, respectively. (M) and (S) are the monomer and solvent concentrations and A and B constants to convert the G values into appropriate units, i.e., moles, liters, seconds.

The method of Chapiro (1) using known $k_p/k_t^{1/2}$ values for the pure monomer can then be used to calculate the overall R_i if the radical yields are purely additive and ideal kinetics apply. The latter implies that all the solvent radicals react immediately with monomer and the termination step is by the recombination of the growing chains. If the purely additive effect, sometimes called the simple dilution effect, applies, then a plot of the overall rates of initiation versus m, the mole fraction of monomer, gives a straight line. If the plot curves, it indicates that energy transfer is taking place as will be discussed later. Curvature above the line shows sensitization and below indicates deactivation by the solvent.

A more useful analysis has been made by Chapiro (1) who showed that:

$$\frac{R_p}{R_{po}} = \left[\frac{m}{m + (1-m)(V_s/V_m)}\right]^{3/2} \left[1 + \phi_{rel}\frac{(1-m)}{m}\right]^{1/2} \qquad (3)$$

Where R_p and R_{po} are the rates of polymerization in solution and pure monomer, respectively. $\phi_{rel} = \phi_s/\phi_m$ where ϕ_s and ϕ_m are the molar yield constants for free radical production per unit exposure dose and unit volume from the solvent and monomer, respectively; (these are almost proportional to the G values). V_s and V_m are the respective molar volumes. Plots of R_p/R_{po} versus m give a family of curves according to the value of ϕ_{rel} as shown in Figure 1. When ϕ_{rel}=1.0, as obtained with styrene in benzene or methyl methacrylate in ethyl acetate, for example, the kinetics are similar to those obtained with catalyst-initiated polymerization systems for which there is no effect of the solvent on the initiation rate.

The curves can usually be fitted with a constant value of ϕ_{rel}; a number of examples may be found in reference 1. This is not always the case, however. Values of ϕ_{rel} may be calculated from equation 3 at various monomer concentrations. If these are not constant it is usually assumed that energy transfer processes are operative. The term energy transfer process in this work is used in its most general sense. The usual meaning implies rapid transfer of excitation from an absorbing site (such as a solvent molecule) over several molecular lengths to an energy sink which becomes reactive (such as a solute which becomes an effective producer of radicals). We use the term to also include fast, but ordinary, chemical reactions which effectively transfer the reactive site. Because these two mechanisms are equivalent, we shall describe the solvent effect in terms of neutral excited species arising from the direct absorption of radiation by the monomer and solvent and the transfer of excitation by exchange processes. Regardless of the details of the mechanism, the processes are equivalent and are described below in the conventional sense of radiation-produced excited species that

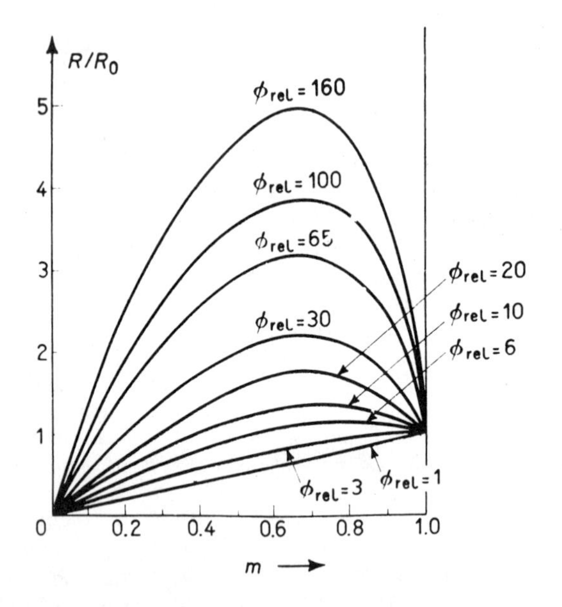

Figure 1. Relative rates of radiation-induced polymerization in bulk and in solution (Eq. 3) at various Ø values. (Reprinted with permission from Ref. 1. Copyright 1962, John Wiley & Sons.)

interact with monomer and solvent thereby producing free radicals or an exchange of energy, thus

$$M^* + S \rightarrow S^* + M \rightarrow S\bullet + M \quad (4)$$

$$S^* + M \rightarrow M^* + S \rightarrow M\bullet + S \quad (5)$$

An extreme example of energy transfer behavior was reported by Miller and Stannett (2) with styrene in n-dibutyl disulfide. The results are presented in Table I and show a progressive decrease in ϕ_{rel} with increasing amounts of disulfide. This implies that energy transfer processes, predominately from styrene to the disulfide, are taking place. The overall kinetics showed that classic kinetics were otherwise normal. A number of other examples are presented in reference (1) and by Stannett et al.(4).

The kinetics of radiation induced polymerization in solution in which energy transfer is operative have been developed by Nikitina and Bagdassarian (3). A term was introduced, P_{rel}, representing the relative probability of energy transfer from M to S versus transfer from S to M. This leads to a modified expression for R_p/R_{po}:

$$R_p/R_{po} = \left[\frac{1}{m + (1-m)\, V_s/V_m}\right]^{3/2} \left[\frac{1 + \phi_{rel}\, P_{rel}\, \frac{(1-m)}{m}}{1 + P_{rel}\, \frac{(1-m)}{m}}\right]^{1/2} m \quad (6)$$

The development and limitations of this equation have been discussed by Chapiro (1). The data with styrene and dibutyl disulfide were analyzed leading to single values of $\phi_{rel}=8.5$ and $P_{rel}=3.1$. The degree of fit of the data is shown in Figure 2. Using the value obtained for ϕ_{rel} after correcting for energy transfer, G(radical) value (strictly speaking, the G value for chain initiation for dibutyl disulfide was estimated to be 3.1. The behavior and values were found to be in good agreement with other data published for various disulfides.

When the ϕ_{rel} values are less than unity and plots of R_i versus m curve below the ϕ=1.0 line, there is deactivation by the solvent whereas curvature above the line signifies $\phi_{rel} > 1.0$, indicating sensitization of the polymerization by the solvent.

The effects of solvents on the rates of initiation can have considerable implications for radiation grafting. This was clearly shown for the grafting of styrene to cellulose acetate in solution (4). The use of pyridine with a $\phi_{rel} \sim 0$ compared with dimethyl formamide with $\phi_{rel} \sim 5.0$ was found to give similar yields of graft copolymer. However, the accompanying homopolymer was only 8% in the case of pyridine compared with 32% with dimethyl formamide as the solvent under comparable conditions.

While the energy transfer model is remarkably successful in summarizing the polymerization kinetics of vinyl monomer solutions, there may be other explanations with certain monomer-

Table I: The Variation of ϕ_{rel} with Percent Solvent for the Radiation Initiated Free Radical Polymerization of Styrene in n-Dibutyl Disulfide at 25°C. Dose Rate 0.005 Mrad per hour (ref. 2)

Vol Percent DBS	5	10	20	40	60	80
ϕ_{rel}	44.0	23.8	16.6	17.3	13.6	6.9

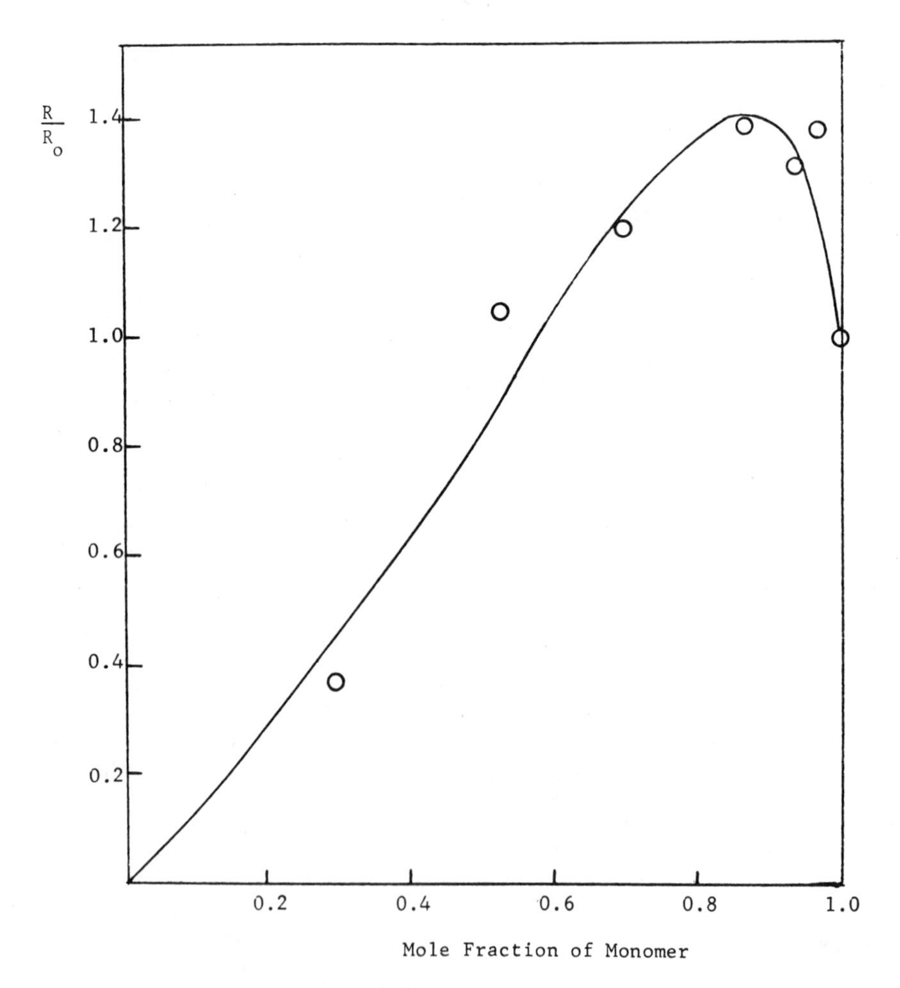

Figure 2. Relative rates for radiation-induced polymerization of styrene in dibutyl disulfide at 25 °C. Key: ———, equation 6 with values; ϕ_{rel} = 8.5; P_{rel} = 3.1. (Reprinted with permission from Ref. 2. Copyright 1969, John Wiley & Sons.)

solvent systems (5). The principal observations leading to an alternative explanation are these: (1) the G value for ion pair formation in gaseous hydrocarbon monomers is ~ 4; (2) G (free ions) in neat hydrocarbon liquid monomers is ~ 0.1; (3) the G value for propagating radical chains in neat hydrocarbon liquid monomers is < 1.0; (4) we have yet to find a value greater than 7 for G (propagating radicals) in liquid solutions of hydrocarbon monomers. This information is the basis of a model in which the sole function of the additive is a fast reaction which converts into propagating free radicals the primordial charged species that would otherwise recombine without propagation.

Consider the radiation chemistry of methanol-styrene solutions. From the data of Huang and Chandramouli (6) we calculate G(chains)=0.7 in neat styrene; it rises to 5.7 in 5 vol % methanol in styrene and levels off at 6.5 at higher methanol concentrations. According to the energy transfer model, these are the steps that take place in styrene containing a small amount of methanol. Energy absorbed in the styrene is efficiently transferred to methanol in quanta well matched to the C-H bond energy. By the alternative model (5), the principal effect of ionizing radiation is to produce cation and anion radicals. In neat solutions most of them recombine in intraspur reactions before they can diffuse away from each other or before they can add more than a few units by propagation. Neutralization provides the energy for the formation of the small fraction of reactive units that survive and propagate as neutral free radicals. When a small concentration of methanol is present, however, the charge carriers could undergo rapid diffusion-controlled reactions with the additive that convert them into initiators of the radical polymerization. Further information concerning this approach may be found in reference (5).

The zero activation energy for radiation initiation, E_i, leads to some interesting practical consequences. The overall activation for the rate, $E = E_p + 1/2\ E_i - 1/2\ E_t$, where E_p and E_t are the activation energies for propagation and termination, respectively. Since E_i for most catalytic initiation is about 30 and E_t close to zero, this leads to a value of about 22 kcal per mole compared with about 7 kcal per mole for radiation for which E_i=0. The practical advantages of this were referred to earlier.

On the other hand, the overall activation energy for the degree of polymerization is equal to $E_p - 1/2\ E_i - 1/2\ E_t$. This leads to an approximate value of -8 kcal per mole for chemical and +7 kcal per mole for radiation initiation. If one wishes to increase the rate by raising the reaction temperature without changing the initiator concentration, this leads to a decrease in molecular weight for chemical but with a corresponding increase with radiation initiation. The practical advantages of this reversal are quite clear.

Radiation Initiated Ionic Polymerization

The radiation initiated polymerization of vinyl monomers in pure liquid or in solution was initially thought to be exclusively free radical in nature. In 1957 Davison et al. (7) demonstrated that isobutylene could be polymerized both with high energy electrons and with cobalt 60 radiation. Because this monomer polymerizes only by a cationic mechanism it was clear that radiation could, indeed, initiate ionic polymerization in the liquid state. Since then there has been considerable research with a number of monomers. The field has, however, continued to be dominated by studies of free radical processes. Almost all the ionic studies have been concerned with cationic processes although a few anionic polymerizations have been reported, notably with nitroethylene (8).

The simplified steps leading to cationic and anionic initiation are as follows:

$$\mathrm{M\ (or\ S)} \rightarrow \mathrm{M}^{+}\ (\mathrm{or\ S}\overset{+}{\cdot}) + e_{free} \quad (7)$$

$$e_{free} \rightarrow e_{thermal} \quad (8)$$

$$e_{thermal} + \mathrm{M\ (or\ S)} \rightarrow \mathrm{M}\overset{-}{\cdot}\ (\mathrm{or\ S}\overset{-}{\cdot}) \quad (9)$$

The exact nature of the initiating species is not known in most cases. In principle they could be positive or negative ion radicals as shown. It seems more likely however that normal carbonium ions or carbanions are formed and that these are the true initiators. This is implied, however, by gas phase or solid state studies. Pulse radiolysis in the liquid state is now beginning to give more direct evidence of the exact nature of the processes.

Typical reactions of the primary reactions which can occur would be:

$$\text{Isobutylene: } C_4H_8\overset{+}{:} + C_4H_8 \rightarrow (CH_3)_3C^{+} + \underset{\text{(allylic radical)}}{CH_2{=}C(CH_3)\dot{C}H_2} \quad (10)$$

$$\text{Nitroethylene: } C_2\bar{H}_3NO_2 + \underset{NO_2}{CH_2{=}CH} \rightarrow \underset{NO_2}{CH_3CH} + \underset{NO_2}{CH_2 = C\cdot} \quad (11)$$

The isobutylene mechanism appears plausible and is strongly supported by mass spectrometer studies (9). The nitroethylene mechanism remains however highly speculative. Considerable relevant information of electron transfer polymerizations is however available from the work of Szwarc and others (10).

In the case of pure styrene, the mechanism for formation of the propagating carbonium ion is not clear. Recent results with picosecond pulse radiolysis has, however, clarified the situation. The principal early effect appears to be ionization leading to the cation radical and the anion radical (by electron capture). Within 10 ps, the cation radical adds a monomer to become the dimerized cationic species, with an estimated first order rate constant of 8×10^9 $M^{-1}sec^{-1}$ (5). The dimer cation disappears by a first order process with a lifetime of 20 ns (5). This is presumably the trimerization and thus the first step of the cationic polymerization, the calculated rate constant is 4×10^6 $M^{-1}sec^{-1}$. This is in excellent agreement with the values estimated from a combination of conductivity and polymerization rate studies with radiation initiation (11,12). Chemically initiated polymerizations are somewhat lower but these are not strictly relevant since they were conducted in chlorinated solvents. It has been well established that solvation and other effects reduce the rate constants with free cationic polymerizations.

These results are consistent with the following mechanisms:

Ionization

$$\phi\,CH{=}CH_2 \rightsquigarrow \phi\,CH\dot{-}CH_2^{+} + e^- \tag{12}$$

Dimerization of cation radical

$$\phi\,CH\dot{-}CH_2^{+} + \phi CH{=}CH_2 \xrightarrow[10ps]{8\times10^{9}M^{-1}s^{-1}} H\dot{C}(\phi)H_2C{-}CH_2\overset{+}{C}H(\phi) \tag{13}$$

Trimerization by cationic addition

$$H\dot{C}(\phi)H_2C{-}CH_2\overset{+}{C}H(\phi) + \phi CH{=}CH_2 \xrightarrow[20\ ns]{4\times10^{6}M^{-1}s^{-1}} H\dot{C}(\phi)H_2C{-}CH_2CH(\phi){-}CH_2\overset{+}{C}H(\phi) \tag{14}$$

Formation of anion radical

$$\phi CH{=}CH_2 + e^- \rightarrow \phi CH{=}CH_2^{\overline{\cdot}} \tag{15}$$

The close proximity of the dimer cation and the anion radical in the spur could lead to some of the high yield of dimers and trimers first reported by Machi, Silverman and Metz (13).

Pulse radiolysis results were also obtained (5) suggesting that the role of methanol is to serve as a rapid proton donor to the anion radical converting the latter to a neutral propagating free radical. The resulting methoxide anion could neutralize the cation radical converting it to an additional neutral propagating free radical. Methanol has, indeed, been observed to scavenge both the cationic and anionic species (5) and to increase the rate of free radical polymerization (11).

The G(i) values for the yield of positive or negative initiating species, which lead to estimates of the rates of initiation, are difficult to determine. The methods available are the scavenging of ions with various compounds and the electrical conductivity method using steady state and pulsed charge collection techniques. They do not always show good agreement; however the range of values is not too large in most cases. Since the R_i values enter into the rate expression as a square root, the errors become even less significant in evaluating the other kinetic parameters. The free ion yields of only a few monomers have been measured directly (11,14-15). However those for a large number of other organic liquids have been determined and have been collected together in an important and useful publication (15). Apart from the values themselves, interesting trends are presented in reference 16. The yields increase with temperature to a limited extent, for example, by about six-fold from -78°C to +110°C for a number of dienes. The free ion yields also increase with increasing dielectric constant, but with some important exceptions, notably the alkanes; for the latter, free ion yields are greater for branched alkanes than for normal alkanes although there are no significant differences in the dielectric constant. With polar organic of compounds the correlation is often good. [See Figure 3 (16).] This enables one to make an estimate for, say, methylene chloride where no values have been determined. It could be asked whether these free ion yields are really equal to the yield of chain initiation. Only one direct comparison has been made, with ethyl vinyl ether in n-pentane and neopentane solutions where G (free ion) values range 0.16 to 1.0 (18). The rates of polymerization of styrene were proportional to the square root of the computed overall G(i) values. In this case, at least, it appears that all the free ions do indeed participate in the initiation process.

It is believed that the free positive and negative species annihilate each other immediately on contact. This is borne out by the strict square root relationship which is found between the rates of polymerization and the dose rate of the radiation. The growing chain ends are therefore free in nature, i.e., with no ion-pair component. This makes radiation initiated ionic polymerization an excellent method for studying free ion polymerization: examples of the power of this method have been presented for p-methoxy styrene (19) and the vinyl ethers (16,20,21).

The work of Hayashi et al. (22), however, has indicated that an ion pairs component can be introduced by adding strong electrons acceptors such as pyromellitic anhydride.

Finally, it is clear that certain monomers, notably styrene, can polymerize simultaneously by free radical, free anion, and free cation mechanisms. After the extremely rapid reactions leading to the production of the chain initiators, interactions between these different mechanisms appear to be minimal but are unknown. However it is logical to believe that the anion and cation chains will terminate each other.

There are sufficient quantitative data for neat styrene at room temperature to estimate the relative rates at which each process will proceed.

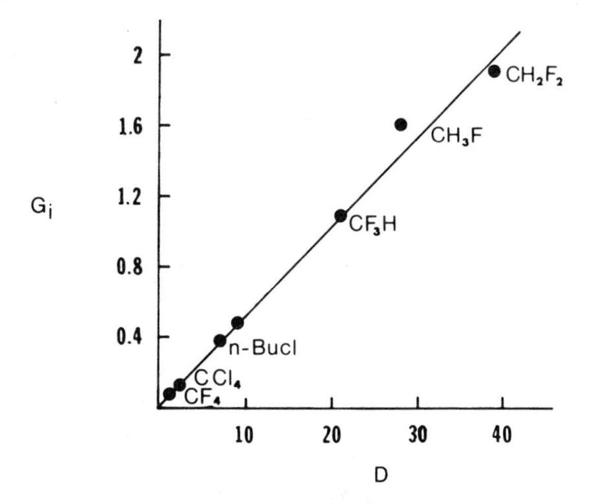

Figure 3. Correlation between the G (free ions) and dielectric constant, D, values for a series of halogenated compounds used to estimate methylene chloride values. (Reprinted with permission from Ref. 17. Copyright 1982, Butterworth & Co. (Publishers, Ltd.)

As indicated earlier, G=0.7 for free radical initiation and about 0.1 for the free ions. The termination rate constants are 3×10^7 $M^{-1}sec^{-1}$ for free radicals and 2×10^{11} $M^{-1}sec^{-1}$ for the free ions. The propagation rate constants have been determined to be 30, 4×10^6 and about 10^5 $M^{-1}sec^{-1}$ for free radical, cationic and anionic polymerization, respectively.

The steady state rates of polymerization, R_p, are given by

$$R_p = k_p \ (M) \ (R_i/k_t)^{\frac{1}{2}} \tag{16}$$

using the correct rate constants for each case.

$$R_i = 1.04 \times 10^{-9} \ G(i) \ \rho \ \dot{D} \tag{17}$$

where R_i is the initiation rate in $M^{-1}sec^{-1}$, ρ is the specific gravity, and $\dot{D}$ is the dose rate in rads per second.

For a typical case of gamma polymerization in which $\dot{D}$=100, R_i is 6.6×10^{-8} and 0.943×10^{-8} $M^{-1}sec^{-1}$ for radicals and ions, respectively.

The rates of polymerization ($M^{-1}sec^{-1}$) calculated by means of equation 17 are as follows.

radical	1.23×10^{-5}
cation	7.6×10^{-3}
anion	1.9×10^{-4}

The ratio is 1:618:15.4 for the free radical, cationic and anionic respectively. The overwhelming mechanism is clearly cationic under super dry conditions with about 2.5% anionic and a negligible radical contribution.

Literature Cited

1. Chapiro, A., "The Radiation Chemistry of Polymeric Systems", Interscience Publishers, New York, NY 1962.
2. Miller, L. A. and Stannett, V., J. Poly. Sci., A1, 7, 3159 (1969).
3. Nikitina, T. and Bagdassarian, Sbornik Rabot po Radiatsionnoi Khimii, Moscow Academy of Sciences, U.S.S.R., p. 183 (1955) see also Ref. 1 pages 266-269.
4. Stannett, V., Wellons, J. D., and Yasuda, H., J. Poly. Sci., C4, 551 (1964).
5. Tagawa, S., Silverman, J., Kobayashi, H., Katasumura, Y., Washio, M., and Tabata, Y., Radiation Phys. & Chem., in press (1982).
6. Huang, R. Y. M. and Chandramouli, P., J. Poly. Sci., B, Polymer 7 , 245 (1969).

7. Davison, W. H. T., Pinner, S. H., and Worrall, R., Chem. and Ind., 1274 (1957).
8. Yamaoka, H., Williams, F., and Hayashi, K., Trans. Farad. Soc., 63, 376 (1967).
9. Lampe, F. W., J. Phys. Chem., 63, 1986 (1959).
10. Szwarc, M., "Carbanions, Living Polymers and Electron Transfer Processes", Wiley, N.Y. (1968).
11. Williams, F., Hayashi, Ka., Ueno, K., Hayashi, K., and Okamura, S., Trans. Farad. Soc. 63, 1501 (1967).
12. Hayashi, Ka., Hayashi, K., and Okamura, S., Polym. J. (Japan) 4, 426 (1973).
13. Machi, S., Silverman, J., and Metz, D. J., J. Phys. Chem, 76, 930 (1972).
14. Hayashi, Ka., Yamazawa, Y., Takagaki, T., Williams, F., Hayashi, K., and Okamura, S., Trans. Farad. Soc. 63, 1489 (1967).
15. Hayashi, Ka., Hayashi, K., and Okamura, S., J. Poly. Sci., A1, 9, 2305 (1971).
16. Allen, A. O., "Yields of Free Ions Formed in Liquids by Radiation", N.B.S. Reference Data System NSRD-NBS 57 (1976).
17. Deffieux, A., Hsieh, W. C., Squire, D. R. and Stannett, V., Polymer, 23, 65 (1982).
18. Kubota, H., Kabanov, Ya., Squire, D. R. and Stannett, V., J. Macromol. Sci. - Chem. A12, 1299 (1978).
19. Deffieux, A., Squire, D. R. and Stannett, V., Polymer Bull., 2, 469 (1981).
20. Hsieh, W. C., Kubota, H., Squire, D. R., and Stannett, V., J. Poly. Sci. - Chem. 18, 2773 (1980).
21. Deffieux, A., Hsieh, W. C., Squire, D. R. and Stannett, V., Polymer 22, 1575 (1981).
22. Hayashi, K., Irie, M., and Yamamoto, Y., J. Poly. Sci., Sym. Series 56, 173 (1977).

RECEIVED January 10, 1983

Temperature and Dilution Effects in Polymerization Induced by Peroxides Undergoing Rapid Decomposition ($t_{1/2}$ <60 Minutes)

NORMAN G. GAYLORD

Gaylord Research Institute, Inc., New Providence, NJ 07974

The polymerization of maleic anhydride, norbornene and various norbornene derivatives, including the Diels-Alder adducts of cyclopentadiene with maleic anhydride, N-phenylmaleimide, acrylonitrile and methyl acrylate, is initiated by the addition of a peroxide or peroxyester having a short half life and undergoing rapid decomposition at the reaction temperature. The higher the temperature the higher the conversion with catalysts having the same half-life, i.e. less than 60 min, at temperature. The polymerizations of MAH, norbornene and the CPD-MAH and CPD-NPMI Diels-Alder adducts are subject to a dilution effect and are suppressed when the solvent concentration is greater than 25%. The polymerizability of these monomers, "unreactive" under normal conditions, in the presence of peroxides undergoing rapid decomposition, the condition wherein chemically induced dynamic nuclear polarization is observed, is consistent with their tendency to undergo photosensitized reactions under ultraviolet light.

The initiation of the radical polymerization of reactive unsaturated monomers, e.g. methyl methacrylate, styrene and vinyl chloride, generally requires the use of a catalyst with a half life of 3-10 hrs at the reaction temperature. The polymerization may be carried out in bulk or in dilute or concentrated solutions.

Various monomers which fail to undergo polymerization under these conditions are designated "unreactive". However, effective polymerization of these monomers, e.g. maleic anhydride (MAH) and norbornene derivatives, is initiated in bulk or in concentrated solutions, at temperatures where a radical precursor has a short half life, i.e. less than 60 min. It is apparent that these polymerizations require more than the availability of free radicals and that other factors prevail under these conditions.

0097-6156/83/0212-0449$06.00/0

Evidence suggesting a change in the mode of peroxide decomposition under conditions of rapid thermolysis is provided by the decomposition of dicumyl peroxide in mineral oil (1). Thus, although the yield of cumyl alcohol is the same whether the decomposition occurs at 135° or 180°C, the yield of acetophenone and methane, the products of ketonic scission of the cumyloxy radical, essentially doubles at the higher temperature.

An examination of the behavior of peroxides in the course of their rapid thermal decomposition, provides some insight into the nature of the decomposition intermediates and/or products. NMR spectra taken during the thermolysis of various peroxides and peroxyesters in solution at temperatures where they have short half lives, e.g. benzoyl peroxide and t-butyl peroxypivalate at 110°C, contain emission peaks and/or enhanced absorption peaks. These peaks are presumed to be those of the reaction products resulting from interactions involving the transient decomposition products, e.g. interactions between pairs of caged radicals or from radical pair encounters. The NMR spectra are indicative of the occurrence of "chemically induced dynamic nuclear polarization" (CIDNP).

During thermal decomposition of peroxides, two radicals with antiparallel electron spins are simultaneously generated. These radicals consist of both triplets and singlets. When singlets, having paired electron spins, couple, no CIDNP is observed. However, when radicals diffuse away from each other and undergo addition and abstraction reactions, new radicals with differing populations of the electron-nuclear spin states are formed. When such radicals undergo coupling or termination reactions, CIDNP is observed.

Although NMR spectra without emission peaks are obtained during thermal decomposition of benzoyl peroxide in CCl_4 at 87°C, CIDNP spectra are obtained during thermolysis in cyclohexanone at 110°C or in CCl_4 at 87°C in the presence of tetramethyldioxetane. 1,2-Dioxetanes, which are cyclic peroxides and appear to be critical intermediates in many reactions characterized by chemiluminescence, undergo thermal decomposition to generate excited triplet carbonyl compounds capable of transferring their excitation energy to acceptors which subsequently undergo a photochemical change. Thus, the decomposition of benzoyl peroxide in CCl_4 at 87°C in the presence of tetramethyldioxetane, which is concurrently undergoing decomposition to triplet acetone, results in CIDNP spectra.

The implied relationship between the rapid thermal decomposition of peroxides and photochemical reactions is supported by the generation of CIDNP spectra when benzoyl peroxide is exposed to ultraviolet light in CCl_4 at 25°C. CIDNP spectra are also obtained when di(4-chlorobenzoyl) peroxide and other peroxidic compounds either undergo rapid decomposition at temperatures where they have short half lives or are exposed to ultraviolet light at ambient temperatures.

The significant point in the rapid thermal decomposition is not the occurrence of CIDNP but the formation of the precursors whose presence leads to CIDNP, i.e. the singlet and triplet radicals. The triplets are excited species and it has been proposed that the role they play in the polymerization of "unreactive" monomers is to transfer their excitation energy to the monomers. The resultant excimers or dimers are the true polymerizable species (2).

The distinction suggested by the terms "reactive" monomer and "unreactive" monomer is not supported by the results obtained in the homopolymerization of styrene and of maleic anhydride (MAH) in the presence of a peroxyester undergoing rapid decomposition. Thus, the addition of 0.5 mmole di-sec-butyl peroxydicarbonate ($t_{1/2}$ 1.3 hr) in 4 portions at 2 min intervals (total reaction time 15 min) to "unreactive" MAH and "reactive" styrene at 53°C indicates that, under these conditions, the labels are reversed.

	mmoles	Maximum Reaction Temp, °C	Polymer Yield, %
MAH	10	60	92.8
Styrene	10	55	3.8

The results suggest that MAH undergoes excitation and polymerization while styrene acts as a quencher.

Polymerization of Maleic Anhydride

The polymerization of MAH does not occur under normal conditions but is readily initiated under gamma or ultraviolet radiation and by the use of radical catalysts at high concentrations or having a short half life at the reaction temperature. The radical initiated homopolymerization is promoted by the presence of photosensitizers in the absence of light (2, 3). It has been proposed that under these conditions MAH undergoes excitation and the excited monomer, actually an excited dimer or charge transfer complex, polymerizes. The participation of the excimer or excited complex and the cationic character of the propagating chain has been confirmed by the total inhibition of MAH polymerization in the presence of small amounts of dimethylformamide which has no effect on the polymerization of "reactive" acrylic monomers (4).

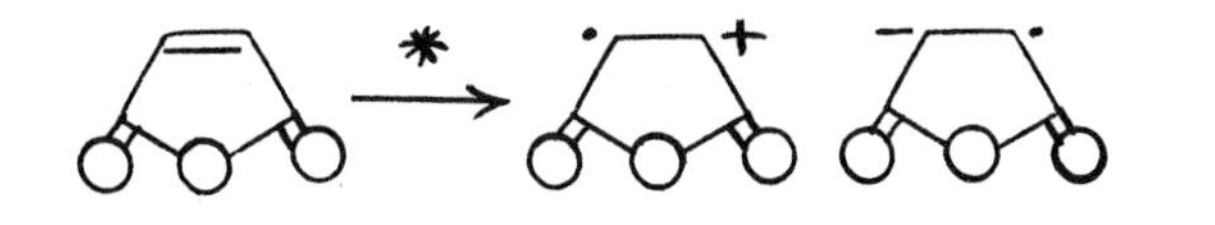

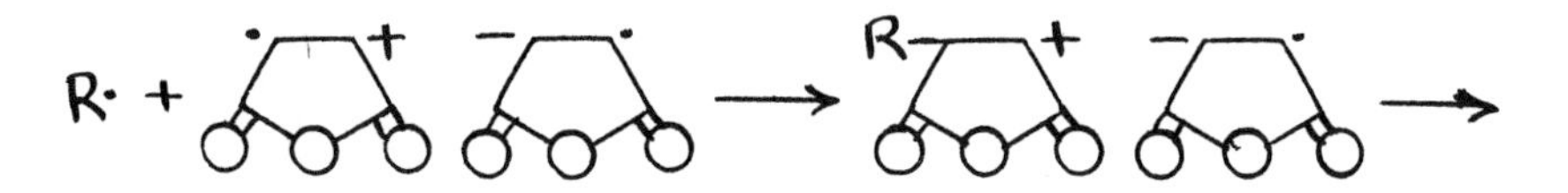

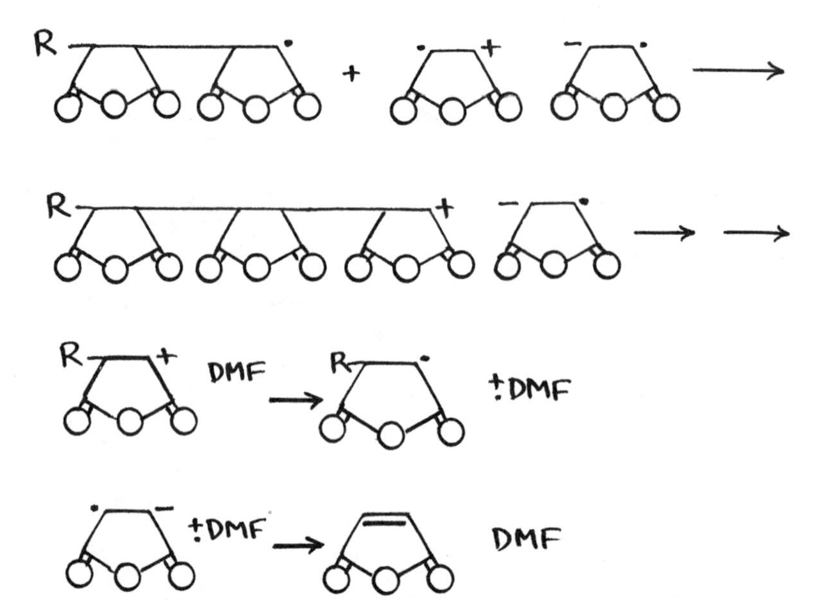

The solution polymerization of MAH is subject to a dilution effect, i.e. the yield of poly-MAH is dramatically reduced when the solvent concentration is at levels which are completely satisfactory for the polymerization of reactive monomers. This may be attributed to the short half life of the MAH excimer and/or complex or the interaction of the excited species with the solvent.

Table I. Homopolymerization of Maleic Anhydride[a]

Temp, °C	tBPP $t_{1/2}$,min	Solvent	ml	PMAH Yield,%
60	345	-	-	30
80	30	-	-	40
100	2.5	-	-	62
80	30	acetone	1	57
		benzophenone	1	62
		benzene	5	5
		xylene	5	16
			50	8 + 45[b]

[a] 1 mmole t-butyl peroxypivalate (tBPP) added in 4 portions over 20 min to 4.9g (50 mmoles) MAH; total reaction time 60 min

[b] Waxy solid; infrared spectra indicate presence of aromatic structures

Polymerization of Norbornene

Analogous to the situation with MAH, the polymerization of norbornene shows a strong dependence on the catalyst half life, i.e. bulk polymerization does not occur unless the catalyst half life is less than 2 hrs at the reaction temperature. Further, even when the catalyst half life is 30 min, the higher the temperature the higher the yield of polynorbornene, probably reflecting a strong dependence of the concentration of polymerizable species and/or the propagation rate on the temperature (5). NMR spectra of the polymer indicate that a rearrangement of the structure occurs during propagation.

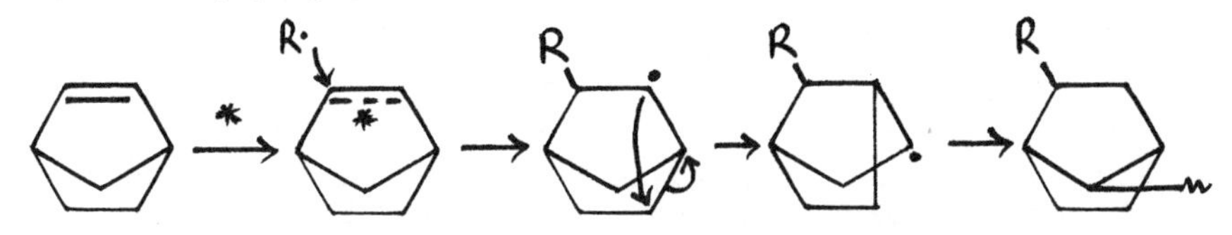

The solution polymerization of norbornene occurs only when low concentrations of solvents are present (6).

Table II. Homopolymerization of Norbornene[a]

Catalyst[b]	$t_{1/2}$ min	Temp, °C	Solvent	ml	Polymer Yield,%
DsBPDC	30	70	-	-	12
tBPP	30	80	-	-	12
tBPA	30	130	-	-	58
DsBPDC	90	50	acetone	0.3	2
AIBN	60	70	benzene	2.0	0
				0.05	11
			dioxane	1.85	3
				10.0	0
tBPA	75	120	benzene	0.045	20

[a] 1 mmole catalyst added in 4 portions over 20 min to 1.84g (20 mmoles) norbornene; total reaction time 60 min

[b] DsBPDC = di-sec-butyl peroxydicarbonate; tBPA = t-butyl peroxyacetate; AIBN = azobisisobutyronitrile

Polymerization of Substituted Norbornenes

The Diels-Alder adducts of cyclopentadiene (CPD) with MAH and with N-phenylmaleimide (NPMI) contain the norbornene moiety and undergo polymerization under the same conditions as MAH and norbornene, i.e. in the presence of radical catalysts undergoing rapid decomposition at the reaction temperature and subject to a significant dilution effect (7, 8).

Table III. Homopolymerization of CPD-MAH Adduct[a]

Adduct	Catalyst[b]	Temp, °C	Solvent[d]	ml	Polymer Yield,%
endo	BPO[c]	80	dioxane	5	0
	tBPP	80	dioxane	20	0
	tBPA	120	CB	2	50
		147	xylene	0.3	50
	tBHP-70	170	-	-	47
		240	-	-	80
exo	tBPA	120	CB	2	72
		147	xylene	0.3	65
	tBHP-70	240	-	-	35

[a] 1.25 mmoles catalyst added in 4 portions over 20 min to 2g (12 mmoles) adduct; total reaction time 60 min
[b] BPO = benzoyl peroxide; tBHP-70 = 70% t-butyl hydroperoxide with 20% di-t-butyl peroxide
[c] Reaction time 5 hrs
[d] CB = chlorobenzene

Table IV. Homopolymerization of CPD-NPMI Adduct[a]

Adduct	Catalyst[b]	Temp, °C	Solvent[d]	ml	Polymer Yield,%
endo	BPO[c]	80	CB	20	0
	tBPP	80	dioxane	20	0
		85	DCE	10	9
	tBPA	120	CB	3	60
	tBPB	150	CB	2	53
	tBHP-70	150	-	-	0
		210	-	-	20
exo	tBPP	85	DCE	16	0
	tBPA	120	CB	3	0
		155	CB	3	60
	tBHP-90	260	-	-	30

[a] 1.25 mmoles catalyst added in 4 portions over 20 min to 2g (8.4 mmoles) adduct; total reaction time 60 min
[b] tBPB = t-butyl perbenzoate; tBHP-90 = 90% t-butyl hydroperoxide
[c] Reaction time 5 hrs
[d] DCE = 1,2-dichloroethane

The dilution effect is very significant in some conjugated diene-MAH copolymerizations which also require the use of catalysts at temperatures where they have a short half life. Thus, whereas the copolymerization of butadiene and MAH occurs readily at 80°C on the addition of tBPP to a solution containing as much as 90% dioxane, the copolymerization of CPD and MAH in the presence of tBPP at 80°C requires that the dioxane concentration be less than 25% (7). The copolymerization of CPD and NPMI also requires the use of catalysts which have short half lives at the reaction temperature and takes place in bulk or in the presence of low concentrations of solvents (8).

The copolymerization of butadiene with MAH yields an unsaturated 1:1 alternating copolymer while the copolymerizations of CPD with MAH and with NPMI yield saturated 1:2 copolymers. The copolymerization of the CPD-MAH adduct with MAH and the copolymerization of the CPD-NPMI adduct with NPMI yield the same copolymers as are obtained from CPD and MAH and CPD and NPMI, respectively. The CPD copolymerizations may actually proceed through the rapid formation of the adducts, followed by their copolymerization with MAH and NPMI, respectively.

The homopolymerizations of the CPD-MAH and CPD-NPMI adducts, as well as the copolymerizations of the adducts with MAH and NPMI, respectively, at temperatures above the exo-endo isomerization temperature of the adducts, yield products containing both endo and exo structures, indicating that retrograde dissociation of the adducts has regenerated the charge transfer complex which under the influence of the rapidly decomposing catalyst undergoes excitation and polymerization. The dilution effects in these polymerizations at temperatures below and above the isomerization temperature, may be attributed to the short half life of the excited adducts and the excited complexes, respectively.

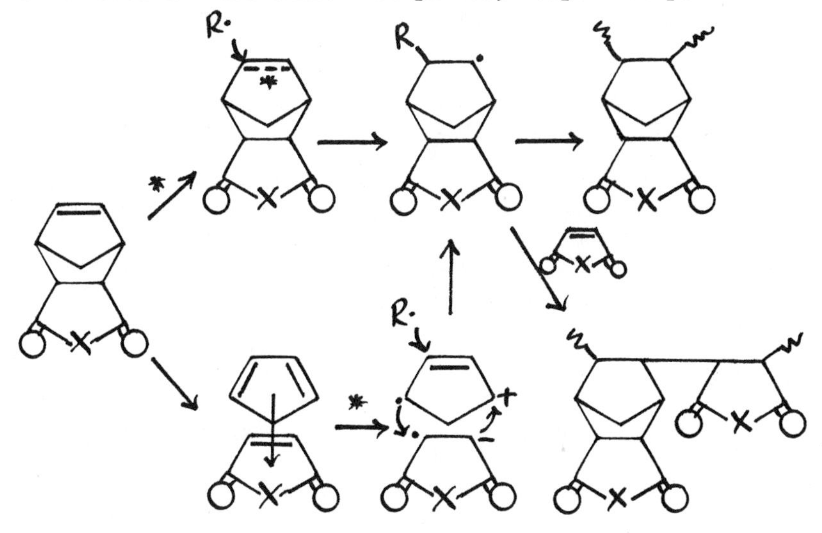

X = N-Ø or O

Mechanisms involving rearrangements in the norbornene moiety to yield products with 2,7 linkages, analogous to the rearrangement noted in the homopolymerization of norbornene, have been proposed previously (7, 8). However, there is as yet no direct evidence of rearrangement in the products of the adduct homopolymerizations and copolymerizations.

A significant temperature effect is noted in the bulk polymerizations of the liquid Diels-Alder adducts of CPD with acrylonitrile (AN) and methyl acrylate (MA) (9).

Table V. Homopolymerization of CPD-AN and CPD-MA Adducts[a]

Catalyst	mole-%	$t_{1/2}$ min	Temp °C	Time hr	Yield %	Mp, °C	MW (vpo)
			CPD-AN Adduct				
tBPP	6	30	80	1	10		
BPO	5	30	100	39	13		
tBPA	10	3.5	150	1	16		
tBHP-90	2	2	220	2	27	138-150	425
	10	2	220	2	54	193-198	892
			CPD-MA Adduct				
tBPP	10	30	80	1	0		
tBPO	5	30	100	39	0		
tBPA	10	3.5	150	1	3		
tBHP-90	2	2	220	2	11	110-113	1640
	10	2	220	2	32	139-144	2015

[a] Catalyst added over 20 min to 25 mmoles adduct; total reaction time as indicated

The yield of homopolymer increases with temperature. The increase in the molecular weight of the polymer with increasing catalyst concentration at the same temperature may be attributed to an increased concentration of polymerizable, i.e. excited species, as the catalyst concentration increases and rapid decomposition occurs.

Literature Cited

1. Dannenberg, E. M.; Jordan, M. E.; Cole, H. M. J. Polym. Sci. 1958, 11, 253.
2. Gaylord, N. G. Applied Polymer Symposium 1975, 26, 197.
3. Gaylord, N. G.; Maiti, S. J. Polym. Sci., Polym. Lett. Ed. 1973, 11, 253.
4. Gaylord, N. G.; Koo, J. Y. J. Polym. Sci., Polym. Lett. Ed. 1981, 19, 107.

5. Gaylord, N. G.; Mandal, B. M.; Martan, M. J. Polym. Sci., Polym. Lett. Ed. 1976, 14, 555.
6. Gaylord, N. G.; Deshpande, A. B.; Mandal, B. M.; Martan, M. J. Macromol. Sci.-Chem. 1977, A11, 1053.
7. Gaylord, N. G. Polymer Preprints 1976, 17, 666.
8. Gaylord, N. G.; Martan, M. Polymer Preprints 1981, 22, 11.
9. Gaylord, N. G.; Schildknecht, E. A.; unpublished results.

RECEIVED January 10, 1983

Homogeneous Lanthanide Complexes as Polymerization and Oligomerization Catalysts: Mechanistic Studies

P. L. WATSON and T. HERSKOVITZ

Central Research and Development Department, E. I. du Pont de Nemours and Company, Experimental Station, Wilmington, DE 19898

The trivalent lanthanide complexes $M(\eta^5\text{-}C_5Me_5)_2CH_3 \cdot L$ (M = Yb, Lu; L = diethyl ether, tetrahydrofuran or trimethylaluminum) and divalent $M(\eta^5\text{-}C_5Me_5)_2 \cdot L$ complexes (M = Yb, Eu, Sm; L = diethyl ether or tetrahydrofuran) are catalysts for polymerization of ethylene (hexane, 30-100°C, 60 psig; M_v product $\sim 10^5$) and oligomerization of propene. The initial insertion reactions of propene with $Lu(\eta^5\text{-}C_5Me_5)_2\text{-}CH_3 \cdot L$, studied by NMR, give well characterized $Lu(\eta^5\text{-}C_5Me_5)_2$-isobutyl and -2,4-dimethylpentyl complexes. Prior dissociation of ligand L is necessary. Constants for the dissociative pre-equilibria are determined from the rates of the propene insertion reactions and ligand exchange rates. Rates of ethylene polymerization in the presence of various ligands L are modified qualitatively as for the propene insertion reactions viz. L = no ligand > diethyl ether > THF > $Al(CH_3)_3$ (slowest rate). Ethylene propagation is faster than propene propagation by $\sim 10^4$.

Much effort has been devoted during the last 30 years toward understanding the mechanisms operative in the coordination catalysis of ethylene and α-olefin polymerization using Ziegler-Natta systems (metal halide and aluminum alkyl, sometimes with Lewis base modifiers). Aspects of the complex heterogeneous reactions have been elucidated (1-5) but the intimate mechanistic detail - for example the role of inhibitors and promoters, kinetics and thermodynamics of chain growth, modes of chain transfer and termination - comes primarily from studies of homogeneous catalysts (5-7).

Ethylene polymerization catalyzed by the well-characterized homogeneous lanthanide complexes $[M(C_5H_4R)_2R']_2$ has been studied,

0097-6156/83/0212-0459$06.25/0

in addition to some model reactions with substituted olefins. (8,9) Mechanistic details of the actual olefin insertion reactions were obscured in these systems by the stability of the alkyl-bridged dimeric structures and by the slow rate of α-olefin insertion relative to further insertion of a second olefin and/or β-hydrogen elimination.

Stereospecific polymerization of 1,3-dienes (10-18) (to butadiene) and isoprene homo- and copolymers), dimerization of propene (19) and recently stereospecific polymerization of acetylene (20) to high cis-content polyacetylene have all been reported using lanthanide catalysts. Sen (21) has reported the preparation of cationic europium systems (which perhaps function as cationic initiators) for polymerization of norbornadiene and 1,3-cyclohexadiene.

Described in this paper is a model system - one in which well-characterized lanthanide complexes exhibit high catalyst activities for ethylene polymerization but where the corresponding oligomerization of propene is sufficiently slowed so that stepwise insertion of the olefin can be studied quantitatively and all important intermediates observed or isolated. Emphasized in this paper is the effect of added Lewis acids and bases on the rate of olefin insertions, and comparison between ethylene and propene reactions. The catalysts, of general structure $M(\eta^5\text{-}Cp^*)_2CH_3 \cdot L$ (M = Yb, Lu; $Cp^* = C_5(CH_3)_5$; L = CH_3Li, ether, THF or $Al(CH_3)_3$), polymerize ethylene at rates comparable to $Zr(benzyl)_4/Al_2O_3$. Thus they are less active by two orders of magnitude than the zirconium-aluminoxane systems reported recently by Kaminsky and Sinn, (22) but much more active than most other reported homogeneous and heterogeneous ethylene polymerization catalysts. (23, 24)

Within the context of lanthanide chemistry it is interesting to note reports by Chinese (10,15) and Russian (12) scientists that activities of lanthanide-based Ziegler-Natta catalysts for polymerization of 1,3-dienes show much higher activity for early metals than for the end-row metals Er-Lu. The potential for higher activity analogs of the complexes described here is thus evident.

Experimental

All synthetic manipulations were carried out in a Vacuum Atmospheres HE-63 Drybox under a very slow continuous purge of dry N_2 using predried glassware. Solvents (THF, ether, toluene, pentane) were dried by distillation from Na/benzophenone under N_2. Ethylene and propene were purchased from Matheson (Research grade, 99.98°/min and 99.7% min respectively).

^1H NMR measurements were made on a Nicolet 360 MHz spectrometer. Both ^1H and ^{13}C chemical shifts were referenced to TMS by correcting the shifts relative to residual deuterated solvent peaks. Paramagnetic materials were referenced (in Hz) to the highest field solvent peak (+ = down field). Samples for NMR were approximately 0.05 M in lanthanide complex. Olefin was condensed into the sample on the vacuum line and then the sample was sealed under vacuum. Elemental analyses were done by Franz Pascher, Microanalytisches Lab., Bonn, West Germany.

Ethylene polymerizations were carried out in a glass 500cc batch reactor with high speed (1000 rpm) stirring and a thermocouple for internal temperature sensing. The stainless steel head also had an inlet for gases and septum port for catalyst injection and sampling. Absence of all poisons was ensured in each polymerization by either: a) titration with sacrificial catalyst under 10 psig C_2H_4 until polyethylene formation ensued; or b) addition of 10^{-4} moles of tetraneophylzirconium. Typical runs involved preequilibration of solvent (cyclohexane, 150 mL) with C_2H_4 (60 psig) in the reactor, followed by injection of catalyst (10^{-3} to 10^{-5} mol). Reaction times were usually 60-120 sec and were terminated with ethanol/acetic acid quench. Average polymerization rates were calculated from the total polymer yield obtained during the reaction. Ethylene uptake profiles were always monitored using a mass flowmeter (Brooks 5810) on the ethylene feed line, and recorded for analysis.

It should be noted that the abbreviation Cp* = η^5-$C_5(CH_3)_5$ is used in the experimental section below.

Li[MCp*$_2$(CH$_3$)$_2$](THF)$_{1-3}$, **2A** (M = Lu), **2B** (M = Yb). (25) To a 0°C solution of Li[MCp*$_2$Cl$_2$](THF)$_3$ (26) was added 2 eq. CH_3Li. After 30 min the solution was evaporated to dryness. Ether extracts of the solid residues were filtered and the filtrate cooled, yielding microcrystalline white (**2A**) or yellow (**2B**) product. For **2A**: ^1H NMR (glyme-d$_{10}$) δ-1.77 (s, 2CH$_3$, Lu-CH$_3$), 1.86 (m, 12H, THF), 2.28 (s, 30H, ring CH$_3$, overlapping THF) and 3.68 (m, 12H, THF). Anal. Calcd. for [$C_{26}H_{46}LiLuO$, i.e. 1 THF): C, 56.11; H, 8.33; Li, 1.25; Lu, 31.44. Found: C, 56.59; H, 8.40; Li, 1.17; Lu, 31.20. Less than 0.15% Cl found. For **2B**: Anal. Calcd. for [$C_{34}H_{60}LiO_3Yb$]: C, 58.6; H, 8.60; Li, 0.90; Yb, 24.83. Found: C, 58.53; H, 8.83; Li, 1.00; Yb, 24.82. [N.B. Samples for elemental analysis were stored under vacuum for several weeks resulting in some slow loss of THF].

Li[MCp*$_2$(CH$_3$)$_2$], **3A** (M = Lu), **3B** (M = Yb). Solid Li[MCp*$_2$(CH$_3$)$_2$](THF)$_3$ was heated under vacuum at 75° for 1-2 days. ^1H NMR in glyme-d$_{10}$ showed no THF or ether.

MCp*$_2$Al(CH$_3$)$_4$, **4A** (M = Lu), **4B** (M = Yb). To a stirred slurry of Li[MCp*$_2$(CH$_3$)$_2$] (solvent-free) in toluene was added 2 eq. $Al(CH_3)_3$. After most of the solids were taken into solution the mixture was filtered. The product was crystallized (-40°C) from the concentrated filtrate (M = Lu, white; M = Yb, purple). For **4A**: ^1H NMR (toluene-d$_8$) δ2.06 (s, 30H, Cp*), -0.11 (s, 6H, bridging CH$_3$) and -0.21 (s, 6H, terminal CH$_3$). Anal. Calcd. for [$C_{24}H_{42}AlLu$]: C, 54.13; H, 7.95; Al, 5.07; Lu, 32.86.

Found: C, 54.16; H, 7.94; Al, 4.86; Lu, 32.70. For 4B: ^{1}H NMR (toluene-d_8) overlapping peaks at +92 ($W_{\frac{1}{2}}$ ∿30 Hz) and +62 Hz (est. $W_{\frac{1}{2}}$ ∿40 Hz), in ratio 10:4. Anal. Calcd. for [$C_{24}H_{42}AlYb$]: C, 54.32; H, 7.98; Al, 5.09; Yb, 32.61. Found: C, 53.90; H, 7.95; Al, 4.88; Yb, 32.50.

MCp*$_2$CH$_3$·ether, 5A (M = Lu), 5B, (M = Yb). (27) MCp*$_2$Al(CH$_3$)$_4$ was dissolved in diethyl ether. On cooling the solution to -30°C crystals of the product formed (5A white; 5B orange). For 5A: ^{1}H NMR (toluene-d_8) δ-0.05 (s, 3H, Lu-CH$_3$), 1.09 (t, 6H, ether), 2.19 (s, 30H, Cp*), 3.49 (q, 4H, ether); (cyclohexane-d_{12}): -0.95 (Lu-CH$_3$), 1.20 (ether), 1.95 (Cp*), 3.56 (ether). ^{1}H NMR (-90° 0.04 M LuCp*$_2$CH$_3$·ether and 0.4 M ether in toluene-d_8): δ-0.55 (s, Lu-Me), 0.41 (t, CH$_3$ coord. ether), 0.60 (t, CH$_3$ coord. ether), 1.17 (t, CH$_3$ free ether), 2.04 (s, Cp*), 2.79 (q, CH$_2$ coord. ether), 2.86 (q, coord. ether), and 3.16 (q, CH$_2$ free ether). Anal. Calcd. for [$C_{25}H_{43}OLu$]: C, 56.17; H, 8.11, O, 2.99; Lu, 32.73. Found: C, 56.04; H, 8.11, O, 2.90; Lu, 33.20. For 5B: ^{1}H NMR (toluene-d_8) J = +9 Hz, $W_{\frac{1}{2}}$ = 90 Hz. Anal. Calcd. for [$C_{25}H_{43}OYb$]: C, 56.37; H, 8.14; O, 3.00; Yb, 32.49. Found: C, 56.05; H, 8.13; O, 2.90; Yb, 32.35.

[MCp*$_2$CH$_3$]$_2$, 7A (M = Lu). To a stirred solution of MCp*$_2$CH$_3$·ether in toluene was added 1-3 eq. NEt$_3$. Solvent was removed under vacuum. The resulting residues were washed with n-pentane, collected by filtration and dried. ^{1}H NMR (cyclohexane-d_{12}, 0.05 M, 25°C): δ-1.10 (s, 3H, Lu-CH$_3$) and 2.01 (s, 30H, Cp*); (toluene-d_8, 0.05 M, 25°C): δ-0.5 (s, 3H, Lu-CH$_3$) and 2.13 (s, 30H, Cp*). Anal. Calcd. for [$C_{21}H_{33}Lu$]: C, 54.77; H, 7.22; Lu, 38.00. Found: C, 54.44; H, 7.19; Lu, 37.90.

MCp*$_2$CH$_3$·THF, 6A (M = Lu), 6B (M = Yb). MCp*$_2$CH$_3$·THF (M = Yb, Lu) was prepared by dissolving MCp*$_2$CH$_3$·ether in THF. Evaporation of solvent followed by recrystallization from n-pentane (-40°) gave the white (Lu) or orange (Yb) product. For 6A: ^{1}H NMR (glyme-d_{10}) δ-1.18 (s, 1CH$_3$, Lu-CH$_3$), 1.80 (m, 4H, THF overlapping Cp*), 1.89 (s, 10CH$_3$, Cp*) and 3.64 (m, 4H, THF) (toluene-d_8): δ-0.79 (s, 1CH$_3$, Lu-CH$_3$), 1.21 (m, 4H, THF), 1.93 (s, 10CH$_3$, Cp*) and 3.28 (m, 4H, THF). ^{1}H NMR (-90°, 0.04 M LuCp*$_2$CH$_3$·THF and 0.02 M THF in toluene-d_8): δ-0.32 (s, Lu-CH$_3$), 1.13 (unres., β and β′ CH$_2$ of coord. THF), 1.57 (m, free THF), 2.27 (s, Cp*), 3.04 (unres., α CH$_2$ of coord. THF), 3.38 (unres., α′ CH$_2$ of coord. THF), 3.81 (m, free THF). Anal. Calcd. for [$C_{25}H_{41}LuO$]: C, 56.38; H, 7.76; Lu, 32,86. Found: C, 56.51; H, 7.81; Lu, 32.75. Anal. Calcd. for [$C_{25}H_{41}OYb$]: C, 56.58; H, 7.79; Yb, 32.61. Found: C, 56.61; H, 7.83; Yb, 32.75.

Results and Discussion

Synthesis and Characterization of Lutetium- and Ytterbium-Methyl Complexes. The synthetic strategy outlined in Scheme 1 allows the preparation of a family of methyl complexes having the general structure $MCp^*_2CH_3 \cdot L$ (M = Lu, Yb; Cp* = η^5-$C_5(CH_3)_5$; L = Lewis bases: ether, THF, CH_3Li; and Lewis acids: $AlMe_3$, $MCp^*_2CH_3$). (25, 26, 27)

Static structures of several complexes - $[YbCp^*_2(CH_3)_2]$-Li(THF)(ether)$_2$, $YbCp^*_2CH_3 \cdot THF$ and $YbCp^*_2CH_3 \cdot$ether - have been confirmed by X-ray crystallography. Ligand arrangements are represented adequately by the drawings in Scheme 1. All these structures contain a basic sandwich arrangement of metal atom between η^5-$C_5(CH_3)_5$ rings (mutually staggered) similar to structures reported for related halogen complexes. (26) The methyl group and ligand L are coordinated in the plane between the η^5-$C_5(CH_3)_5$ rings. An ORTEP drawing of $YbCp^*_2CH_3 \cdot$ether is provided in the abstract, (28) and of $YbCp^*_2CH_3 \cdot THF$ in Figure 1.

Variable temperature ^1H NMR provides information about the lability of ligands L in $LuCp^*_2CH_3 \cdot L$ complexes (Eq. 1). For all ligands L considered here Eq. 1 lies strongly to the left. Dissociative rather than associative exchange is confirmed

$$LuCp^*_2CH_3 \cdot L \underset{k_1}{\overset{k_1}{\rightleftharpoons}} LuCp^*_2CH_3 + L \qquad (1)$$

by the independence of ^1H NMR coalescence temperatures on concentration of excess L. Exchange of coordinated ether or THF (L) in **5A** or **6A** with uncoordinated L is extremely rapid at 25°C in toluene or cyclohexane. At -90°C, limiting spectra show sharp separate resonances for coordinated L and added free L. Intramolecular exchange - presumably rotational - of the chemically equivalent α, α' sites and β, β' sites of coordinated L is fast on the NMR time scale above -90°C for ether and above -65°C for THF. Coalescence measurements show that intermolecular ligand exchanges then become rapid at higher temperatures. Dissociative ligand exchange rates at 15°C (Eq. 1 forward, extrapolated from data in Figure 2) are 4.5×10^5 s^{-1} for ether and 1.2×10^3 s^{-1} for THF in toluene-d_8. Data shown in Figure 2 indicate that these ligand dissociation processes are 2-3 times faster in cyclohexane than in toluene.

The structure of $MCp^*_2Al(CH_3)_4$ (**4**, see Scheme 1) is inferred from the NMR spectrum, and by analogy with the structures of $YbCp^*_2AlCl_4$ (26) and $Y(C_5H_5)_2Al(CH_3)_4$. (29) Exchange with $Al(CD_3)_3$ occurs within minutes at 15°C. Bridge-terminal methyl group site exchange (extrapolated from coalescence temperatures in Figure 2) at 15°C is $\sim$0.3 s^{-1}. If this exchange results from dissociation of $Al(CH_3)_3$, then 0.3 s^{-1} is the rate of Eq. 1 forward. However, if the exchange is intermolecular then 0.3 s^{-1} is an upper limit for the dissociative process. Since the resonance for free, added $Al(CH_3)_3$ (at δ-0.3, superimposed on the lower-field half of the coalescenced bridge-terminal resonance) is also broadened it is likely that intermolecular exchange is

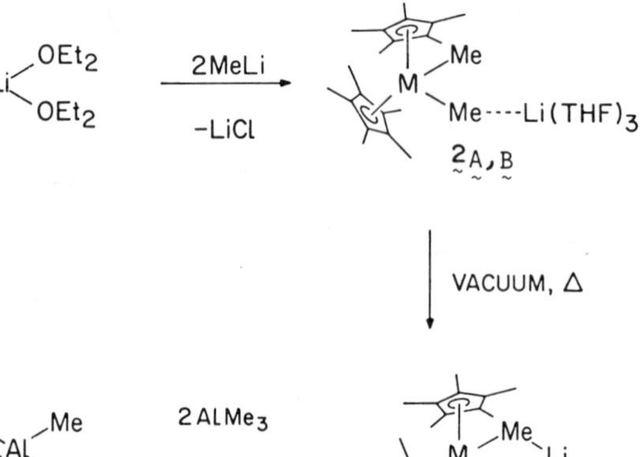

VACUUM, Δ

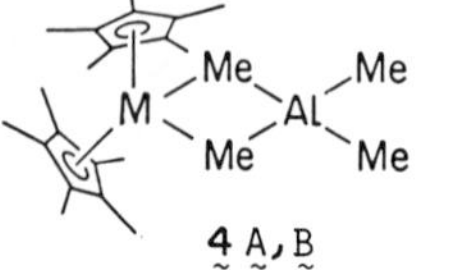

2AlMe3

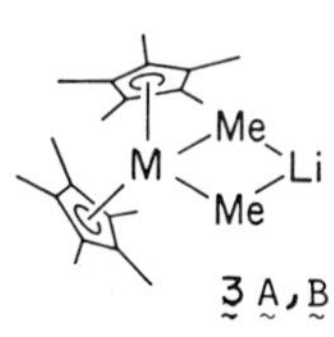

ether
−40°

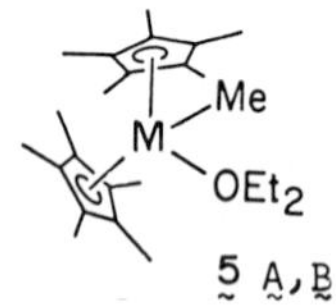

NEt3
VACUUM

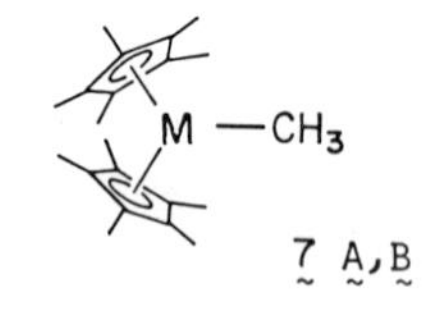

THF

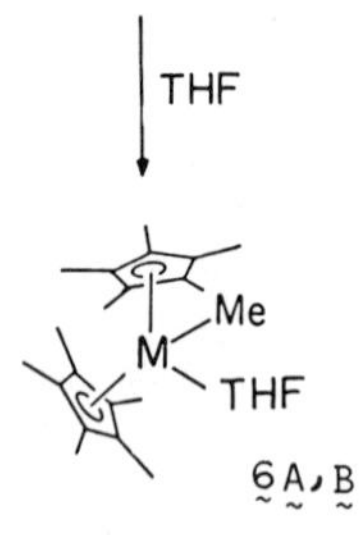

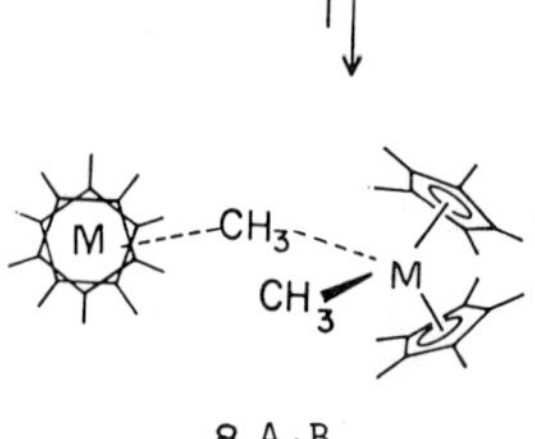

Scheme 1.

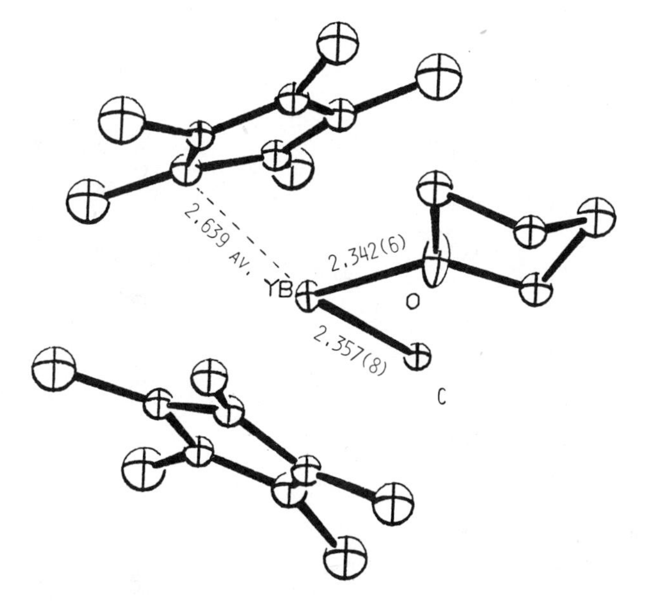

Figure 1. Structure of $Lu(C_5Me_5)_2CH_3$ in THF.

Figure 2. Coalescence measurements[a] for ligand exchange.

Complex	Ligand	Solvent	$\Delta\nu^b$ 80MHz	k^c (s^{-1})	T (°C)	$\Delta\nu$ 360MHz	k (s^{-1})	T (°C)	k_1^d (s^{-1})
5A	ether	tol-d_8[g]	27.08[e]	120.3	-65	122.8[e]	545.6	-55	~4.5×10^5
5A	ether	tol-d_8	52.75[f]	234.4	-59	237.4[f]	1054.8	-48	
5A	ether	MCH-d_{14}[g]				100.1	448.0	-63	
6A	THF	tol-d_8	42.4[e]	188.4	-5.5	190.8	847.7	11	~1.2×10^3
6A	THF	tol-d_8	32.9[f]	146.2	-10	148.0	657.5	7	
6A	THF	MCH-d_{14}				19.1	84.8	-10	
4A	$Al(CH_3)_3$	tol-d_8				46.4	206.3	107	
4A	$Al(CH_3)_3$	MCH-d_{14}	38.3	170.1	94.5	172.3	765.5	120	≤ 0.3

a. Measurements were made on solutions 0.5M in complex and 0.5M in added ligand L. Changes of the coordinated-ligand line shapes with temperature were independent of concentration of added L. *b*. $\Delta\nu = \nu(\text{free}) - \nu(\text{coord})$. *c*. Calculated for two site model, $k = \sqrt{2}\pi\Delta\nu$ at coalescence. *d*. By extrapolation. *e*. Protons α to oxygen. *f*. Protons β to oxygen. *g*. Toluene-d_8 and methylcyclohexane-d_{14}.

occurring. Again the exchange rate is independent of the amount of added ligand, $Al(CH_3)_3$.

The dimer $[MCp*_2CH_3]_2$, **8**, can also be viewed as a Lewis acid adduct. Monomer and dimer are in rapid equilibrium down to -60°C, giving rise to a single $Lu-CH_3$ resonance in the ^1H NMR spectrum of **8A**. The limiting spectrum of **8A** at -90°C shows a 1:1:2 pattern for the Cp* peaks and a 1:1 set for the $Lu-CH_3$ groups. The structure shown in Scheme 1 is proposed for this dimer. Presumably steric bulk constrains the two $LuCp*_2$ fragments to be mutually orthogonal and prevents bridging of both methyl groups.

The methyl complexes are thus well-characterized with respect to structure and lability of ligand L. Comparatively, lability of L decreases in the series $MCp*_2CH_3$ ~ether>THF>$AlMe_3$. As will be seen, this is also approximately the ordering of equilibrium constants K for Eq. 1.

Reaction of Methyl Complexes with Propene. Products. Reaction of $[MCp*_2CH_3]_2$, **8**, with propene initially yields isobutyl complexes $MCp*_2CH_2CH(CH_3)_2$, **9**, as shown in Scheme 2. Secondary, slower reaction of propene with **9** gives the 2,4-dimethylpentyl species **10**, and subsequently higher oligomers.

Confirmation of the structures comes from several sources. GC-MS of hydrolyzed samples of **9** and **10** show isobutane and 2,4-dimethylpentane (respectively) as the only C_4 or C_7 isomers formed, confirming the regiospecificity of the insertions. ^1H NMR spectra of **9A** and **10A** (and of ^2H specifically-labelled samples of **9A** and **10A**) are fully assigned (30) and consistent with proposed structures. Additionally, 9A is also formed from the reaction of $LuCp*_2H$,(31) **11A**, with isobutene (reaction 5, Scheme 2), and **10A** is formed by addition of either 2,4-dimethylpentene to $LuCp*_2H$ or by addition of 4-methylpentene to $LuCp*_2CH_3$.

Polypropene is not formed in these systems. Analysis (by hydrolysis then GC) of mixtures of propene (60 psi) and **8A** (0.1 M in cyclohexane) reveals that oligomers at least up to C_{24} are formed. However the product distribution above C_{10} becomes progressively more complex, indicating chain transfer and chain termination processes which compete with propagation.

Both **9** and **10** are thermally unstable with half-lives at 25°C of ~3 hr. The thermal decomposition of **9A** in cyclohexane is under study as a model for chain termination for these systems. Both β-hydrogen elimination (giving **11A** and isobutene) and β-alkyl elimination (giving **8A** and propene) are important kinetically accessible processes, both for **8A** and for the longer chain $LuCp*_2(CH_2CHCH_3)_nCH_3$ complexes.(31)

Kinetics. Kinetic analysis of the reaction of the dimer **8A** with propene confirms that dissociation to the active monomer **7A** is necessary and that the rate-determining step is insertion of propene into the $Lu-CH_3$ bond of **7A** (reactions 1 and 2 in Scheme 2). At 15°C $K = k_1/k_{-1} = 4.1 \times 10^{-3}\ M^{-1}$ and $k_2 = 1.22 \times 10^{-1}\ M^{-1}s^{-1}$

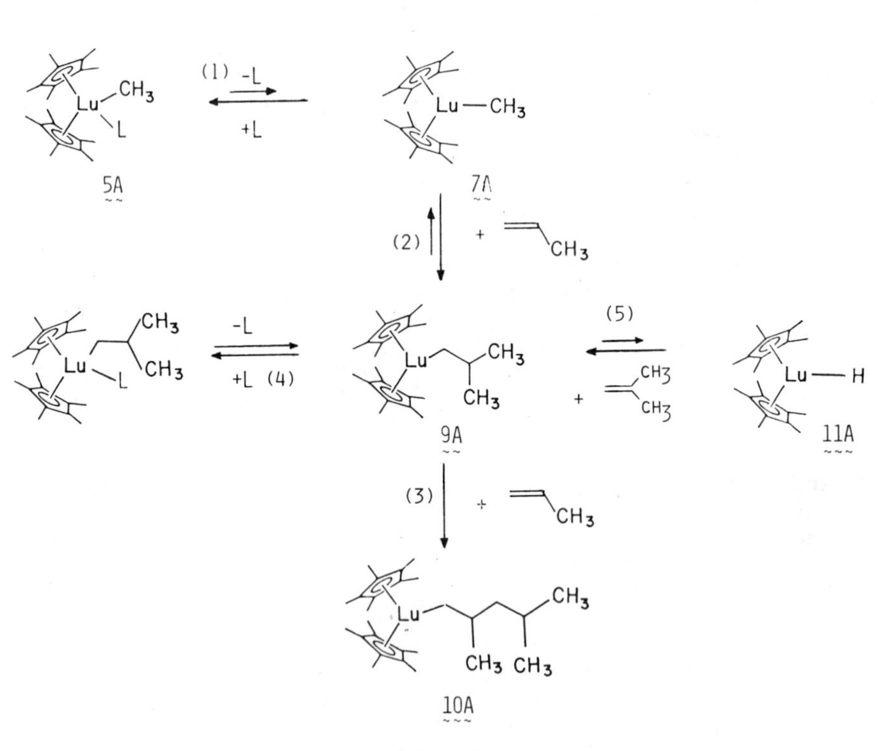
(1) -L
+L
CH3
Lu
L
5A
Lu—CH3
7A
(2)
+
CH3
-L
+L (4)
(5)
Lu
CH3
CH3
9A
+
Lu—H
11A
(3)
10A
CH3 CH3

Scheme 2.

in cyclohexane. Previous kinetic studies (27) of the reaction of $LuCp^*_2CH_3$·ether, 5, with propene also showed the presence of a dissociative preequilibrium giving 7A and ether followed by a slower olefin insertion. The rate expression derived (Eq. 2) is consistent with Scheme 2. Knowing k_2 exactly, values for

$$\frac{-d[7A\cdot L]}{dt} = \frac{k_1 k_2 [7A\cdot L][propene]}{k_{-1}[L]} \quad (2)$$

the preequilibrium constants $K = k_1/k_{-1}$ can be derived from the observed rates of reaction of the complexes with propene under pseudo-first order conditions. These are shown in Figure 3 for various ligands L.

Inhibition of propene insertion by ether or monomer 7A is mild but kinetically important at the concentrations used for NMR experiments (∿0.05 M), slowing the rates by a factor of 20-30. Inhibition is very pronounced with the stronger ligands $Al(CH_3)_3$ and THF. This kinetic inhibition is due to the thermodynamic stability of the adducts (see Figure 4). In fact, complexation of 7L by both $Al(CH_3)_3$ and THF slows the propene insertion reaction sufficiently that only very low steady-state concentrations of the initial Lu-isobutyl product are observed. Observed rates for reaction of propene with $LuCp^*_2CH_3$·L (L = $LuCp^*_2CH_3$, ether, THF, $AlMe_3$) are all dependent on propene concentration. The highest energy transition state is therefore always that for the olefin insertion, not for ligand dissociation.

Inhibition of the reaction of isobutyl complex 9A with propene does not occur with added ether, indicating that formation of 9A·ether is not significant. However, THF does inhibit this reaction and a value of $K = k_4/k_{-4}$ (Scheme 2) = $\sim 1 \times 10^{-2}$ M^{-1} is obtained. The relative ability of 7A and 9A to coordinate THF is probably largely steric in origin. Finally, the rate of propene insertion into the Lu-C bond of isobutyl complex 9A, $\sim 1.1 \times 10^{-3}$ M^{-1} s^{-1} is a good estimate of the rate of propagation during further oligomerization of propene.

Polymerization of Ethylene by Methyl-Lutetium and -Ytterbium Complexes. Products. Ethylene (60 psi, 30-100°C) is polymerized rapidly by 10^{-4} to 10^{-6} M cyclohexane solutions of the methyl complexes 3, 4, 5, 6, and 8 (Table). Low Mw oligomers were not formed (analysis by GC) in experiments 1a or 3a, Table. At 160°C both the Lu and Yb etherates produced very little polyethylene ($\sim 10^{-3}$ of the Table yields at 40-100°) reflecting thermal decomposition of the active species.

Inherent viscosities of the polymer (Table) show that $M_v \sim 10^5$ and that polymerization at higher temperatures reduces the molecular weight (Figure 5). Infrared spectral analysis of the polyethylenes obtained in the Table showed no $H(R)C{=}CR_2$ groups that would result from α-olefin formation then incorporation into growing polymer then chain termination *via*

Figure 3. Equilibrium constants for Equation 1, from kinetics of the reaction of propene with lutetium complexes.[a]

Complex	Ligand	$k_{observed}$ (s^{-1})	K^b (M)	$k_{-1}=k_1/K$ [c]
5A	ether	2.3×10^{-4}	1.9×10^{-3}	$\sim 2.4 \times 10^{8}$
6A	THF	1.7×10^{-5}	1.4×10^{-4}	$\sim 8.5 \times 10^{6}$
4A	$Al(CH_3)_3$	3.0×10^{-6}	2.4×10^{-5}	$< 1.2 \times 10^{4}$
8A	$LuCp^*_2CH_3$	-	4.1×10^{-3}	-
9A.THF[d]	THF	1.2×10^{-5}	1.0×10^{-2}	-

a. In cyclohexane-d_{12} at 15°C. *b*. At 15°C, calculated from the rate = $k_1k_2[\text{propene}]/k_{-1}[L]$, with $k(\text{observed}) = Kk_2 = k_1k_2/k_{-1}$. Rates were obtained from plots of -ln (5A, 6A or 4A) versus time for reactions having 5-20 fold excess propene. Values for dissociation constant K(8A) $=k_1/k_{-1}$ (Scheme 2) $=4.1 \times 10^{-3}$ M; k_2 (Scheme 2) $=1.22 \times 10^{-1} M^{-1}s^{-1}$; and k_3 (Scheme 2) $=1.1 \times 10^{-3} M^{-1}s^{-1}$ were determined independently.
c. From Figure 2. *d*. Using k_3 in place of k_2.

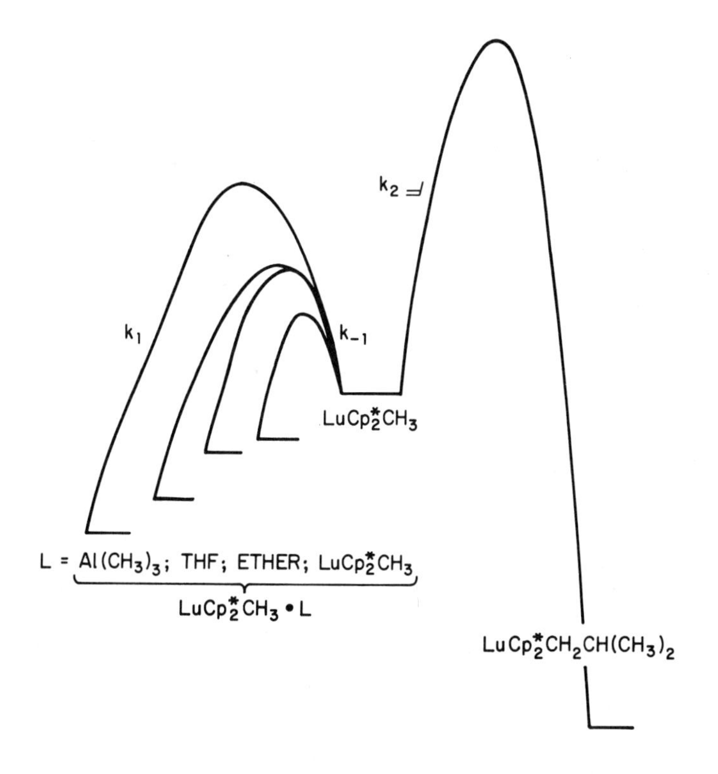

Figure 4. Thermodynamic stability of several adducts.

Table: ETHYLENE POLYMERIZATIONS[a]

CATALYST			CATALYST EFFICIENCY[b]	PEAK TURNOVER NUMBER[b]	YIELD	RUN TIME	BATH TEMP	REACTOR TEMP. °C	
	Type	Am't. (mmol)	g PE/mmol M/min/ $Kgcm^{-2}$ C_2H_4	moles C_2H_4/cat/sec ($X10^3$)	g PE	sec.	°C	start	end
1. a.	$Lu(Cp^*)_2CH_3 \cdot OEt_2$	0.01	115±8	1.00	4.0	60	60	52	80
b.	$Lu(Cp^*)_2CH_3 \cdot OEt_2$	0.01	55±5	.62	1.8	60	100	88	97
2. a.	$Lu(Cp^*)_2CH_3 \cdot AlMe_3$	0.01	45±5[c]	0.54	3.6	120	60	52	80
b.	$Lu(Cp^*)_2CH_3 \cdot AlMe_3$[b]	0.005	140±15	1.01	2.4	60	100	86	101
3. a.	$Yb(Cp^*)_2CH_3 \cdot OEt_2$	0.01	70±8	0.52	2.6	60	60	53	72
b.	$Yb(Cp^*)_2CH_3 \cdot OEt_2$	0.005	90±9	.68	1.5	60	100	86	94
4. a.	$Yb(Cp^*)_2CH_3 \cdot AlMe_3$	0.01	45±3[c]	0.33	2.6	90	60	53	71
b.	$Yb(Cp^*)_2CH_3 \cdot AlMe_3$	0.005	105±10	.68	1.8	60	100	86	98
5.	$Lu(Cp^*)_2(CH_3)_2Li$[d]	1.0	∿0	-	<0.4	∿600	40	34	-
6.	$Yb(Cp^*)_2(CH_3)_2Li \cdot OEt_2$[d]	0.8	0.6±0.1	-	26.2	900	40	33	89
7.	1. + $1AlMe_3$	0.01	30±8[c]	0.24	2.9	120	40	37	61
8.	1. + $2AlMe_3$	0.03	∿0[c]	-	0	$<10^3$	RT	RT	-

9.	2. + $nAlMe_3$ (n = 1,4)	0.01	$\sim 0^c$	-	0	$<10^3$	RT	RT	-
10.	2. + $1AlMe_3$	0.14	$\sim 0^c$	-	0	$<10^3$	RT	RT	-
11.	1. + $10Et_2O$	0.01	120±10	1.05	2.7	60	60	52	71
12.	1. + 10THF	0.01	60±5	0.52	2.3	60	60	52	69
13.	$Sm(Cp^*)_2 \cdot OEt_2$	0.05	$2\text{-}11^e$	-	0.5	60	40	37	41
14.	$Eu(Cp^*)_2 \cdot 2THF$	0.09	$<10^{-2}$	-	$<10^{-1}$	$\sim 10^3$	40	∿37	-
15.	$Yb(Cp^*)_2 \cdot OEt_2$	0.05	$<10^{-1}$	-	$<10^{-1}$	300	40	∿37	-

a) Batch, 500 cc cyclohexane, 1000 rpm stirring, 60 psi (4.2 $Kgcm^{-2}$) C_2H_4, 5 L/min maximum C_2H_4 flow. *b)* See Figure 6. *c)* Induction period noted (seconds) (excluded from the catalyst efficiency calculation). *d)* Partially soluble. *e)* Catalyst efficiency range due to poor reproducibility.

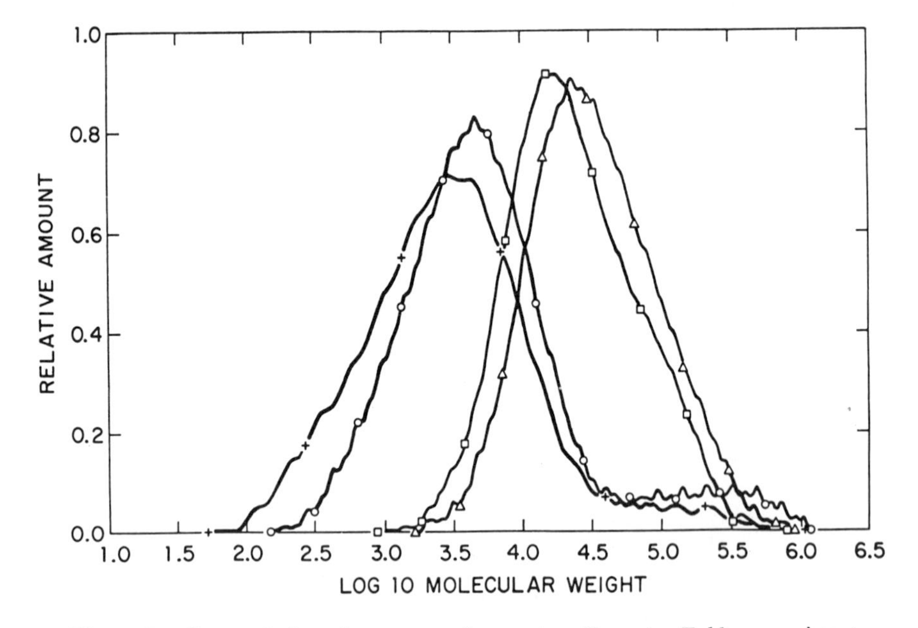

Figure 5. Size exclusion chromatography spectra. Key: +, Table, experiment 2.b.; ○, Table, experiment 1.b.; △, Table, experiment 2.a.; □, Table, experiment 1.a.

β-hydride transfer. $H_2C=C(H)R$ groups were always evident and increase in concentration by the higher temperature runs (e.g. 1b vs. 1a and 2b vs. 2a) indicating relatively more β-hydride transfer. This concurs with the observed lower PE Mw at higher polymerization temperatures. Size exclusion chromatography (SEC) (Figure 5) confirms the shift to lower Mw by the Lu^{III} (Figure 5) and the Yb^{III} complexes. Figure 6 shows that the Lu^{III} etherate and $AlMe_3$ adduct afford the same polyethylene Mw distributions, supporting the presence of the same catalyst in both cases. Further polymer characterizations are pending.

Activity. Calculated catalyst efficiencies (g PE/mmol catalyst/min/$Kgcm^{-2}$ C_2H_4) and peak turnover numbers (moles C_2H_4/catalyst/sec) for the various catalysts (or more accurately, catalyst precursors) are shown in the Table. In the batch reactor used it is not possible to reduce the reaction rate sufficiently to effect isothermal runs with these very active catalysts, so the catalyst efficiencies are measured over a temperature range. Activities reported for the different catalysts are, however, reproducible. The calculated catalyst efficiencies in the Table suggested that ethylene diffusion was rate limiting at 60 psi. Indeed, experiments at higher (100 psi) and lower (40 psi) confirmed this fact. This, combined with very low catalyst concentrations (i.e. all adducts largely dissociated), made quantitative comparisons of the ligand effects for propene vs. ethylene insertion rates difficult. Qualitatively however, the rate of ethylene polymerization decreases with complexation by ligand L, with $Al(CH_3)_3$ (most potent inhibitor) > THF > ether. Experiments 7-12 in the Table give the appropriate data. $Al(CH_3)_3$ in low excess completely stops polymerization which probably reflects strong binding to the propagating species $LuCp^*_2(CH_2)_nCH_3$. An alternative explanation of the $AlMe_3$ rate depression is the formation of more stable organometallic intermediates (for example $MCp^*_2(H)(R)AlR_2$ from β-hydrogen elimination) which could slow the rate of chain transfer. Excess ether (10 equiv.) has no effect on the propagation rate (Experiment 10, Table) while THF (10 equiv.) does depress the rate of polymerization but not dramatically (Experiment 11). All of the complexes readily polymerize ethylene in diethyl ether solution, but not at all in THF solution, indicating much stronger complexation of $LuCp^*_2(CH_2)_nCH_3$ species by THF. These results are consistent with the trend of equilibrium constants in Figure 3 for ligand binding to methyl complex **7A**, which modifies the rate of propene insertion.

The dimethyl anion **3A** shows very little activity in cyclohexane (Experiment 5). However, this is due to insolubility. This complex readily polymerizes ethylene in diethyl ether solution.

A representative mass flowmeter trace is shown in Figure 6. In all cases polymerization ensued immediately on addition of

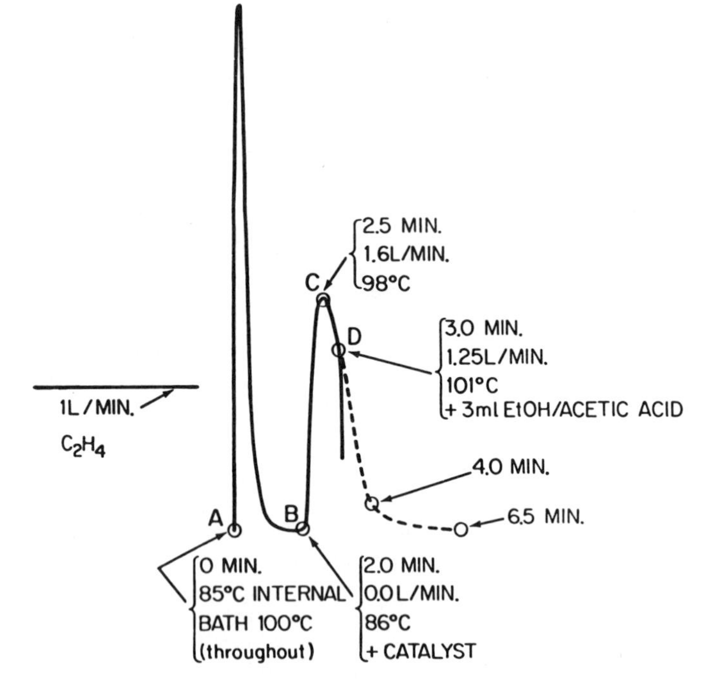

Figure 6. Mass flowmeter trace of run 2.b. in Table. Key: A to B, solvent saturation with 60 psig C_2H_4; B to D, polymerization period affording the catalyst efficiency value; C, point used to obtain the peak turnover number; - - -, C_2H_4 uptake profile without EtOH/acetic acid addition.

catalyst to ethylene-saturated solvent, except for $MCp^*{}_2Al(CH_3)_4$ complexes which showed an induction period. Since ethylene insertion is so rapid ($k_{av} > 10\ M^{-1}\ S^{-1}$) the first insertion may well appear to have a rate-determining dissociation of $Al(CH_3)_3$.

All the catalysts had limited lifetimes (minutes) under the exothermic conditions of these polymerizations. Complete details of chain termination proceeses are not known. Within experimental error analogous lutetium and ytterbium complexes had very similar activities and lifetimes. It can be estimated from the data that the intrinsic ethylene propagation rate is faster than that for propene by a factor of $>10^4$.

Divalent Complexes as Catalysts. The divalent complexes $M(C_5Me_5)_2 \cdot$ether (M = Yb, Eu, Sm) (32, 33) were also examined as ethylene polymerization catalysts (Table). Activity of Eu and Yb complexes (which have relatively stable divalent states) was low, whereas the Sm complex was considerably more active. Since activity of the last complex was quite variable we suspect the true catalyst is formed in situ by some process such as oxidation. Since the color (green) of the catalyst does not change the level of active catalyst formed may be low and variable, indicating the true activity to be much greater than for the lutetium or ytterbium complexes. This would be consistent with the Chinese observations (10, 15) mentioned in the introduction.

Acknowledgments

The excellent technical assistance of J. T. Corle, Jr. and G. Krupa is gratefully acknowledged.

Literature Cited

1. Boor, J., Jr. "Ziegler-Natta Catalysts and Polymerizations"; Academic Press: New York, 1979.
2. Pino, P.; Mülhaupt, R. Angew. Chemie. Int. Ed. Engl. 1980, 19, 857-975.
3. Langer, A. W.; Burkhardt, T. J.; Steger, J. J. Proc. 28th IUPAC Makromol Symp. Amherst, MA. 1982, 244 and references 8 and 10 therein. Langer, A. W. Ann. N. Y. Acad. Sci. 1977, 295, 110.
4. Reichert, K. K. Angew. Makromol. Chemie. 1981, 94, 1-23.
5. Sinn, H.; Kaminsky, W. Adv. Organometal. Chem. 1980, 18, 99-151.
6. Fink, G.; Rottler, R. Angew. Makromol. Chemie. 1981, 94, 25-47. Fink, G.; Rottler, R.; Kreiter, C. G. Angew. Makromol. Chemie. 1981, 96, 1-20.
7. Henrici-Olivé, G.; Olivé, S. Chem. Tech. 1981, 11, 746-752.

8. Ballard, D. G. H.; Courtis, A.; Holton, J.; McMeeking, J.; Pearce, R. J.C.S. Chem. Commun. 1978, 994-995.
9. Ballard, D. G. H.; Courtis, A.; Holton, J.; McMeeking, J.; Pearce, R. Conf. Eur. Plast. Caoutch. [C.R.], 5th. 1978, 1, A4/1-A4/7.
10. Ouyang, J.; Wang, F.; Shen, Z. Proc. China-U.S. Bilateral Symp. Polym. Chem. Phys. 1979 (Publ. 1981), 382-398. Science Press: Peking, People's Republic of China.
11. Mazzei, A. NATO Adv. Study Inst. Ser., Ser. C Math and Phys. Sci. Dordrecht. Neth. 1979, 44 (Organomet. f-element) 379-393.
12. Rafikov, S. R.; Monakov, Yu. B.; Bieshev, Ya. Kh.; Vashtov, I. F.; Murinov, Yu. I.; Toestikov, G. A.; Nikitin, Yu. E. Doklad. Akad. Nauk. SSR 1976, 229, 1174.
13. Yang, J.; Hu, J.; Feng, S.; Pan, E.; Xie, D.; Zhong, C.; Ouyang, J. Sci. Sin., Engl. Ed. 1980, 23, 734-743.
14. Rafikov, S. R.; Monakov, Y. B.; Marina, N. G.; Puvakina, N. V.; Tolstikov, G. A.; Krivonogov, V. P.; Nurmukhametov, F. N.; Kovalev, N.; Tikhomirova, G. A. USSR Pat. No. 730710 (CA 93:96152v).
15. Wang, F.; Sha, R.; Jin, Y.; Wang, Y.; Zheng, Y. Sci. Sin. Engl. Ed. 1980, 23, 172-179.
16. Bieshev, Ya. Kh.; Kozlova, O. I.; Minikhanova, A. M. Issled. Obl. Khim. Vysokomol. Soedin, Neftekhim. 1977, 74, (CA 92:164934h).
17. Shen, Z.; Ouyang, J.; Wang, F.; Hu, Z.; Yu, F.; Qian, B. Hua Hseuh Tung Pao 1979, 5, 426-430 (CA 92:77000g).
18. Tzu, Jan Tsa Chih Rare Earths and Rubbers 1980, 3, (9), 658-660, 711.
19. Yoo, J. S.; Koncos, R. U.S. Pat. No. 3,803,053 1974, (CA 81:50335f).
20. Shen, Z.; Yang, M.; Cai, Y. Proc. 28th IUPAC Makromol. Symp. Amherst, MA. 1982, 431.
21. Sen, A. Abstracts 183rd ACS National Meeting Las Vegas, Nevada March, 1982, Inor. Section 191.
22. Kaminsky, W.; Sinn, H. Proc. 28th IUPAC Makromol. Symp. Amherst, MA. 1982, 247. Sinn, H.; Kaminsky, W.; Woldt, R. Angew. Chemie. Int. Ed. Engl. 1980, 19, 390.
23. Zakharov, V. A.; Yermakov, Y. I. Catal. Rev.-Sci. Eng. 1979, 19, 67-103.
24. Ballard, D. G. H. J. Polymer Sci., Pol. Chem. Ed. 1975, 13, 2191-2212.
25. Watson, P. L. J.C.S. Chem. Commun. 1980, 652-653.
26. Watson, P. L.; Whitney, J. F.; Harlow, R. L. Inorg. Chem. 1981, 20, 3271-3278.
27. Watson, P. L. J. Am. Chem. Soc. 1982, 104, 337-339.
28. Watson, P. L. Abstracts 183rd ACS National Meeting Las Vegas, Nevada March, 1982, Macr. Section 008.
29. Holton, J.; Lappert, M. F.; Ballard, D. G. H.; Pearce, R.; Atwood, J.; Hunter, W. E. J. C. S. Dalton Trans. 1979, 45-53.

30. ^{1}H NMR spectrum of **9A** is described in ref. 27. Complete assignment of the ^{1}H NMR spectrum of **10A** is in ref. 28.
31. Watson, P. L.; Roe, D. C. J. Am. Chem. Soc. 1982, 104, 6471-6473.
32. Watson, P. L.; Harlow, R. L.; Whitney, J. F.; Tilley, T. D.; Andersen, R. A. submitted to Organometallics 1982.
33. Evans, W. J.; Bloom, I.; Hunter, W. E.; Atwood, J. L. J. Am. Chem. Soc. 1981, 103, 6507-6508.

RECEIVED January 10, 1983

INDEX

INDEX

E

Q

R

Jacket design by Kathleen Schaner
Indexing by L. Luan Corrigan
Production by Frances Reed

Elements typeset by Service Composition Co., Baltimore, MD
Printed and bound by Maple Press Company, York, PA